An Introduction to Lightning

Vernon Cooray

An Introduction to Lightning

Springer

Vernon Cooray
Department of Engineering Sciences
Division for Electricity
Uppsala University
Uppsala, Sweden

ISBN 978-94-017-8937-0   ISBN 978-94-017-8938-7 (eBook)
DOI 10.1007/978-94-017-8938-7
Springer Dordrecht Heidelberg New York London

Library of Congress Control Number: 2014951208

© Springer Science+Business Media Dordrecht 2015
This work is subject to copyright. All rights are reserved by the Publisher, whether the whole or part of the material is concerned, specifically the rights of translation, reprinting, reuse of illustrations, recitation, broadcasting, reproduction on microfilms or in any other physical way, and transmission or information storage and retrieval, electronic adaptation, computer software, or by similar or dissimilar methodology now known or hereafter developed. Exempted from this legal reservation are brief excerpts in connection with reviews or scholarly analysis or material supplied specifically for the purpose of being entered and executed on a computer system, for exclusive use by the purchaser of the work. Duplication of this publication or parts thereof is permitted only under the provisions of the Copyright Law of the Publisher's location, in its current version, and permission for use must always be obtained from Springer. Permissions for use may be obtained through RightsLink at the Copyright Clearance Center. Violations are liable to prosecution under the respective Copyright Law.
The use of general descriptive names, registered names, trademarks, service marks, etc. in this publication does not imply, even in the absence of a specific statement, that such names are exempt from the relevant protective laws and regulations and therefore free for general use.
While the advice and information in this book are believed to be true and accurate at the date of publication, neither the authors nor the editors nor the publisher can accept any legal responsibility for any errors or omissions that may be made. The publisher makes no warranty, express or implied, with respect to the material contained herein.

Cover image: © yevgeniy11/fotolia

Printed on acid-free paper

Springer is part of Springer Science+Business Media (www.springer.com)

This book is dedicated to the memory of my late mother, Dorothy Cooray, who taught me that education is a treasure that cannot be stolen and an inheritance that cannot be lost; to my wife, Ruby Cooray, who has the love, consideration, understanding, and patience to live with a person who finds it nearly impossible to decouple his mind from research; and to my 4-year-old grand daughters Sanduni Cooray and Ella Cooray, who think their grandfather is working with Bamse's (a popular Swedish cartoon bear) grandmother to make Thunder Honey (Dunder Honung).

Preface

The material presented in this book is based on a lecture course I gave to postgraduate and undergraduate students at Uppsala University. Over the course of my lectures I realized that most of the available books on lightning were written mainly for experts in the field and, perhaps, for advanced postgraduate students. I came to recognize that there was a need for a book that introduces undergraduate and beginning postgraduate students to the subject of lightning and prepare them for more advanced texts aimed at experts. I hope that this book, which explains various aspects of lightning in a manner suitable for undergraduate and postgraduate students, will satisfy this need. To understand the mechanism of the lightning flash and the features of its electromagnetic fields, some knowledge of the physics of electrical discharges and electromagnetic field theory is necessary. To this end, I have added one chapter on the physics of discharges and another on electromagnetic field theory. Whenever necessary, I provided the mathematical details requisite for an understanding of the effects of lightning and the various theories associated with them. I believe, however, that even a layman can read and understand this book, of course skipping the mathematical details, and still gain appreciable knowledge on the physics and effects of lightning flashes.

This book resulted from my long-term involvement in teaching and researching the effects and physics of lightning flashes. Research work is a group effort, and I wish to thank all my former and current students who keep me constantly busy thinking about lightning and its effects. Furthermore, I would especially like to thank Riduan Ahmed, Mona Riza, Pasan Hettiarachchi, and Muzafar Ismail for providing helpful feedback on the draft version of this book. I also thank Dr. Mahendra Fernando with whom I collaborate closely in my lightning-related research. I appreciate the help of Muditha and Pasan Hettiarachchi in creating the figures in the book. My thanks go to Prof. Farhad Rachidi and Prof. Carlo Alberto for various international collaborations in lightning-related research work.

Uppsala, Sweden
03 Jan 2014

Vernon Cooray

Contents

1 Lightning and Humans: The Early Days 1
   1.1   Possible Connection of Lightning to Early
           Human Development 1
   1.2   Ancient Myths 1
   1.3   First Attempts at a Natural Explanation of Lightning 3
   1.4   Final Breakthrough 3
   1.5   Modern Day Myths 4
   References 5

2 Basic Physics of Electrical Discharges 7
   2.1   Introduction 7
   2.2   Beginning of an Electrical Discharge – Electron
           Avalanche 7
   2.3   Electrical Discharges Consist Only of Electron Avalanches ... 10
   2.4   Streamer Discharges – Avalanche to Streamer Transition 11
   2.5   Corona Discharges Consisting of Both Avalanches
           and Streamer Discharges 16
   2.6   Extension and Charge of Streamer Discharge 17
   2.7   Leader Discharge 18
   2.8   Propagation of Leader Discharge 19
   2.9   Potential of Leader Channel 21
   2.10   Mechanism of Stepped Leader 22
   2.11   Low-Pressure Electrical Discharges 24
   2.12   A Summary of Mechanism of Lightning Flashes 24
   References 26

**3 Basic Electromagnetic Theory with Special Attention to Lightning
Electromagnetics** 29
   3.1   Electric Field Generated by a Point Charge 29
   3.2   Electric Potential of Point Charge 30
   3.3   Gauss's Law 30

	3.4	Electric Field Inside and on Surface of Perfect Conductor	32
	3.5	Electric Field of a Point Charge Over a Perfect Conductor	32
	3.6	Ampere's Law and the Magnetic Field due to a Long Conductor	35
	3.7	Magnetic Field Produced by Current Element	36
	3.8	Faraday's Law and the Voltage Induced in a Loop in the Vicinity of a Current-Carrying Conductor	36
	3.9	Force Between Two Current-Carrying Conductors	38
	3.10	Electric Fields Generated by a Tripolar Thundercloud	40
	3.11	Electric Field Change Due to Cloud Flash	42
	3.12	Electric Field Change Due to Ground Flash	43
	3.13	Electric Field Change Caused by Stepped Leader	44
	3.14	Electric Field Change Caused by Leader Return Stroke Combination	48
	3.15	Time-Varying Electromagnetic Fields	49
	3.16	Relaxation Time of a Conducting Medium	51
	3.17	Electromagnetic Fields of a Dipole	53
	3.18	Electromagnetic Fields of a Dipole Over a Perfectly Conducting Ground Plane	54
	3.19	Electromagnetic Fields of a Current Element in Time Domain Over a Perfectly Conducting Ground Plane	56
	3.20	Electromagnetic Field of a Return Stroke	56
	References		58
4	**Earth's Atmosphere and Its Electrical Characteristics**		**59**
	4.1	Variation of Atmospheric Temperature with Height	59
	4.2	Variation in Air Pressure and Air Density with Height	60
	4.3	Electrical Characteristics of Earth's Atmosphere	61
		4.3.1 Electron Density Profile of Atmosphere	61
		4.3.2 Electric Conductivity of Atmosphere	63
	4.4	Fair-Weather Electric Field in the Atmosphere	66
	4.5	Global Electrical Circuit	68
	References		70
5	**Formation of Thunderclouds**		**71**
	5.1	Cumulus Cloud	71
	5.2	Formation of a Thundercloud	71
	References		77
6	**Charge Generation in Thunderclouds and Different Forms of Lightning Flashes**		**79**
	6.1	Charge Generation in Thunderclouds	79
		6.1.1 Ice–Graupel Collision Mechanism	79
		6.1.2 Consequences of Ice–Graupel Collision Charging Mechanism	81
	6.2	Different Types of Lightning Flash	85

	6.3	Global Distribution of Lightning Flashes	86
	6.4	Density of Lightning Flashes Striking Earth	87
	References		89

7 Mechanism of Lightning Flashes 91
7.1 Initiation of Lightning Flashes Inside a Cloud 91
- 7.1.1 Conditions Necessary for Initiation of Lightning Flashes 91
- 7.1.2 Classical Explanation 92
- 7.1.3 Electron Runaway Breakdown Mechanism 93

7.2 Mechanism of Lightning Flash 94
- 7.2.1 Bidirectional Propagation of a Leader 95
- 7.2.2 Concept of Recoil Leaders 96
- 7.2.3 Mechanism of Ground Flashes 97
- 7.2.4 Mechanism of Cloud Flashes 110
- 7.2.5 Mechanism of Upward Initiated Lightning Flashes 112
- 7.2.6 Mechanism of Positive Ground Flashes and Their Main Difference with Negative Ground Flashes 112

7.3 Nomenclature of Ground Lightning Flashes 114
References 115

8 Electric Currents in Lightning Flashes 117
8.1 Measurement of Lightning Currents at Strike Point 117
- 8.1.1 Measurement of Lightning Currents Using Instrumented Towers 117
- 8.1.2 Measurements of Lightning Currents Using Triggered Lightning 119

8.2 General Features of Currents Measured in Triggered Lightning and Towers 120
- 8.2.1 General Features of Upward Initiated Lightning Flashes 121
- 8.2.2 General Features of Downward Lightning Flashes 124

8.3 Specific Features of Lightning Currents 124
- 8.3.1 Features of First and Subsequent Return-Stroke Currents 125
- 8.3.2 Features of M-Component Current 130
- 8.3.3 Features of Continuing Currents 132

References 132

9 Electromagnetic Fields of Lightning Flashes 135
9.1 Introduction 135
9.2 Measurement of Electric Fields 135
- 9.2.1 Principle of the Vertical Antenna 135
- 9.2.2 Principle of Electric Field Mill 138
- 9.2.3 Measurement of Magnetic Fields 140

	9.3	Electric Field Close to Lightning Channel	142
	9.4	Fine Structure of Radiation Fields from Stepped Leaders and Return Strokes	148
		9.4.1 Slow Front and Fast Transition	152
		9.4.2 Distribution of Peak Radiation Fields of Return Strokes	153
		9.4.3 Time Derivative of Radiation Field	153
		9.4.4 Subsidiary Peaks	154
		9.4.5 Zero Crossing Time	154
		9.4.6 Ionospheric Reflections	154
	9.5	Electromagnetic Fields from Cloud Flashes	155
		9.5.1 Radiation Fields of Cloud Flashes	157
	9.6	Compact Cloud Flashes	160
	9.7	Electromagnetic Spectrum	162
	References		164
10	**The Return Stroke – How to Model It**		**167**
	10.1	Introduction	167
	10.2	Return-Stroke Process	171
	10.3	Current-Generation-Type Return-Stroke Model	172
		10.3.1 Selection of Input Parameters in Current-Generation-Type Return-Stroke Models	178
	10.4	Current-Propagation-Type Return-Stroke Models	179
	10.5	Analytical Expression for Channel Base Current	183
	References		184
11	**Experimental Data and Theories on Return Stroke Speed**		**187**
	11.1	Introduction	187
	11.2	Experimental Observations	187
	11.3	Theories of Return Stroke Speed	189
		11.3.1 Theories of Lundholm and Wagner	189
		11.3.2 Rai's Theory	192
		11.3.3 Cooray – Model I	194
		11.3.4 Cooray – Model II	196
	References		201
12	**Propagation Effects Caused by Finitely Conducting Ground on Lightning Return Stroke Electromagnetic Fields**		**203**
	12.1	Introduction	203
	12.2	Theory	205
	12.3	Illustration of Propagation Effects	210
	References		215
13	**Localization of Lightning Flashes**		**217**
	13.1	Magnetic Direction Finding	217
	13.2	Time-of-Arrival Method	220
	13.3	Interferometric Method	224

	13.4	Thunder Ranging	227
	13.5	Observations from Satellites	227
	13.6	Other Methods	227
		13.6.1 Peak Electric Field [11]	228
		13.6.2 Time-Reversal Method [12]	229
	13.7	Lightning Localization Systems in Practice	229
	References		231
14	Potential of a Cloud and Its Relationship to Charge Distribution on Stepped Leader and Dart Leader Channels		233
	14.1	The Concept of Cloud Potential	233
	14.2	Cloud Potential as a Function of the Return Stroke Peak Current	234
	14.3	Analytical Expression for EGM Striking Distance	239
	14.4	Distribution of Negative Charge on Stepped Leader and Dart Leader Channel	239
	14.5	Distribution of Positive Charge Deposited by Return Stroke on Leader Channel	241
	14.6	Energy Dissipation in First Return Strokes	243
	14.7	Energy Dissipated by Subsequent Return Strokes	244
	References		245
15	Direct and Indirect Effects of Lightning Flashes		247
	15.1	Introduction	247
	15.2	Generation of Thunder	247
	15.3	Temperature of a Lightning Channel and How It Is Measured	249
	15.4	Thickness of a Lightning Channel	250
	15.5	Melting of Material	251
	15.6	Heating of Material as Lightning Current Passes Through an Object	252
	15.7	Attraction of Two Current-Carrying Conductors	252
	15.8	Induction of Voltages Due to Lightning-Generated Magnetic Fields	254
	15.9	Induction Due to Electric Field	255
	15.10	Energy Dissipation During a Lightning Strike	257
	15.11	Effects of Direct Lightning Strikes on Structures	258
		15.11.1 Direct Lightning Strike to a House	258
		15.11.2 Direct Strike to Trees	259
		15.11.3 Direct Strike to an Airplane	263
		15.11.4 Direct Strike to a Windmill	268
		15.11.5 Effects of Lightning on Power Lines	271
	References		279
16	Interaction of Lightning Flashes with Humans		281
	16.1	Lightning Causes Human Casualties	281

16.2		Different Ways in Which Lightning Can Interact with Humans	281
	16.2.1	Direct Strike	283
	16.2.2	Side Flash	284
	16.2.3	Touch Voltage	286
	16.2.4	Step Voltage	287
	16.2.5	Connecting Leader	292
	16.2.6	Shock Waves	292
	16.2.7	A Single Lightning Strike Can Cause Injuries to Many People	293
16.3		Different Types of Injury	293
	16.3.1	Cardiac and Respiratory Arrest	295
	16.3.2	Injuries to the Eye	296
	16.3.3	Injuries to the Ear	296
	16.3.4	Injuries to the Nervous System	296
	16.3.5	Paralysis During Lightning Strikes	297
	16.3.6	Burns Caused by Lightning	297
	16.3.7	Psychological Injuries	298
	16.3.8	Injuries Caused by Shock Waves	299
	16.3.9	Disability Caused by Lightning	299
References			300

17 Basic Principles of Lightning Protection ... 301

17.1	Function of a Lightning Conductor		301
17.2	Attractive Range of a Lightning Conductor – Electro-Geometrical Method		302
17.3	Estimating the Location of Lightning Conductors Using EGM – Rolling Sphere Method		306
17.4	Angle of Protection		308
17.5	Volume of Protection Provided by Horizontal and Vertical Conductors		311
17.6	Estimating the Number of Lightning Flashes Striking a Structure Over a Given Period of Time		312
17.7	Basic Features of External Lightning Protection System		314
17.8	Mesh Protection Method		317
17.9	Summary of External Lightning Protection System		319
17.10	Internal Lightning Protection System		320
	17.10.1	Surge Protective Devices	321
	17.10.2	Electromagnetic Zoning	324
17.11	Inadequacies of EGM		327
17.12	Attachment Models More Advanced Than EGM		328
17.13	Unconventional Lightning Protection Systems		328
	17.13.1	Early Streamer Emission Principle	328
	17.13.2	Lightning Dissipaters	329
References			330

Contents · xv

18 Advanced Procedures to Estimate the Point of Location of Lightning Flashes on a Grounded Structure ... 331
- 18.1 Introduction ... 331
- 18.2 Conditions Necessary for the Inception of a Connecting Leader According to Different Models ... 332
  - 18.2.1 Critical Radius Concept ... 332
  - 18.2.2 Rizk Criterion ... 333
  - 18.2.3 Becerra and Cooray Criterion ... 333
- 18.3 Leader Propagation Models ... 333
  - 18.3.1 Eriksson's Model ... 334
  - 18.3.2 Dellera and Garbagnati Model ... 335
  - 18.3.3 Rizk's Model ... 336
  - 18.3.4 Becerra and Cooray Model (SLIM) ... 337
- 18.4 Comparison of Different Models ... 339
- References ... 340

19 Interaction of Lightning Flashes with the Earth's Atmosphere ... 341
- 19.1 Introduction ... 341
- 19.2 Production of Nitrogen Oxides in the Atmosphere ... 341
- 19.3 Schumann Resonances ... 346
- 19.4 Global Warming and Lightning ... 348
- 19.5 Upper Atmospheric Electrical Discharges ... 349
  - 19.5.1 Sprites ... 351
  - 19.5.2 Blue Jets and Gigantic Jets ... 354
  - 19.5.3 Elves ... 355
- 19.6 Energetic Radiation Production from Lightning and Thunderstorms ... 356
- References ... 357

20 Unusual Forms of Lightning Flashes ... 359
- 20.1 Introduction ... 359
- 20.2 Forked Lightning ... 359
- 20.3 Bead Lightning ... 361
- 20.4 Ribbon Lightning ... 364
- 20.5 Lightning without Thunder ... 364
- 20.6 Ball Lightning ... 365
- References ... 369

Appendices ... 371
- Appendix A: How to Avoid Lightning Injuries ... 371
  - A.1 How to Avoid Lightning Injuries ... 371
  - A.2 Most Commonly Asked Questions ... 375
- Appendix B: Advanced Books on Lightning for Further Reading ... 379

Index ... 381

Chapter 1
Lightning and Humans: The Early Days

1.1 Possible Connection of Lightning to Early Human Development

Lightning has been striking the Earth since before humans first set foot on the planet. At any given instant approximately 2,000 thunderstorms are roaming the Earth's atmosphere and they strike the Earth with lightning flashes about 100 times per second.

A lightning flash may have caused the first chemical reactions that led to the creation of the first building blocks of life in the primordial Earth's atmosphere, which, after being processed and reprocessed by the principles of evolution, led to the transformation of the barren Earth into a living planet. Lightning flashes also played a significant role in the evolutionary process that raised humans above all other animals by introducing to them their first tool: fire.

1.2 Ancient Myths

Some of us are fascinated and impressed by lightning flashes, while others tremble in fear in the presence of lightning. Still others indulge themselves in myths and ascribe to lightning a supernatural origin. The gloomy darkness preceding a lightning flash, the eerie light, and the Earth-shattering explosion instilled so much fear in the hearts of early humans that they associated it with the wrath of god. It is no wonder that Frankenstein's monster received his first breath from a lightning flash and the witches in Shakespeare's *Macbeth* plotted their evil deeds amid lightning flashes. Indeed, lightning is the natural phenomenon most associated with myth.

Before the emergence of modern religions, each culture had a pagan god associated with lightning. For example, in Greek mythology, lightning was created by Cyclops, a giant with a single eye in the middle of his forehead, and it was the weapon of Zeus, the father of all the gods. His favorite winged horse, Pegasus, was the carrier of thunderbolts. In Roman mythology, Jupiter is feared for his thunderbolts, and it was Vulcan who forged thunderbolts for Jupiter. He presented Jupiter with two types of thunderbolt, one with a deadly accuracy and another that scattered out before it hit its object. Scandinavian mythology attributed powers of thunderbolts to Thor, who threw his hammer, *mjöllnir*, at his enemies. After delivering the thunderbolt, the hammer returned to Thor's hand like a boomerang. In Hindu mythology, Indra is the god of lightning. Some Native Americans believed that thunder and lightning were created by a mythological thunderbird. Thunder was created by the clapping of the bird's wings, and lightning flashes came out of its eyes when it blinked.

The Greeks and Romans respected lightning so much that they considered any point struck by lightning as holy and built their temples on these sites. While believing that lightning was associated with various gods, these early cultures also found ways to avoid the wrath of lightning. Emperor Tiberius wore a laurel wreath around his neck during thunderstorms to prevent lightning strikes from hitting him. Early Europeans kept a log of oak under their bed to repel lightning flashes.

It is interesting that these early civilisations, though they regarded lightning as a supernatural phenomenon, made correct observations about it. Herodotus warned humanity that god used his thunderbolts to destroy anything that rose too high above the earth. Humans, humble creatures that they were, were not worthy of creating grand structures that projected into the heavens. The Mongols were in the habit of jumping into rivers and other bodies of water to escape the wrath of lightning. Most likely to discourage this habit, Genghis Khan forbade the Mongols from taking a bath in open waters during thunderstorms.

The power of lightning is associated with God even in the early stages of modern religions. In the Hebrew Bible lightning struck Mount Sinai when God gave to Moses the Ten Commandments. Psalms 18:14 says: '*Yea,* he sent out *his* arrows, and scattered them; *and he shot out lightning, and* discomfited them.' As the concept of God changed from a tyranny to a protector following the advent of Christ, the myths changed accordingly. In medieval Christian church bells one finds the inscription: 'I am the repeller of lightning'. Satan seems to have become the bearer of lightning, and the sounds of the church bells represent a call for help from God. Even atheists of this time accepted the power of ringing bells over lightning. Of course, they believed, incorrectly, that the sound of bells broke the path of lightning channels.

1.3 First Attempts at a Natural Explanation of Lightning

While many people believed in the myths evoked to explain lightning, some attempted to explain lightning rationally. Although some of these explanations are not very different, from a scientific perspective, from the myths described earlier, they did attempt to provide a natural account of it. For example, Aristotle believed that some invisible physical substance rose up from the earth as a result of heating and movement. A lightning flash represented a burning wind caused by this substance in the sky. Socrates believed that lightning was a natural phenomenon. He rejected the Zeus hypothesis by asking: "If Zeus is the bearer of lightning, why does he strike his own temples with it?" Socrates explained lightning and thunder by asserting that a dry gust of wind rose up and got trapped inside a thundercloud. From time to time, the gust was ejected from the cloud with such force that it created thunder and lightning. Lacerates in the first century BC suggested that thunder was caused by the collision of two clouds, whereas Descartes believed that thunder was caused by the vibration of air in a huge organ pipe formed between two clouds.

1.4 Final Breakthrough

The first person to make the connection between electricity and lightning was Benjamin Franklin [1]. Franklin saw the similarities between small sparks and powerful lightning. Both sparks and lightning produce light and a cracking noise, and they both follow a zigzag path. Other scientists also saw these similarities, but they never tried to prove the connection between electric sparks and lightning. In order to pursue these questions further Franklin retired from his printing business and devoted more time to research. He made a detailed comparison between sparks and lightning. The more similarities he saw, the more he became convinced that lightning was simply a long spark. Finally, he said, "Let the experiment be made." He wrote a letter to the Royal Society of England, saying, "I would propose an experiment to be tried where it may be done conveniently. On top of some high tower or steeple place a kind of sentry box, big enough to contain a man and an electrical stand (insulating stool or table). From the middle of the stand let an iron rod rise and pass bending out of the door, and then upright 20 or 40 feet, pointed very sharp at the end. If the electrical stand is kept clean and dry, a man standing on it when such clouds are passing low might be electrified and afford sparks, the rod drawing fire to him from a cloud." This idea of Franklin is depicted in Fig. 1.1. English scientists did not appreciate the idea of Franklin at the time. However, French scientists carried out the experiment upon receiving the proposal. In the meantime, Franklin thought of a much simpler way to prove that lightning was an electric spark. In a flash of insight he realized that a simple kite could show him the connection between electricity and lightning.

It was a wet and humid day, and Benjamin Franklin saw the signs of thunderstorms brewing in the air. He brought out a kite that he had prepared

Fig. 1.1 Benjamin Franklin's sentry box experiment. Under the influence of the electric field of a thundercloud, the tall rod goes into the corona, and since it is resting on a table insulated from ground, an electric charge accumulates on the rod. A spark can be driven into the hand of a person or any other grounded object by bringing it closer to the rod (Figure created by author)

beforehand. It was made of two crossed sticks of cedar across which was drawn a silk handkerchief. To the twine at the bottom of the kite he tied a key and a small piece of silk ribbon. He went into a nearby barn and flew the kite in the direction of a thundercloud, being very careful not to wet the silk ribbon he used to hold the twine. When the kite was a few hundred feet in the air, he slowly moved his knuckles toward the key tied to the twine on the kite. To his great delight and satisfaction, a spark jumped from the key to his hand. At last, he drew an electrical fire from the cloud to his hand, and in that moment humanity made a giant leap in its endeavor to understand nature. Lightning and electricity are the same. Franklin proved that clouds store electricity, and lightning is a manifestation of that electricity. No longer did humans have to hide from the wrath of god. They acquired knowledge and, with that knowledge, ways to protect themselves from lightning flashes.

1.5 Modern Day Myths

Many modern myths related to lightning are the result of a misinterpretation of experimental observations or scientific facts. Unfortunately, some of these myths are propagated by companies peddling lightning protection equipment [2]. Lightning protection engineers must guard against these modern-day myths and to introduce only the scientifically proven lightning protection procedures into the society. Some of these myths are addressed in subsequent chapters describing lightning protection techniques.

References

1. Krider E (2010) Benjamin Franklin and lightning rods. In: Cooray V (ed) Lightning protection. IET Publishers, London, 2010
2. Cooray V (2011) Non conventional lightning protection systems. ELECTRA, No. 258, 36–41, October 2011

Chapter 2
Basic Physics of Electrical Discharges

2.1 Introduction

The main constituents of air in the Earth's atmosphere are nitrogen (78 %), oxygen (20 %), noble gases (1 %), carbon dioxide (0.97 %), water vapor (0.03 %), and other trace gases. Because of the ionization of air by the high-energy radiation of cosmic rays and radioactive gases generated from the Earth, each cubic centimeter of air at ground level contains approximately ten free electrons. In general, air is a good insulator and it can retains its insulating properties until the electric field to which it is exposed to exceeds approximately 3×10^6 V/m at standard atmospheric conditions (i.e., $T = 293$ K and $P = 1$ atm). When the electric field exceeds this critical value, air is converted very rapidly into a conducting medium, making it possible for electrical currents to flow through it in the form of sparks. Let us now consider the basic processes that make possible the conversion of air from an insulator into a conductor and the different types of discharge that take place in air under various conditions.

2.2 Beginning of an Electrical Discharge – Electron Avalanche

As mentioned previously, because of cosmic radiation, there are approximately ten free electrons at any given time in a cubic centimeter of air at ground level. These free electrons are generated continuously. As they are generated they remain free for only a very short time (on the order of tens of nanoseconds) because they become attached to oxygen molecules in the air. When an electric field is applied in air, all ions and free electrons experience a force due to the electric field. Because of their small mass, the electrons accelerate rapidly, gaining energy. As they accelerate, they collide with other atoms and molecules, and part of the energy gained is transferred to these atoms and molecules. Of course, during this process some electrons are captured by oxygen atoms. If the electric field is kept constant, then the electrons soon reach a constant speed; this speed is called the *drift speed*. For a given type of charged particle, the drift speed is a function of the electric field. The drift speed increases as the applied electric field increases. As the electric field increases, the energy gained by the electrons increases, and a stage is reached where the electrons gain so much energy that they start ejecting other electrons from atoms during collisions. For example, to eject an electron from an oxygen atom, one needs approximately 13 eV. Let us start with one electron that has enough energy to knock another electron out of an atom. During a collision with another atom this electron generates another electron. Now both electrons start accelerating in the electric field. When they gain sufficient energy from the field, they give rise to two more electrons, bringing the total number of electrons to four. As illustrated in Fig. 2.1, the continuation of this process leads to a stream of electrons moving in the opposite direction of the electric field (electrons, because of their negative electric charge, move in a direction opposite to the direction of the electric field). While this process of electron accumulation is going on, the process of electron attachment, mainly to oxygen molecules in the air, is also taking place. The process of attachment can be considered as a process that impedes the increase of free electrons in the air because it gives rise to slow moving negative ions that are incapable of ionization as electrons.

Now let us describe the foregoing process mathematically. Assume that a free electron is generated at a given point and starts moving in the background electric field, generating more electrons (Fig. 2.1). The electric field is directed along the

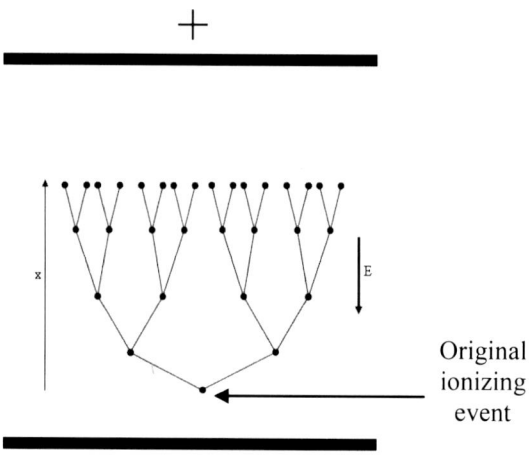

Fig. 2.1 Formation of an electron avalanche. An electron moving in a background electric field gains enough energy to release another electron from an atom during collision. This event is marked as the original ionizing event in the figure. These two electrons accelerate in the background electric field and give rise to two more electrons through collisional ionization. The process repeats itself, giving rise to a cumulative increase in the number of electrons. In the diagram, the locations of electrons are displaced laterally for clarity (Figure created by author)

2.2 Beginning of an Electrical Discharge – Electron Avalanche

negative x-axis. First the electron gains the energy necessary to ionize, and following ionization, there are two electrons. These two electrons now move in the electric field, gaining energy, and once they ionize, they generate four electrons. This shows that the electrons must move a certain distance, say λ, in the electric field before they gain enough energy to ionize. Consequently, when an electron travels a unit distance along the electric field, it gives rise to $1/\lambda$ free electrons. Let us represent this number by the symbol α. Since each new electron created takes part in the ionization process, the number of electrons n present when the first electron travels a distance x is (note that it is not equal to αx !)

$$n = e^{\alpha x}. \tag{2.1}$$

Observe that, according to this equation, the number of electrons increases exponentially with distance. This rapid increase in the number of electrons is called an *electron avalanche*. But this is not the whole story. We must consider the effect of the attachment of electrons to oxygen molecules. Let us denote by η the number of attachments that takes place per unit length traveled by an electron. Then the total number of electrons reaching distance x is given by

$$n = e^{(\alpha - \eta)x}. \tag{2.2}$$

This equation describes completely the growth of electrons in an electron avalanche. This is the first or basic element of an electrical discharge. Now, before proceeding further, we must understand the parameters that control the value of α and η. Experiments and theory show that they are both determined by the electric field (E) and the gas density (δ). Actually, they depend on the ratio of the electric field to the gas density, that is, on E/δ. The value of α increases with increasing E/δ, and η decreases with it. Now, a closer look at the equation shows that the number of electrons increases with distance only when $\alpha > \eta$. For small values of E/δ, $\alpha < \eta$, and the avalanche does not grow. However, for large values of E/δ, $\alpha > \eta$, and the avalanche grows cumulatively, increasing the number of electrons with distance. The critical value of E/δ where the transition takes place is equal to 1.04×10^{-19} Vm². For atmospheric density at sea level (2.7×10^{25} molecules/m³) the value of the electric field corresponding to it is 2.8×10^6 V/m. In practical applications it is common to assume that this critical value is 3.0×10^6 V/m. This is why the electric field in atmospheric air should increase beyond this value for an electrical breakdown to take place. This also shows that with decreasing density, a smaller electric field is needed to cause a breakdown. For example, the critical electric field, E, necessary for an electrical breakdown of air of density δ is given by

$$E = E_o \frac{\delta}{\delta_o}, \tag{2.3}$$

where δ_o is the density of air at sea level under standard atmospheric conditions and E_o is the corresponding critical electric field necessary for an electrical breakdown under the same conditions. Since the density of air in the Earth's atmosphere decreases with height, z, as $\delta = \delta_o e^{-z/\lambda_p}$, with $\lambda_p \approx 7.64 \times 10^3$, the critical electric field necessary to cause an electrical breakdown in the atmosphere decreases with height as

$$E = E_o e^{-z/\lambda_p}. \tag{2.4}$$

2.3 Electrical Discharges Consist Only of Electron Avalanches

Consider a lightning conductor or any other object in the electric field of a thundercloud. When a conducting object is placed in a background electric field, the electric field at the sharp points on the object becomes many times larger than the background electric field because of the separation of electric charge along the object. The longer the conductor or the sharper the pointed tip is, the larger the electric field will be at the tip. Because of this field enhancement of the conductor, the electric field is very high close to the tip of the conductor and decreases rapidly as one moves away from the tip (Fig. 2.2). If the background electric field is large enough, it is possible that over a certain region around the tip of the conductor the electric field will exceed the critical electric field necessary for electrical breakdown. Thus, any free electron generated in this volume gives rise to electron avalanches. As a consequence, the localized space where the electric field exceeds the breakdown value is filled with electron avalanches. In Fig. 2.2 this boundary is located at a distance of 0.1 m (marked x_c in the diagram) from the tip of the conductor. The avalanches do not extend further into the field region, where the electric field is smaller than 3×10^6 V/m, because in this region the coefficient of attachment η is higher than the coefficient of ionization α, leading to their decay.

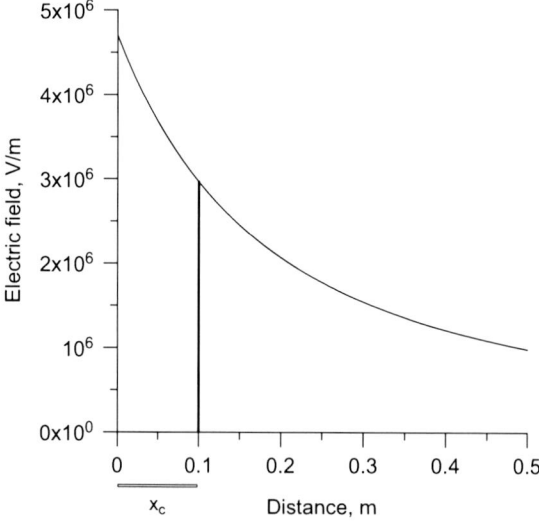

Fig. 2.2 The electric field in front of a 0.2-m-thick, 20-m-long grounded conductor immersed in a background electric field of 10^5 V/m. In the diagram, x_c is the distance from the tip of the conductor within which the electric field exceeds the breakdown electric field in air at standard atmospheric pressure (Figure created by author)

This is the most basic form of electrical discharge that can occur in nature, and it is called a *corona discharge*. Inside the discharge volume, the ionization and deionization of atoms cause light emissions, and in the dark this region can be perceived as a glowing region. When this happens below a thundercloud as a result of the thundercloud electric field, it is called *St. Elmo's fire*. It was observed at the tip of the masts of boats during thundery weather, and sailors interpreted it as a sign from the patron saint of the sailors, St. Elmo.

Now, in the corona discharge described above, the volume of the discharge is limited by the confinement of the high electric field region to a small volume. What happens if the high electric field extends over a much larger space in the air? Does the electron avalanche continue to grow with the number of electrons at its head, reaching an extremely high number? We consider this question in the next section.

2.4 Streamer Discharges – Avalanche to Streamer Transition

The analysis given in Sect. 2.2 shows that as the avalanche increases in length (i.e., increase x in Eq. 2.1), the charge accumulated at the head of the avalanche increases. As a result, the electric field produced by this charge located at the head of the avalanche also increases as the avalanche moves forward. If the background electric field continues to support the growth of the avalanche, a situation arises in which the electric field produced by the charge located at the avalanche head overwhelms the critical electric field necessary for electrical breakdown in the medium. At this stage the electric field produced by the charges located at the avalanche head, i.e., the space charge electric field, starts influencing the ionization processes taking place in the vicinity of the avalanche head. When this stage is reached, the avalanche is converted into a streamer discharge. The exact mechanism of the formation of a streamer discharge from an avalanche depends on the polarity of the source that generates the background electric field. First, consider a source at positive polarity. That is, the electric field generated by the source is such that the electrons are attracted to it. The source could be a charged hail particle in a background electric field of a thundercloud, a Franklin rod exposed to the background electric field of a thundercloud, or a high-voltage electrode. For clarity we refer to it as the *anode*.

If the electric field in front of the anode is high enough, a photoelectron (an electron ejected from an atom by a high-energy photon) generated at a point located in front of the anode initiates an avalanche that propagates toward the anode. The process is depicted in Fig. 2.3. As the electron avalanche propagates toward the anode, an immobile (stationary with respect to the fast moving electrons) positive space charge accumulates at the avalanche head. When the avalanche reaches the anode, the electrons are absorbed into it, leaving behind the net positive space charge. Because of the recombination of positive ions and electrons, the avalanche head is a strong source of high-energy photons. These photons create other avalanches in the

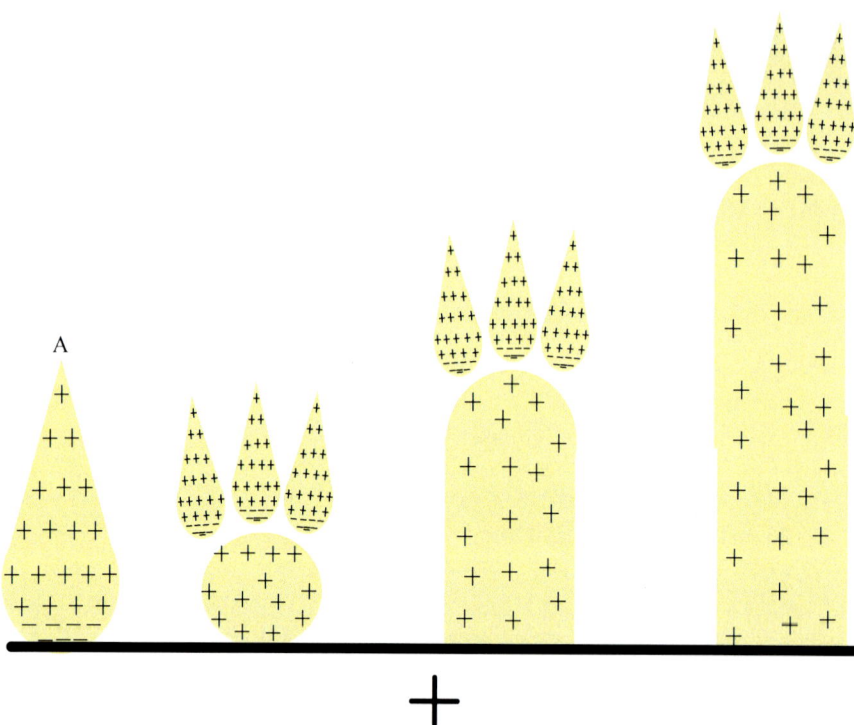

Fig. 2.3 Mechanism of positive streamers. A photoelectron generated at a point located in front of the anode (*point A*) initiates an avalanche that propagates toward the anode. When the avalanche reaches the anode, the electrons are absorbed into it, leaving behind the net positive space charge. If the number of positive ions in the avalanche head is larger than a critical value, secondary avalanches created by the photons are attracted toward the positive space charge. The positive space charge is neutralized by the electrons in the secondary avalanches, creating a weakly conducting channel. Consequently, a part of the anode potential is transferred to the channel, making it positively charged and increasing the electric field at the tip. The high electric field at the tip attracts more electron avalanches to it, and the channel grows as a consequence (Figure created by author)

vicinity of the positive space charge. If the number of positive ions in the avalanche head is larger than a critical value, then the electric field created by the space charge becomes comparable to or overwhelms the critical electric field necessary for breakdown. As a result, the secondary avalanches created by the photons are attracted to the positive space charge. The electrons in the secondary avalanches are neutralized by the positive space charge of the primary avalanche, leaving behind a new positive space charge not too far from the anode. Furthermore, the neutralization process leads to the creation of a weakly conducting channel, and part of the anode potential is transferred to this channel, making it positively charged and increasing the electric field at the tip. The high electric field at the tip of this weakly conducting channel attracts more electron avalanches to it, and the resulting neutralization process causes the weakly conducting channel to extend in a direction away from the anode. This discharge, which travels away from the anode, is called a *positive streamer*.

2.4 Streamer Discharges – Avalanche to Streamer Transition

Now let us consider a source of negative polarity, i.e., a *cathode*. A photoelectron created close to the cathode generates an avalanche (primary avalanche) that moves away from the cathode, leaving behind positive charge close to it. The process is depicted in Fig. 2.4. When the avalanche reaches a critical size, the positive charge of the avalanche starts attracting secondary avalanches to it. As in the case of a positive streamer, the electrons in the secondary avalanches neutralize this positive charge,

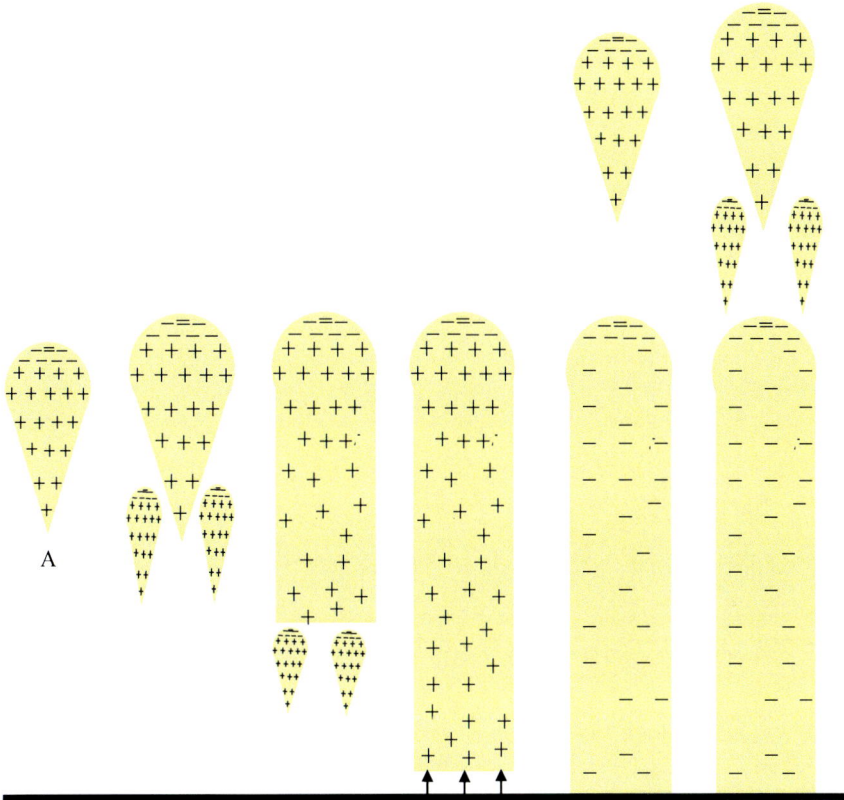

Fig. 2.4 Mechanism of negative streamers. A photoelectron generated close to the cathode (*point A*) generates an avalanche that moves away from the cathode, leaving behind a positive charge close to it. When the avalanche reaches a critical size, the positive charge of the avalanche starts attracting secondary avalanches to it. As in the case of positive streamers, the electrons in the secondary avalanches neutralize this positive charge, effectively moving it toward the cathode. When the positive charge reaches the cathode, the field enhancement associated with the proximity of positive space charge to the cathode leads to the emission of electrons from the latter. These electrons will neutralize the positive space charge, creating a weakly conducting channel that connects the negative head of the electron avalanche to the cathode. Part of the cathode potential will be transferred to the head of this weakly ionized channel (i.e., negative streamer), increasing the electric field at its head. This streamer head now acts as a virtual cathode, and the process is repeated. Repetition of this process leads to the propagation of the negative streamer away from the cathode (Figure created by author)

effectively moving it toward the cathode. When the positive charge reaches the cathode, the field enhancement associated with the proximity of positive space charge to the cathode leads to the emission of electrons from the cathode. These electrons neutralize the positive space charge, creating a weakly conducting channel that connects the negative head of the electron avalanche to the cathode. As a consequence, a part of the cathode potential is transferred to the head of this weakly ionized channel (i.e., *negative streamer*), increasing the electric field at its head. This streamer head now acts as a virtual cathode, and the process is repeated. Repetition of this process leads to the propagation of the negative streamer away from the cathode.

If the background electric field is very high, the positive space charge of the primary avalanche may reach the critical size necessary for streamer formation before reaching the anode. This may lead to the formation of a bidirectional discharge whose two ends travel toward the anode and the cathode, the former as a negative streamer and the latter as a positive streamer. Such a discharge is called a *midgap streamer* (Fig. 2.5).

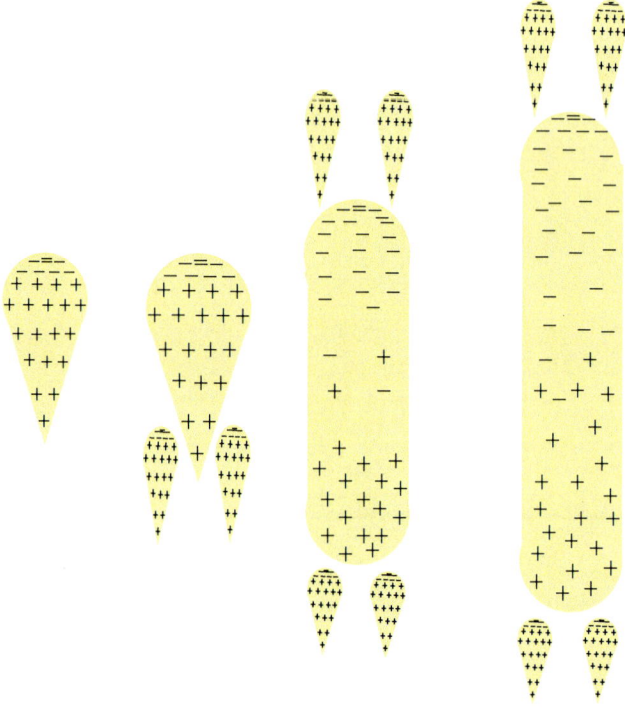

Fig. 2.5 Mechanism of midgap streamers. If the background electric field is very high, then the positive space charge of the primary avalanche may reach the critical size necessary for streamer formation before reaching the anode (compare this picture with that in Fig. 2.3). This may lead to the formation of a bidirectional discharge whose two ends travel toward the anode and the cathode, the former as a negative streamer and the latter as a positive streamer (Figure created by author)

2.4 Streamer Discharges – Avalanche to Streamer Transition

So far we have not discussed the exact conditions under which an avalanche is converted into a streamer. As was mentioned earlier, the avalanche-to-streamer transition takes place when the number of charged particles at the avalanche head exceeds a critical value, N_c. From cloud chamber photographs of avalanches and streamers, Raether [1] estimated that an avalanche converts into a streamer when the number of positive ions in the avalanche head reaches a critical value of approximately 10^8. A similar conclusion was also reached independently by Meek [2]. On the other hand, Bazelian and Raizer [3] suggest 10^9 as a reasonable value for this transformation. Thus, the condition for the transformation of an avalanche into a streamer can be written as

$$e^{\int_0^{x_c} [\alpha(x) - \eta(x)]dx} = 10^8 - 10^9. \tag{2.5}$$

Note that in writing down the preceding equation it is assumed that the electric field is not uniform, and therefore both α and η are functions of distance. Moreover, in the preceding equation, the distance x is measured from the origin of the avalanche, and x_c is the distance from the origin of the avalanche where the background electric field goes below the critical value necessary for electrical breakdown. In the electric field shown in Fig. 2.2, the region where the foregoing integration should be performed is marked x_c.

The advancement of the streamer in a given background electric field is based on the distortion of the electric field at the streamer head and the enhanced production of energetic photons from the streamer head. These energetic photons create secondary electrons in front of the streamer head, and the secondary electrons give rise to secondary avalanches that move, in the case of positive streamers, toward the streamer head. Once initiated, the streamers have been observed to travel in background electric fields that are not large enough to support avalanche formation. Secondary avalanche formation in a streamer is confined to a very small region around the streamer head where the electric field exceeds 2.8×10^6 V/m, which is the minimum electric field required for cumulative ionization in air (or the formation of avalanches) under standard atmospheric conditions. This region is called the *active region*. The dimension of the active region is approximately 200 μm, and the streamer radius was found to be on the order of 10–50 μm [4, 5]. This value, however, may correspond to short streamers. Since electron multiplication in the active region is supported by the space charge electric field of the streamer head, the streamer can propagate in electric fields that are much smaller than the critical electric field necessary for cumulative electron ionization. In air, the background electric field necessary for positive streamer propagation lies in a range of 4.5–6×10^5 V/m [6–8]. For negative streamers it lies in a range of 1–2×10^6 V/m. Any variation in the electron loss processes, such as electron attachment and gas density, can change this electric field. For example, when air is saturated with water vapor, the critical electric field necessary for positive streamer propagation grows from 4.7×10^5 V/m at a humidity of 3 g/m^3 to 5.6×10^5 V/m at a humidity of 18 g/cm^3 [8–10]. The critical electric field necessary for streamer propagation decreases approximately linearly with decreasing air density [3, 10].

In background electric fields close to the critical value necessary for streamer propagation, streamer speed is approximately 10^5 m/s. However, the streamer speed increases with increasing background electric field. Another interesting physical parameter pertinent to streamers is the potential gradient along the streamer channel. No direct measurements are currently available on the potential gradient of streamer channels. Experiments conducted with long sparks show that the average potential gradient of an electrode gap when the positive streamer bridges the gap between the two electrodes is approximately 5×10^5 V/m [11]. This indicates that the potential gradient of positive streamer channels in air at atmospheric pressure is close to this value. Note that this value is approximately the same as the critical electric field necessary for the propagation of positive streamers.

2.5 Corona Discharges Consisting of Both Avalanches and Streamer Discharges

Let us consider again a lightning rod placed in the background electric field of a thundercloud or a stepped leader (described in Chap. 7) coming down from a thundercloud. Assume that at any given instant the electric field exceeds the breakdown electric field in a certain region around the conductor. Let us assume that this high field region extends to a distance of r_0 in front of the conductor tip. Consider a free electron created at the edge of this region, that is, at a distance of r_0 from the conductor. This electron gives rise to an electron avalanche that travels toward the tip of the lightning conductor. As it travels toward the tip of the conductor, the number of electrons at the head of the avalanche increases. When the electrons reach the conductor, they are absorbed into it, leaving behind a blob of positive space charge in front of the conductor. If the background electric field is large enough or the distance to the outer boundary is large enough, this space charge gives rise to a streamer (as explained in the previous section) that starts propagating away from the tip of the conductor. Actually, when this condition is reached, many of the avalanches traveling toward the lightning conductor give rise to streamers, and there is a burst of streamers that start propagating away from the tip of the conductor. These streamers do not stop at the boundary defined by the radius r_0 because the critical electric field they need for their propagation is smaller than the electric field at radius r_0. Thus, they continue to propagate into the low field region where the electric field is less than the breakdown electric field. Actually, they propagate until they completely exhaust the available potential difference. The way to estimate the distance the streamers travel, i.e., the length of the streamers, is described in the next section.

In the situation described previously, both avalanches and streamer discharges are found in the high field region surrounding the conductor. Now, it is possible to understand what happens at the tip of a lightning conductor (or any other structure for that matter) when the background electric field continues to increase with time as happens, for example, just before a lightning strike. As the electric field continues to increase, first electron avalanches are created in the vicinity of the conductor tip. As the electric field increases further, the discharge process changes, and streamers are created. These streamers travel a greater distance than the boundary of the region of space where the electric field is confined to 3×10^6 V/m. In the literature, this process is called *streamer inception* from the tip of the lightning rod. Now, let us see how to evaluate the extension of the streamer region and the charge generated by streamers.

2.6 Extension and Charge of Streamer Discharge

Earlier it was mentioned that streamers propagate until they completely exhaust the available potential difference for propagation. The potential difference in front of a grounded lightning conductor in the presence of a uniform background electric field is shown in Fig. 2.6. The potential is zero at the tip of the conductor and increases as the point of observation moves away from the tip of the conductor. Initially it increases rapidly but then levels off as the electric field becomes smaller. Of course, it will not become constant because there is a background electric field. Now, recall that as a streamer moves in this background potential, it maintains a constant potential difference of 5×10^5 V per meter of its length. Thus, it moves to a point located at a distance, say d, such that the potential at that point is $d \times 5.0 \times 10^5$ V. In the distance–voltage diagram, the extension of the streamer can be obtained by drawing a straight line with a gradient equal to 5×10^5 V/m and locating the point where this straight line crosses the potential curve. This procedure is illustrated in

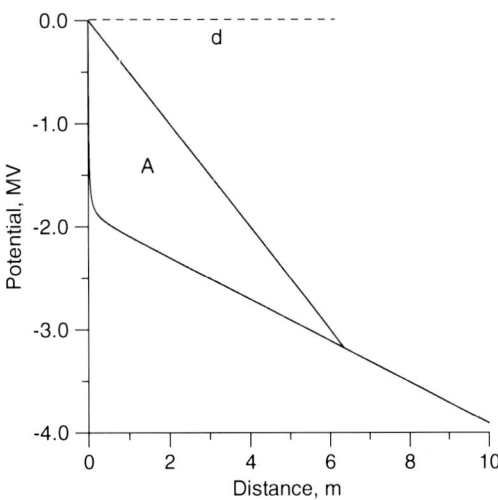

Fig. 2.6 The potential in front of a 0.02-m-thick, 10-m-long grounded conductor immersed in a background electric field of 2×10^5 V/m. In the diagram, d is the distance from the tip of the conductor up to which the streamers generated from the tip of the conductor extend. The *straight line* corresponds to a line with a potential gradient of 5×10^5 V/m (Figure created by author)

Fig. 2.6, and the distance up to which the streamers extend is marked d. If the background potential is known, then one can investigate how far the streamers will travel from the conductor using this procedure. Now, as we will see subsequently, we also wish to evaluate the amount of charge being deposited by the streamers. Analysis shows that this charge is proportional to the area between the potential curve and the straight line with a gradient of 5×10^5 V/m. This area is marked A in Fig. 2.6. Thus, the charge in the streamer region can be written as

$$Q = KA, \qquad (2.6)$$

where K is a constant that depends on the spatial variation of the electric field through which the streamers are propagating. In the case of lightning-related problems, it was shown by Becerra and Cooray [12] that a suitable value for this constant is approximately $(3-4) \times 10^{-11}$ C/Vm. Now, let us see what happens as the electric field continues to increase with time.

2.7 Leader Discharge

If we look carefully at the streamer bursts generated by conductors raised to a high potential or by conductors at zero potential but placed in a background electric field, we see that, like branches coming out from a tree stem, the streamer burst emanates from a small bright region attached to the conductor. This region is called the *stem of the streamers* (Fig. 2.7). Now, as the streamers move out from the conductor, the current associated with these streamers is forced to move through the stem of the

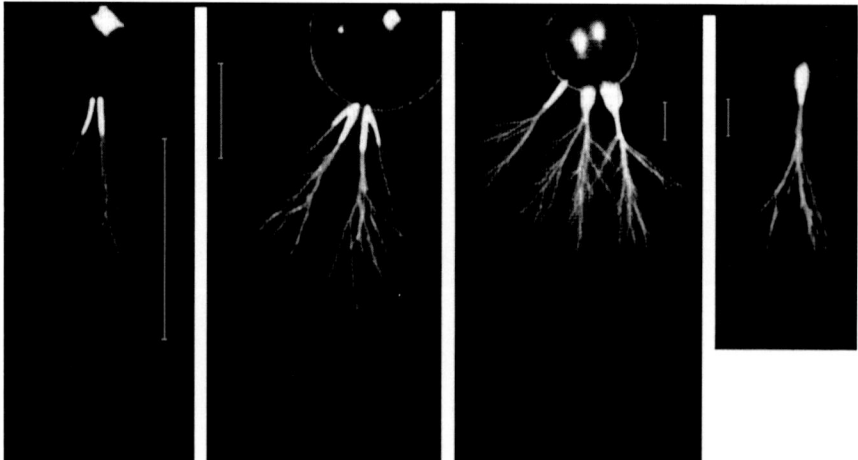

Fig. 2.7 The bright regions close to the electrode are the stem of streamer channels. Note that several streamer channels could originate from a single stem. Thus, the current generated by all these streamers goes through the stem (Figure courtesy of Division for Electricity, Uppsala University, Sweden)

streamer burst. This causes the streamer stem to heat up. Theory and experiment show that if the charge in the streamer burst is greater than approximately 1 µC, then the stem is heated to a temperature exceeding 1,600 K.

In the streamer phase of the electric discharge, many free electrons are lost because of attachment to electronegative (having an tendency to become attached to electrons) oxygen in air. Furthermore, a considerable amount of energy gained by electrons from the electric field is used in exciting molecular vibrations (especially of nitrogen molecules). Since electrons can transfer only a small fraction of their energy to atoms during elastic collisions (one can show this very easily using Newtonian mechanics and the fact that atoms are much heavier than electrons), electrons have a higher temperature than atoms. That is, the gas and the electrons are not in thermal equilibrium. As the gas temperature in the streamer stem rises to approximately 1,600–2,000 K, rapid detachment of the electrons from oxygen negative ions takes place, and this supplies the discharge with a copious amount of electrons, thereby enhancing the ionization [13]. Moreover, as the temperature rises, the time required to convert the energy stored in the molecules as vibrational energy into thermal or translational energy (i.e., kinetic energy) decreases and the vibrational energy converts back into translational energy, thereby accelerating the heating process. As the ionization process continues, the electron density in the channel continues to increase. When the electron density increases to approximately 10^{17} cm^{-3}, a new process starts in the discharge channel. This is the strong interaction of electrons with each other and with positive ions through long-range Coulomb forces [13]. This leads to a rapid transfer of the energy of electrons to positive ions, causing the electron temperature to decrease while the ion temperature increases. The positive ions, having the same mass as the neutrals, transfer their energy very quickly, in a time on the order of 10^{-8} s, to neutral atoms. This results in a rapid heating of the gas. At this stage the energy of the ions and neutral atoms becomes so large that they also start ionizing during collisions. This form of ionization is called *thermal ionization*. Once thermal ionization sets in, the electron density in the channel increases rapidly, leading to an increase in the conductivity of the streamer stem. This process is called *thermalization*. During thermalization, as the electron temperature decreases (because the electrons transfer a large fraction of their energy to ions and neutrals), the gas temperature increases, and very quickly all the components of the discharge, namely electrons, ions, and neutrals, achieve the same temperature and the discharge reaches a local thermodynamic equilibrium. The result is the conversion of the streamer stem into a hot conducting channel. This hot channel is called a *leader* or *leader discharge*.

2.8 Propagation of Leader Discharge

When a small channel section (i.e. streamer stem) attached to a grounded conductor (in the case of a lightning rod) is heated to a high temperature, the channel section becomes highly conducting, and as a consequence, its potential becomes approximately that of the grounded conductor (i.e., ground potential). This change in the potential causes an increase in the electric field at the tip of this conducting section (or the leader channel). This increase in the electric field at the tip of the leader channel gives rise to a streamer burst now emanating from a stem attached to the tip of the leader. As before, the current flowing in the streamer burst heats this new streamer stem, converting it into a hot channel, thereby extending the hot channel section or the leader. In this manner, the leader discharge, with the aid of streamer discharges, propagates away from the conductor (Fig. 2.8). As we will see in Chap. 7, a ground flash is initiated by a discharge that travels from cloud to ground. This discharge is called a *stepped leader*. The stepped leader carries a negative charge to ground, and as the leader extends toward the ground, the electric field at ground increases. When a leader is generated from a lightning conductor because of the influence of the increasing electric field caused by the stepped leader, it is called a *connecting leader*. Initially, the speed of the connecting leader is on the order of 10^4 m/s but increases as it extends toward the negative stepped leader. Theory shows that the amount of charge necessary to heat a unit length of leader channel,

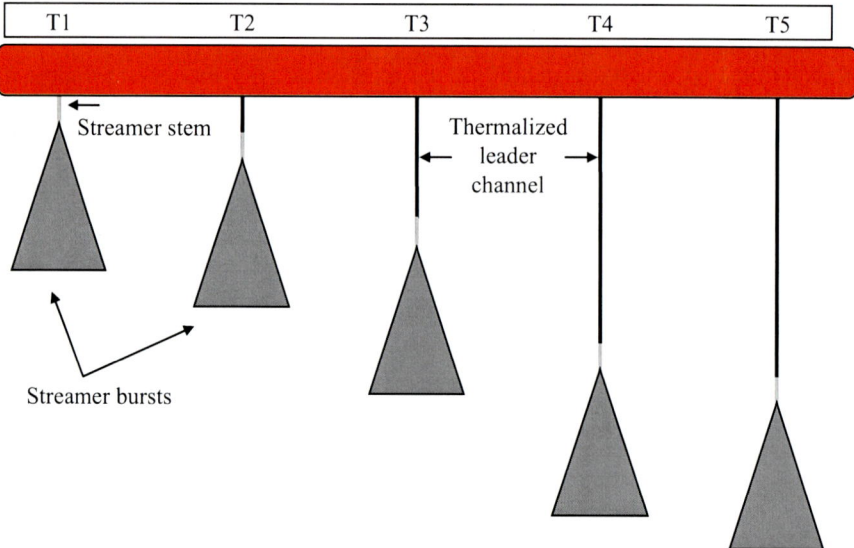

Fig. 2.8 Mechanism of positive leaders. When the electric field at the surface and in the vicinity of the anode increases to a value large enough to convert avalanches into streamers, a burst of streamers is generated from the anode (*T1*). Many of these streamers have their origin in a common channel called the *streamer stem*. The combined current of all streamers flowing through the stem causes this common region to heat up, and as a result, the stem is transformed into a hot and conducting channel called the *leader* (*T2*). Because of its high conductivity, most of the voltage of the anode is transferred to the head of the leader channel, resulting in a high electric field there. This high electric field now leads to the production of streamer discharges from a common stem located at the head of the leader channel (*T2*). With the aid of cumulative streamer currents the new stem is gradually transformed into a newly created leader section with the streamer process now repeating at the new leader head (*T3*, *T4*, *T5*) (Figure created by author)

say q_l, is approximately 60 μC. Thus a streamer burst associated with a charge Q will extend the leader by an amount equal to Q/q_l. However, theory also shows that this value depends additionally on the speed of the leader, and it increases as the speed of the connecting leader increases [12].

The preceding description is for a discharge that results in response to the electric field generated by the downward moving stepped leader carrying negative charge to ground. The electric field at the tip of the lightning conductor is such that it moves positive charges away from the tip and negative charges toward the tip. This means that in all stages of the discharge (i.e., avalanche, streamer, and leader) electrons travel toward the conductor, while positive charges are moved away from the conductor. This is defined as a *positive polarity discharge* because the charge deposited in the streamer bursts and on the leader channel is positive. Experimental data show that positive leaders travel more or less continuously. On the other hand, experimental data additionally show that negative leaders do not travel continuously but in steps. The reason for this is explained in a section to follow.

2.9 Potential of Leader Channel

The leader channel maintains a certain potential gradient along its channel, and for this reason the potential at the head of the leader is different from the potential at the point of initiation of the leader. Moreover, the potential gradient along the leader channel is not constant. It is smaller in the older sections of the leader channel and larger in newly created sections. The development of the potential gradient along the leader channel can be derived iteratively using the theory developed by Gallimberti [13]. Assume that at any time the current flowing through a section of the leader channel, the electric field in that channel section, and the radius of that channel section are known. Let us represent them by $I(t)$, $E(t)$, and $a(t)$. Then the radius of the leader channel at a time $t + \Delta t$ is given by

$$\pi \cdot a^2(t + \Delta t) = \pi \cdot a^2(t) + \frac{\gamma - 1}{\gamma \cdot p_0} E(t) \cdot I(t) \cdot \Delta t. \tag{2.7}$$

The electric field in the leader channel at that time is given by

$$E(t + \Delta t) = \frac{a^2(t)}{a^2(t + \Delta t)} E(t). \tag{2.8}$$

In the preceding equations, γ is the ratio of specific heats at constant volume and constant pressure, and p_0 is the atmospheric pressure. In writing down the preceding equations it was assumed that the current in the leader channel remained more or less the same when the time changed from t to $t + \Delta t$. Starting from an initial value for the radius of the leader channel and the electric field in the channel one can use the preceding equations to study the development of the potential gradient of the leader channel. It is common practice to assume that the initial radius of the leader channel is approximately 50–100 μm and the initial potential gradient approximately 500 kV/m. In practical applications the leader channel is divided into a large number of small sections, and the development of the potential gradient in each section is calculated using the preceding set of equations. Of course, the calculation assumes that the current flowing along the leader channel as a function of time is known.

The calculation of the leader potential gradient can be simplified if, instead of calculating the time evolution of the leader potential gradient in each segment as previously, one uses the expression derived by Rizk [14] for the potential of the tip of the leader channel, which is given by

$$U_{tip} = lE_\infty + x_o E_\infty \ln \left[\frac{E_{str}}{E_\infty} - \frac{E_{str} - E_\infty}{E_\infty} e^{-\{l/x_0\}} \right].$$

In the preceding equation, l is the total leader length, E_∞ is the final quasistationary leader gradient, and x_0 is a constant parameter. The values of parameters to be used are $E_\infty = 3 \times 10^4$ V/m, $E_{str} = 4.5 \times 10^5$ V/m (potential gradient of streamer channels), and $x_0 = 0.75$ m.

2.10 Mechanism of Stepped Leader

The development of the negative leader discharge is slightly more complicated. It also maintains its propagation with the aid of negative streamers generated from its head. As in the case of positive leaders, a negative leader also originates with a streamer burst issued from the high-voltage electrode, i.e., the cathode in this case. However, the mechanism of its propagation is different from that of positive leaders. Let us consider the events taking place immediately after the generation of a streamer burst from the head of the negative leader. The same set of events take place immediately after the generation of a streamer burst from the cathode with the hot streamer stem acting as the negative leader head during the creation of the negative leader. A simplified schematic diagram giving the main features of propagation of a negative leader is shown in Fig. 2.9. Once a negative streamer burst is generated from the leader head, a unique feature called a *pilot system*, which does not exist in positive leaders, is created. The pilot system consists of a bright spot called a *space stem*, from which streamers of both polarities develop in opposite directions. The location of the space stem is usually at the edge of the negative streamer system. The action of these streamers heats the space stem and converts it into a hot channel. This is called a *space leader*. The positive streamers from the space leader propagate toward the head of the negative leader, and the negative streamers generated from the other end of the space leader propagate in the opposite direction. Indeed, the positive streamers of the space stem propagate in the region previously covered by negative streamers. The space leader lengthens at a higher speed toward the cathode (3×10^4 m/s) than toward the anode (10^4 m/s). As the space leader approaches the main leader, the velocity of both increases exponentially. The connection of the two leaders is accompanied by a simultaneous

2.10 Mechanism of Stepped Leader

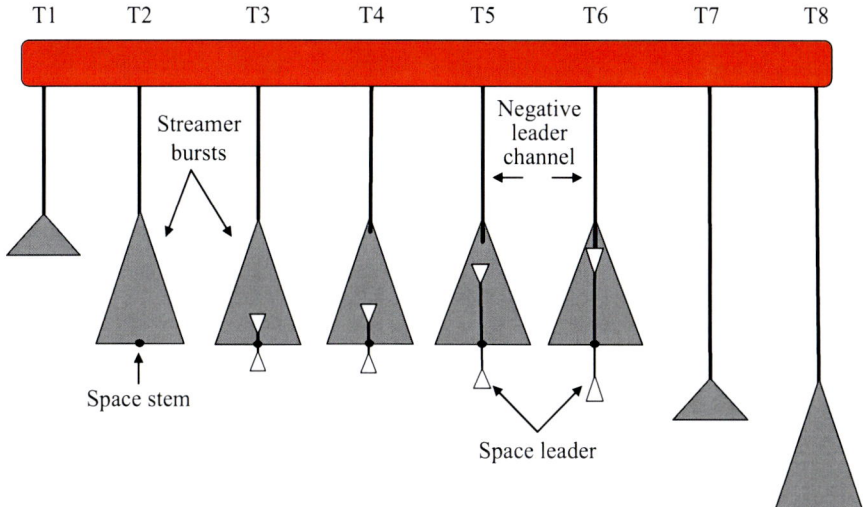

Fig. 2.9 Propagation of negative leaders. Once a negative streamer burst is generated from the leader head, a unique feature, called a *pilot system*, that does not exist in the positive leaders manifests in the system. The pilot system consists of a bright spot called a *space stem*, from which streamers of both polarities develop in opposite directions (*T2–T3*). The space stem is usually located at the edge of the negative streamer system. The action of these streamers heats the space stem and converts it into a hot channel. This is called a *space leader*. The space leader advances in both directions (the speed of extension of the positive end is generally higher than that of the negative end) through the cumulative action of positive streamers (generated from the side facing the negative leader) and negative streamers (generated from the opposite side) (*T4–T6*). The connection of the two leaders is accompanied by a simultaneous illumination of the whole channel starting from the meeting point (*T7*). During this process the space leader acquires the potential of the negative leader, and the negative end of the space leader becomes the new tip of the negative leader. During the formation of the step a new streamer burst is generated from the new leader head and the process is repeated (*T8*). Note that while the space leader travels toward the negative leader, the latter itself may continue to grow in length, as shown in the diagram. (The processes associated with the origin of the leader, which are almost identical to that of positive leaders, are not shown in the diagram.) (Figure created by author)

illumination of the channel of the space leader starting from the meeting point. During this process the space leader acquires the potential of the negative leader, and the negative end of the space leader becomes the new tip of the negative leader. In photographs, the negative leader appears to extend itself abruptly in a leader step. The change in the potential of the previous space leader generates an intense burst of negative corona streamers from its negative end, which has now become the new head of the negative leader. Now a new space stem appears at the edge of the new streamer system, and the process repeats itself. Recent evidence shows that, as in the case of laboratory sparks, repeated interaction of the negative leader with the space leader is the reason for the stepwise elongation of the negative leaders, as observed in negative stepped leaders in lightning flashes [15]. Models that describe the propagation of negative leaders taking into account the space leaders were published by Mazur et al. [16] and Arevalo and Cooray [17].

2.11 Low-Pressure Electrical Discharges

As described previously, an avalanche-to-streamer transition requires that the avalanche grow to approximately 10^8 electrons and the space charge in the avalanche tip create an electric field that is capable of attracting electron avalanches toward it. As the atmospheric pressure (or the density) decreases, the avalanche must grow to an ever greater length before it can accumulate enough space charge at its head to modify the background electric field. The reason for this is that, in comparison to an avalanche at atmospheric pressure, a larger gas volume is needed to create a given amount of charge at low pressure. Thus, with decreasing pressure, the charge density associated with a given amount of space charge decreases. This could be the case in low-pressure discharges taking place in the upper atmosphere (i.e., sprites and elves) during thunderstorms. In these discharges, the length of streamers like discharges may exceed hundreds of meters to kilometers [18]. Another interesting feature of these low-pressure discharges is the absence of the thermalization process. As mentioned earlier, the thermalization requires increasing the electron density beyond a certain limit (i.e., approximately 10^{17} cm^{-3}). In low-pressure discharges the density of molecules and atoms is such that the electron densities never reach the critical values necessary for thermalization. In these discharges, the electron temperature remains very high while the gas temperature remains close to ambient. The ionization by electron impacts is the dominant mechanism of ionization in these discharges. These discharges are also known as *Townsend discharges*.

2.12 A Summary of Mechanism of Lightning Flashes

A full description of the mechanism of lightning flashes is provided in Chap. 7. However, to introduce the reader to the nomenclature associated with the study of lightning flashes, this brief description is presented here. The following description is adapted from [19].

A thundercloud generally contains two main charge centers, one positive and the other negative, and a small positive charge pocket located at the base of the cloud. A ground flash occurs between the charge centers of the cloud and the ground, and the result is a transfer of charge from cloud to ground. When a ground flash brings positive charge down to Earth, it is called a *positive ground flash*; when it brings negative charge it is called a *negative ground flash*.

Electromagnetic field measurements show that a ground flash is initiated by an electrical breakdown process in the cloud called the *preliminary breakdown*. This process leads to the creation of a column of charge called the *stepped leader*, which travels from cloud to ground in a stepped manner. Some researchers use the term *preliminary breakdown* to refer to both the initial electrical activity inside the cloud and the subsequent stepped leader stage.

2.12 A Summary of Mechanism of Lightning Flashes

On its way to ground a stepped leader may give rise to several branches. As the stepped leader approaches ground, the electric field at ground level increases steadily. When the stepped leader reaches a height of approximately a few hundred or fewer meters from ground, the electric field at the tip of grounded structures increases to such a level that electrical discharges are initiated from them. These discharges, called *connecting leaders*, travel toward the downward moving stepped leader. One of the connecting leaders may successfully bridge the gap between the ground and the downward moving stepped leader. The object that initiated the successful connecting leader is the one that will be struck by lightning. Once the connection is made between the stepped leader and ground, a wave of near ground potential travels along the channel toward the cloud and the associated luminosity event that travels upward at a speed close to that of light is called the *return stroke*.

Whenever the upward moving return-stroke front encounters a branch, there is an immediate increase in the luminosity of the channel; such events are called *branch components*. Although the current associated with the return stroke tends to last for a few hundred microseconds, in certain instances the return-stroke current may not go to zero within this time but continue to flow at a low level for a few tens to a few hundreds of milliseconds. Such long-duration currents are called *continuing currents*.

The arrival of the first return-stroke front at the cloud end of the return-stroke channel leads to a change of potential in the vicinity of this point. This change in potential may initiate a positive discharge that travels away from the end of the return-stroke channel. Occasionally, a negative recoil streamer may be initiated at the outer extremity of this positive discharge channel and propagate along it toward the end of the return-stroke channel. Sometimes, discharges originate at a point several kilometers away from the end of the return-stroke channel and travel toward it. On some occasions these discharges may die out before they make contact with the end of the return-stroke channel. Such events are called *K changes*. If these discharges make contact with the previous return-stroke channel, the events that follow may depend on the physical state of the return-stroke channel. If the return stroke channel happens to be carrying a continuing current at the time of the encounter, it results in a discharge that travels to ground. These are called *M-components*. When M-components reach ground, no return strokes are initiated, but recent analyses of the electric fields generated by M-components show that the current wave associated with them may reflect from the ground. If the return-stroke channel happens to be in a partially conducting stage with no current flow during the encounter, it may initiate a dart leader that travels toward the ground. Sometimes the lower part of the channel has decayed to such an extent that the dart leader stops before actually reaching the ground. These are termed *attempted leaders*. In other instances, the dart leader may encounter a channel section whose ionization has decayed to such an extent that it cannot support the continuous propagation of the dart leader. In this case, the dart leader may start to propagate toward the ground as a stepped leader. Such a leader is called a *dart-stepped leader*. If these leaders travel all the way to ground, then another return stroke, called the *subsequent return stroke*, is initiated. In general, dart leaders travel along the residual channel of the first return strokes, but it is not uncommon for the dart leader to take a different path than the first stroke. In this case, it ceases to be a dart leader and travel toward the ground as a stepped leader. The point at which this leader terminates may be different from that of the original first leader. The separation between such subsequent channels was observed to be approximately a few kilometers on average.

Electrical activity similar to that which occurs after the first return strokes may also take place after subsequent return strokes. Note, however, that branch components occur mainly in the first return strokes and occasionally in the first subsequent stroke. This is the case because, in general, dart leaders do not give rise to branches. In the literature on lightning, the electrical activity in the cloud that takes place between the strokes and after the final stroke are called, collectively, *junction processes* or *J processes*.

The previously given description is based on observations of negative ground flashes. The mechanism of positive ground flashes is qualitatively similar to that of negative flashes, with differences in the details. In addition to these typical ground flashes, lightning flashes can also be initiated by tall structures. These are called *upward initiated lightning flashes*.

Cloud flashes normally occur between the main negative and upper positive charge of the cloud. The result is a partial neutralization of the charge centers. Cloud flashes do not contain return strokes, but they do contain discharge processes identical to the K changes described earlier.

References

1. Raether H (1939) Z Phys 112:464
2. Meek JM (1940) Phys Rev 57:722
3. Bazelyan EM, Raizer YP (1997) Spark discharge. CRC Press, New York
4. Marode E (1983) In: Kunhardt E, Larssen L (eds) The glow to arc transition, in electrical breakdown and discharges in gases. Plenum Press, New York
5. Marode E (1975) The Mechanism of Spark Breakdown in Air at Atmospheric Pressure between a Positive Point to Plane. J Appl Phys 46:2005–2020
6. Les Renardiéres Group (1977) Positive discharges in long air gaps at Les Renardiéres-1975 results. Electra 53:31–153
7. Les Renardiéres Group (1981) Negative discharges in long air gaps at Les Renardiéres-1978 results. Electra 74:67–216
8. Gao L, Larsson A, Cooray V, Scuka V (2000) Simulation of streamer discharges as finitely conducting channels. IEEE Trans Dielectr Electr Insul 7(3):458–460
9. Griffiths RF, Phelps CT (1976) The effects of air pressure and water vapour content on the propagation of positive corona streamers. Q J R Meteorol Soc 102:419–426
10. Giffiths RF, Phelps CT (1976) The dependence of positive corona streamer propagation on air pressure and water vapour content. J Appl Phys 47:2929
11. Paris L, Cortina R (1968) Switching and lightning impulse discharge characteristics of large air gaps and long insulation strings. IEEE Trans PAS-98:947–957
12. Becerra M, Cooray V (2006) A self-consistent upward leader propagation model. J Phys D Appl Phys 39:3708–3715

References

13. Gallimberti I (1979) The mechanism of the long spark formation. J Phys 40(C7):193–250
14. Rizk F (1989) A model for switching impulse leader inception and breakdown of long air-gaps. IEEE Trans Power Deliv 4(1):596–603
15. Biagi CJ, Uman MA, Hill JD, Jordan DM, Rakov VA, Dwyer J (2010) Observations of stepping mechanisms in a rocket-and-wire triggered lightning flash. J Geophys Res 115: D23215. doi:10.1029/2010JD014616
16. Mazur V, Ruhnke L, Bondiou-Clergerie A, Lalande P (2000) Computer simulation of a downward negative stepped leader and its interaction with a grounded structure. J Geophys Res 105(D17):22361–22369
17. Arevalo L, Cooray V (2011) Preliminary study on the modeling of negative leader discharges. J Phys D Appl Phys 44(31). doi:10.1088/0022-3727/44/31/315204
18. Winckler JR, Lyons WA, Nelson TE, Nemzek RJ (1996) New high-resolution ground-based studies of sprites. J Geophys Res 101(D3):6997–7004
19. Cooray V (2013) Mechanism of lightning flashes. In: Cooray V (ed) The lightning flash. IET Publishers, London

Chapter 3
Basic Electromagnetic Theory with Special Attention to Lightning Electromagnetics

The goal of this chapter is not to provide a reference for the theory of electricity and magnetism. The study of the physics and effects of lightning flashes entail certain elements of electricity and magnetism that are used to describe various interactions. Some of the equations of electromagnetic theory that will be used either directly or indirectly in the book are presented here. For a complete treatment of the subject the reader is referred to Refs. [1] and [2].

3.1 Electric Field Generated by a Point Charge

Consider a point charge located at point O. Let us denote by P the point of observation where we would like to study the effect of the electric charge (Fig. 3.1). The electric charge gives rise to an electric field in space, and the direction and magnitude of this electric field change from one point to another. According to Coulomb's law, the electric field (measured in volts per meter) produced at P by a point charge q (whose magnitude is measured in Coulombs) located at O is

$$E = \frac{q}{4\pi\varepsilon_0 r^2}\mathbf{a_r}, \tag{3.1}$$

where $\mathbf{a_r}$ is a unit vector directed toward OP. If the charge is positive, then this electric field is directed away from point O. In the preceding equation, the parameter ε_0 is called the *permittivity of free space*. Its value is 8.85×10^{-12} Farads/m (F/m).

Fig. 3.1 A point charge is located at point O. The electric field produced by the point charge at point P is directed toward unit vector $\mathbf{a}_r$ (Figure created by author)

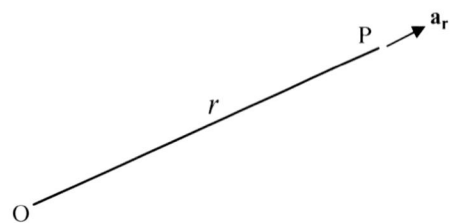

3.2 Electric Potential of Point Charge

Consider a point charge at point O. The potential of point P due to the presence of the point charge is

$$V = \frac{q}{4\pi\varepsilon_0 r}. \tag{3.2}$$

This is the amount of work that must be done (or the energy necessary) to bring a point charge of 1 C from infinity (where the potential is zero) to point P. The potential is measured in volts. Thus, the work necessary (measured in Joules) to bring a charge q_1 from infinity to point P is

$$W = \frac{qq_1}{4\pi\varepsilon_0 r}. \tag{3.3}$$

At the same time, if the charge q_1 is moved from point P to infinity, then the amount of energy released is also equal to the value given by Eq. 3.3. This shows that when we transfer a charge Q across a potential difference of V, the energy released (or the work that needs to be done, depending on the direction of movement of the charge) is equal to QV.

3.3 Gauss's Law

According to Gauss's law, the flux of an electric field coming from a closed volume is related to the total charge located inside that closed volume. Let **E** be the electric field at any point on a closed surface and **ds** a small area vector (the direction of the vector is the outward normal to the surface; Fig. 3.2). Then the flux of the electric field coming from the closed volume is given by $\oint_s \mathbf{E}.\mathbf{ds}$. The 's' sign on the integral shows that it is a surface integral, and the circle shows that it is performed around the closed surface. Gauss's law states that this is equal to

$$\oint_s \mathbf{E}.\mathbf{ds} = \frac{Q}{\varepsilon_0}, \tag{3.4}$$

3.3 Gauss's Law

where Q is the total charge located inside the closed volume. Now let us apply this equation to calculate the electric field in the vicinity of a long charged line or a channel. Let us assume that the charge per unit length of the channel is ρ. Since the channel is very long, the electric field lines by symmetry are radial to the channel. Now consider a closed surface in the form of a cylinder having radius r (Fig. 3.3). Applying Gauss's law to this cylinder one can write directly that

$$2\pi r l E = \frac{\rho l}{\varepsilon_0}. \tag{3.5}$$

Note that since the field lines are radial, the flux coming from the edges of the cylinder is zero (at the edges, $\mathbf{E} \cdot \mathbf{ds} = 0$). Thus, the radius r at which the electric field is equal to E is given by

$$r = \frac{\rho}{2\pi\varepsilon_0 E}. \tag{3.6}$$

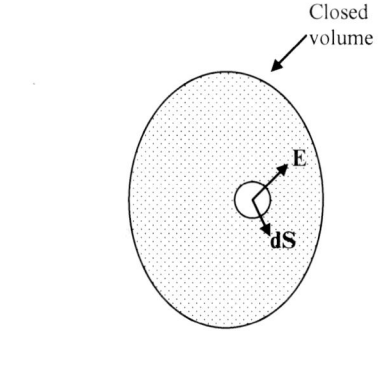

Fig. 3.2 Geometry pertinent to definition of Gauss's law. The *white patch* is a small surface area on the closed surface. The outward normal to the surface area is **dS**, and **E** is the electric field vector at the location of the elementary area (Figure created by author)

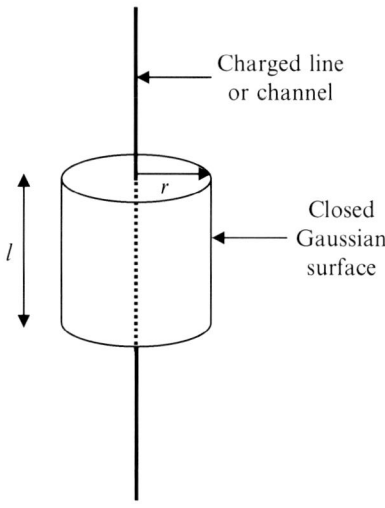

Fig. 3.3 Geometry pertinent to application of Gauss's law to a long charged line or channel. The closed volume is composed of the cylindrical surface together with the two circular surfaces covering the two ends. Since the electric field is radial, all the flux of the electric field passes through the cylindrical surface (Figure created by author)

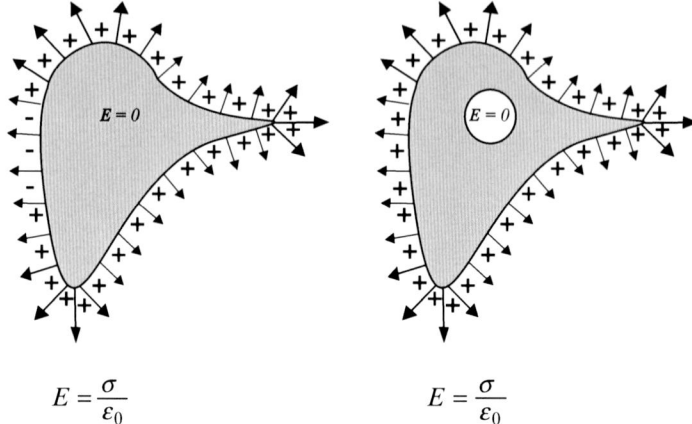

Fig. 3.4 The electric field on the surface of a perfect conductor is always perpendicular to the conductor. The local electric field is proportional to the local surface charge density, σ, and is given by σ/ε_0. The local charge density is proportional to the curvature of the surface. The surface charge and, hence, the electric field is large at sharp points on the conductor. The electric charge resides always on the surface of the conductor, and the electric field is zero inside a closed conductor. For the same reason the electric field inside a cavity of the conductor is also zero. This is the origin of the Faraday cage principle (Figure created by author)

3.4 Electric Field Inside and on Surface of Perfect Conductor

On the surface of a perfect conductor the electric field is always perpendicular to the surface. The component of the electric field parallel to the surface is zero (Sect. 3.16). This can be understood as follows. If there is an electric field parallel to the surface of a good conductor, free charges in the conductor will displace along the conductor and create an electric field opposite to that of the applied electric field. The displacement of the charge continues until the two electric fields cancel each other, making the electric field parallel to the surface zero. For the same reason the electric field inside a perfect conductor is also zero. That is, the potential at any point of the conductor is the same. This also explains why the electric charge of a charged conductor resides on the surface of the conductor. These facts can be used together with Gauss's law to show that the electric field inside the cavity of a perfect conductor is zero irrespective of the amount of charge that resides on the surface. This is the origin of the Faraday cage principle of lightning protection. Some of these points are illustrated in Fig. 3.4.

3.5 Electric Field of a Point Charge Over a Perfect Conductor

Now consider a point charge located over a perfectly conducting plane of zero potential. In problems dealing with electrostatics or slowly changing fields, the Earth's surface can be treated as a perfect conductor. For all practical purposes we can also

3.5 Electric Field of a Point Charge Over a Perfect Conductor

treat the potential of the ground as zero. Now, the presence of charge q (assumed to be positive) repels any positive charge from points directly below the ground while attracting negative charge into that region. This gives rise to a charge distribution (induced) on the surface of the ground (recall that charges reside on the surface of a conductor). The electric field at any point above the ground is the sum of the electric field produced by the electric charge q and this induced charge. Moreover, as discussed earlier, the electric field at the ground plane is perpendicular to it because it is a good conductor. The field configuration of a point charge over a perfectly conducting ground plane is shown in Fig. 3.5a. Note that without affecting the field distribution, one can replace the conducting plane by placing a negative charge of the same magnitude at the mirror point (Fig. 3.5b). That is, the effect of the induced negative charge on the ground can be represented by a negative charge, equal in magnitude to the positive charge, located at the mirror point of the charge. Now consider a point charge q located at height h above a perfectly conducting ground plane of potential zero (see Fig. 3.5c). The electric field at point P located above the ground plane consists of two components, one from the real charge (positive) and the other from the image charge (negative). The component from the real charge is

$$E_r = \frac{q}{4\pi\varepsilon_0 r_r^2} \mathbf{a_{rr}}, \tag{3.7}$$

where $\mathbf{a_{rr}}$ is a unit vector directed toward r_r (see Fig. 3.5c). Separating the electric field components into vertical and horizontal directions we obtain

$$E_{rx} = \frac{q}{4\pi\varepsilon_0 r_r^2} \sin\theta_r, \tag{3.8}$$

$$E_{rz} = -\frac{q}{4\pi\varepsilon_0 r_r^2} \cos\theta_r. \tag{3.9}$$

Similarly, the electric field due to the image charge is

$$E_i = \frac{q}{4\pi\varepsilon_0 r_i^2} \mathbf{a_{ri}}. \tag{3.10}$$

Separating this into z and x components we obtain

$$E_{ix} = -\frac{q}{4\pi\varepsilon_0 r_i^2} \sin\theta_i, \tag{3.11}$$

$$E_{iz} = -\frac{q}{4\pi\varepsilon_0 r_i^2} \cos\theta_i. \tag{3.12}$$

The total x and z components are the sum of these components. These are given by

$$E_x = \frac{q}{4\pi\varepsilon_0 r_r^2} \sin\theta_r - \frac{q}{4\pi\varepsilon_0 r_i^2} \sin\theta_i, \tag{3.13}$$

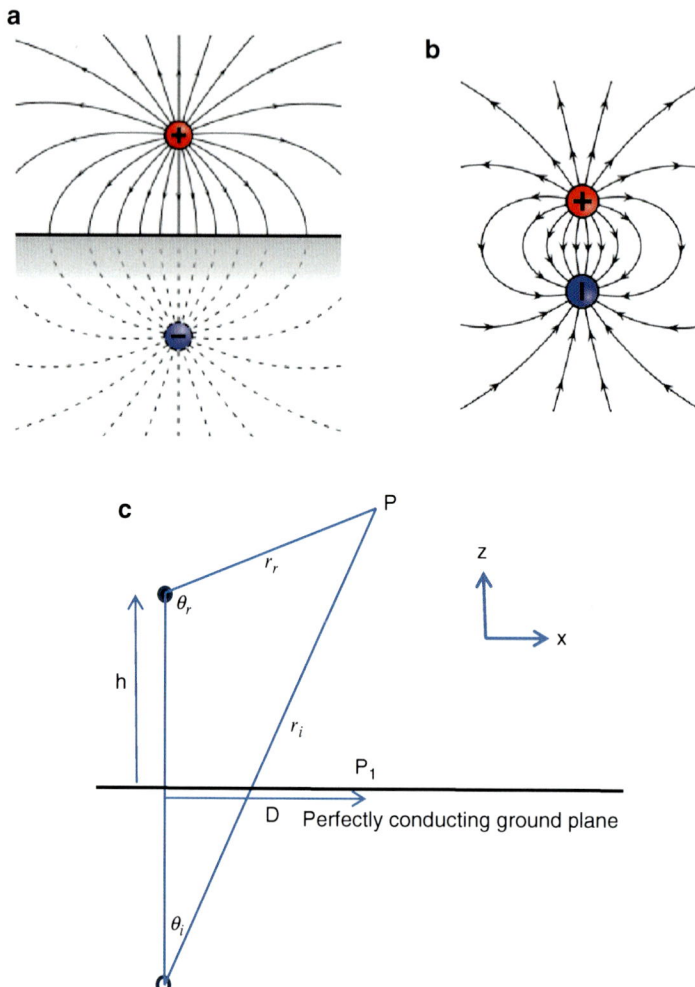

Fig. 3.5 (a) Field configuration of a point charge located over a perfectly conducting ground plane. (b) One can replace the perfect conductor with a charge of opposite polarity located at the image point without changing the field configuration above the conducting plane. (c) Geometry relevant to calculation of electric field from point charge located at height h over perfectly conducting ground plane. The perfectly conducting plane is replaced by a charge of equal magnitude but of opposite polarity located at the mirror image point (Panels **a** and **b** from Wikipedia (http://en.wikipedia.org/wiki/Electric_charge), Panel **c** created by author)

$$E_z = -\frac{q}{4\pi\varepsilon_0 r_r^2}\cos\theta_r - \frac{q}{4\pi\varepsilon_0 r_i^2}\cos\theta_i. \qquad (3.14)$$

One can see directly at ground level, where $r_r = r_i$ and $\theta_r = \theta_i$, the x component, i.e., the component parallel to the surface, goes to zero. The vertical electric field on the surface of the ground at a horizontal distance D from the point charge is (see Fig. 3.5c for the geometry)

$$E_{z,\text{ground}} = -\frac{qh}{2\pi\varepsilon_0 (D^2 + h^2)^{3/2}}. \tag{3.15}$$

3.6 Ampere's Law and the Magnetic Field due to a Long Conductor

Ampere's law relates the line integral of the B-field or the H-field around a closed path to the electric current I passing through the closed path (Fig. 3.6):

$$\oint_l \mathbf{B} \bullet \mathbf{dl} = \mu_0 I. \tag{3.16}$$

In the above equation the letter '*l*' on the integral sign denotes that it is a line integral and the circle indicates that it is performed around a closed path. Applying this to a long current-carrying conductor and using the facts that the magnetic field forms closed loops around the conductor and the magnetic field has the same magnitude along any of these loops, one obtains the B-field at a radial distance r from the conductor as

$$2\pi r B(t) = \mu_0 I(t), \tag{3.17}$$

$$B(t) = \frac{\mu_0 I(t)}{2\pi r}. \tag{3.18}$$

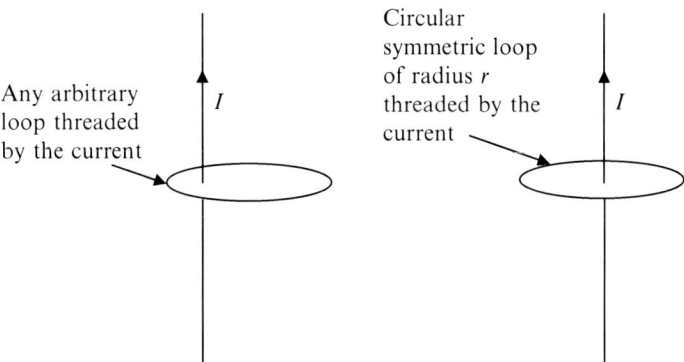

Fig. 3.6 Geometry relevant to definition of Ampere's law, which says that the line integral of the B-field around the loop is equal to $\mu_0 I$, where I is the current passing through the loop. The law is valid irrespective of the orientation of the loop or the point of intersection of the loop and the wire (Figure created by author)

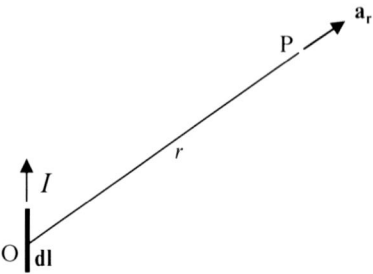

Fig. 3.7 Geometry relevant to definition of magnetic field produced by a current element according to Biot-Savart's law. In the diagram, **dl** is an elementary current element (or a small piece of the conductor carrying a current). The direction of the vector **dl** is the same as the direction of the positive current flow (Figure created by author)

3.7 Magnetic Field Produced by Current Element

Consider a current element dl through which a current I is flowing (Fig. 3.7). The current element can be represented by a vector **dl** with magnitude dl and direction specified by the direction of current flow. According to Biot-Savart's law, the B-field produced by this current element at point P is given by

$$d\mathbf{B} = \mu_o I \cdot \frac{\mathbf{dl} \times \mathbf{a_r}}{4\pi |R|^2}, \qquad (3.19)$$

where $\mathbf{a_r}$ is a unit vector in the direction of OP (Fig. 3.7). Note that the direction of the magnetic field is perpendicular to both the current element **dl** and the vector joining the current element and the point of observation (i.e., $\mathbf{a_r}$). The magnitude of the magnetic field is proportional to the current in the current element, and its strength decreases with $1/R^2$.

The magnetic field produced by a conductor of any shape can be calculated by dividing the conductor into elementary sections and summing up the contribution to the B-field from each element using the preceding equation.

3.8 Faraday's Law and the Voltage Induced in a Loop in the Vicinity of a Current-Carrying Conductor

The essence of Faraday's law is that it defines and quantifies the natural law that a changing magnetic field gives rise to an electric field. Consider a closed path in a region where there is a changing magnetic field. According to Faraday's law, the E-field, **E**, generated by this changing magnetic field is such that

$$\oint_l \mathbf{E} \bullet \mathbf{dl} = -\frac{d\psi}{dt}, \qquad (3.20)$$

3.8 Faraday's Law and the Voltage Induced in a Loop in the Vicinity...

where the left-hand side is the line integral of the electric field taken along the closed path and the right-hand side is equal to the negative rate of change of magnetic flux ψ passing through the closed path. The magnetic flux passing through the closed path can be calculated as

$$\psi = \int_S \mathbf{B} \bullet \mathbf{ds}, \qquad (3.21)$$

where the surface integral is carried out over a surface bounded by the closed path (Fig. 3.8a). The positive direction of **ds** can be decided as follows. Place a right handed screw inside the closed path and rotate it in a circular direction in which the line integral is performed. If the screw moves out of the surface then the positive direction of **ds** is the outward normal to the surface. If the direction of motion of the screw is into the surface the positive direction of **ds** is the inward normal to the surface. Since the electromotive force, *emf*, generated around the closed path under consideration is given by the line integral of the E-field along that path, we can write

$$emf = -\frac{d\psi}{dt}. \qquad (3.22)$$

Let us consider a square loop located in the vicinity of a current-carrying conductor, as shown in Fig. 3.8b. The magnetic field produced by this conductor at a radial distance r from the conductor is given by

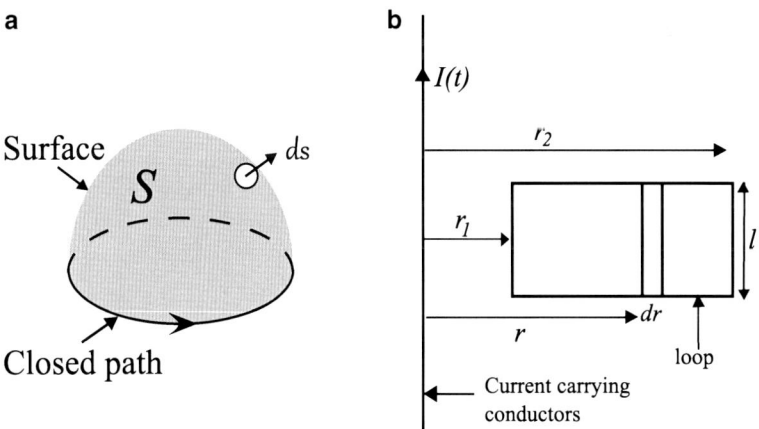

Fig. 3.8 (**a**) The definition of a surface bounded by a closed path. The white patch is a small area element on the surface. The magnitude of the vector **ds** is equal to the area of the element and the direction of the vector is decided by the right hand screw rule (see the text). (**b**) Geometry relevant to application of Faraday's law in case of a conducting loop located in vicinity of current-carrying conductor (Figure created by author)

$$B(t) = \frac{\mu_0 I(t)}{2\pi r}. \tag{3.23}$$

Now, let us divide the square loop into small elementary rectangular sections, as shown in Fig. 3.8b. The flux through the small rectangular element located at distance r is

$$d\psi = \frac{I(t)\mu_0}{2\pi r} l\, dr. \tag{3.24}$$

The total flux is obtained by integrating this result from distance r_1 to r_2. The result is

$$\psi = \int_{r_1}^{r_2} \frac{I(t)\mu_0 l}{2\pi} \frac{dr}{r}. \tag{3.25}$$

This reduces to

$$\psi = \frac{I(t)\mu_0 l}{2\pi} \ln \frac{r_2}{r_1}. \tag{3.26}$$

Thus, the rate of change of magnetic flux through the loop is

$$\frac{d\psi}{dt} = \frac{dI(t)}{dt} \frac{\mu_0 l}{2\pi} \ln \frac{r_2}{r_1}. \tag{3.27}$$

The induced *emf* in the loop according to Faraday's law is

$$emf = -\frac{dI(t)}{dt} \frac{\mu_0 l}{2\pi} \ln \frac{r_2}{r_1}. \tag{3.28}$$

Note that the *emf* is proportional to the rate of change of the current passing through the conductor.

3.9 Force Between Two Current-Carrying Conductors

Consider a conductor carrying a current I located in a magnetic field **B** (Fig. 3.9a). Consider a small element **dl** on this conductor. The element is represented by a vector, with a magnitude dl and the positive direction the same as that of the direction of positive current flow. The force acting on the element **dl** due to the magnetic field is given by

3.9 Force Between Two Current-Carrying Conductors

$$d\mathbf{F} = I d\mathbf{l} \times \mathbf{B}. \qquad (3.29)$$

This can be expanded as

$$dF = B I dl \sin\theta. \qquad (3.30)$$

Now consider a situation where two parallel current-carrying conductors are separated by a distance a (Fig. 3.9b). We represent the current in the two conductors as $I_1(t)$ and $I_2(t)$. Assuming that the radii of the conductors are much smaller than the separation a, the magnetic field generated by conductor 1 at the location of conductor 2 is

$$B(t) = \frac{\mu_0 I_1(t)}{2\pi a}. \qquad (3.31)$$

Since this magnetic field is perpendicular to conductor 2 (i.e. $\sin\theta = 1$ in equation 3.30), the force per unit length (in N/m) on conductor 2 due to the current in conductor 1 is (using Eq. 3.30)

$$F(t) = \left(\frac{\mu_0 I_1(t)}{2\pi a}\right) I_2(t). \qquad (3.32)$$

The same force also acts on conductor 1. If the two currents are equal [say $I(t)$], then the force between the two conductors is

$$F(t) = \left(\frac{\mu_0}{2\pi a}\right) I^2(t). \qquad (3.33)$$

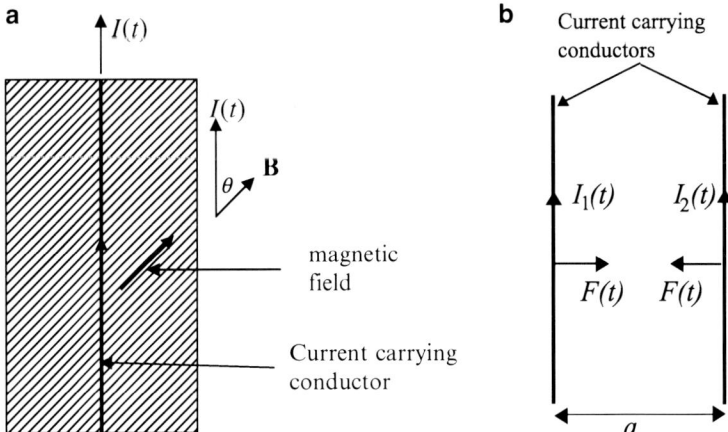

Fig. 3.9 (a) Current-carrying conductor located in magnetic field. (b) Two parallel current-carrying conductors separated by a distance a (Figure created by author)

If a transient current flows along the conductors, the change in momentum of the conductors (i.e., the relative movement due to the force) is given by the impulse, S, due to the force. This is given by

$$S = \left(\frac{\mu_0}{2\pi a}\right) \int_0^\infty I^2(t) dt. \quad (3.34)$$

In lightning research, the parameter $\int_0^\infty I^2(t) dt$ is called the *action integral*.

3.10 Electric Fields Generated by a Tripolar Thundercloud

Consider a tripolar cloud with three charge centers. The heights of the charge centers are h_1, h_2, and h_3. The amount of charge in the respective charge centers is q, Q, and $-Q$, respectively. In the literature, the charge center closest to the ground and located close to the base of the cloud is called the *positive charge pocket*. The charge center in the middle, which carries a negative charge, is called the *negative charge center*. The charge center at the top is called the *positive charge center*. The geometry is depicted in Fig. 3.10. The z-axis is directed out of the conducting plane (or the ground plane), as shown in the figure. In presenting the results of the calculations we treat the electric field directed along the positive z-direction (i.e., a vector directed away from the ground surface) as positive. This is called the *physics sign convention*. The opposite sign convention, where the electric field directed into the ground is assumed to be positive, is called the *atmospheric sign convention*. The vertical electric fields at ground level generated by individual

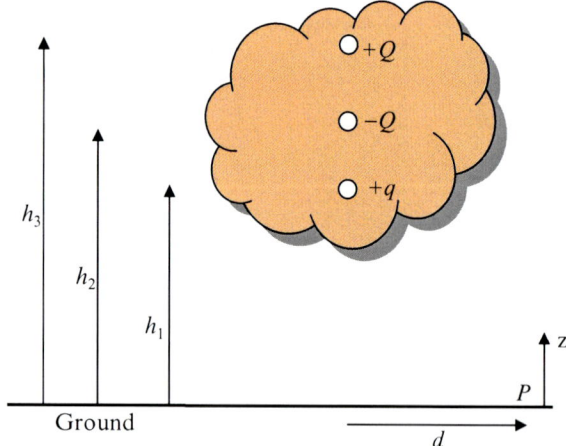

Fig. 3.10 Idealized tripolar cloud with three charge centers located over ground. The amount of charge in the respective charge centers are denoted by q, Q, and $-Q$. In the calculations the positive direction of the electric field is in the z-direction (Figure created by author)

3.10 Electric Fields Generated by a Tripolar Thundercloud

charges at a point of observation located at a horizontal distance d from the axis of the charge centers are given by

$$E_1 = -\frac{q}{2\pi\varepsilon_0} \frac{h_1}{\left(h_1^2 + d^2\right)^{3/2}}, \qquad (3.35)$$

$$E_2 = \frac{Q}{2\pi\varepsilon_0} \frac{h_2}{\left(h_2^2 + d^2\right)^{3/2}}, \qquad (3.36)$$

$$E_3 = -\frac{Q}{2\pi\varepsilon_0} \frac{h_3}{\left(h_3^2 + d^2\right)^{3/2}}. \qquad (3.37)$$

The total vertical electric field E at distance d is then given by

$$E = E_1 + E_2 + E_3. \qquad (3.38)$$

Figure 3.11 depicts the electric field at a point of observation as a function of the distance to the thundercloud. In the calculations, we assume $Q = 100$ C, $h_1 = 4$ km, $h_2 = 6$ km, and $h_3 = 9$ km. Results are shown for three values of q equal to 0, 5, and 10 C to show the effect of the positive charge pocket. This diagram also shows how the electric field at ground level varies as the thundercloud approaches a point

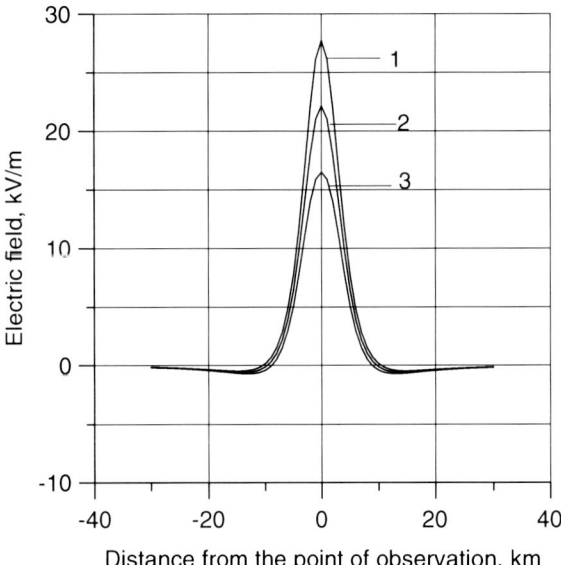

Fig. 3.11 Electric field at point P (Fig. 3.10) located at surface of ground as function of distance d to tripolar thundercloud (distance to vertical axis where charges are located). In the calculation, $Q = 100$ C. Results are shown for three values of q: (1) 0, (2) 5 C, and (3) 10 C. In the calculation, $h_1 = 4$ km, $h_2 = 6$ km, and $h_3 = 9$ km. In the presentation the electric field produced by a negative charge in the cloud is assumed to be positive. If the charge q is significantly larger than the preceding values, the electric field at 0 km could also become negative (Figure created by author)

of observation, passes over it, and then recedes from the point of observation. Note also how the polarity of the electric field changes as a thundercloud approaches the point of observation.

3.11 Electric Field Change Due to Cloud Flash

Let us consider a tripolar cloud. The heights of the charge centers are h_1, h_2, and h_3. The amount of charge in the respective charge centers before a cloud lightning flash is q, Q, and $-Q$, respectively. The cloud lightning flash leads to the neutralization of the ΔQ charge from the negative and positive charge centers (Fig. 3.12). Let $Q' = Q - \Delta Q$. The electric field at ground level at a horizontal distance d from the cloud just before the lightning flash is

$$E_i = -\frac{q}{2\pi\varepsilon_0} \frac{h_1}{\left(h_1^2 + d^2\right)^{3/2}} + \frac{Q}{2\pi\varepsilon_0} \frac{h_2}{\left(h_2^2 + d^2\right)^{3/2}} - \frac{Q}{2\pi\varepsilon_0} \frac{h_3}{\left(h_3^2 + d^2\right)^{3/2}}. \qquad (3.39)$$

After the lightning flash the field reduces to

$$E_f = -\frac{q}{2\pi\varepsilon_0} \frac{h_1}{\left(h_1^2 + d^2\right)^{3/2}} + \frac{Q'}{2\pi\varepsilon_0} \frac{h_2}{\left(h_2^2 + d^2\right)^{3/2}} - \frac{Q'}{2\pi\varepsilon_0} \frac{h_3}{\left(h_3^2 + d^2\right)^{3/2}}. \qquad (3.40)$$

The field change ΔE caused by the cloud lightning flash is

$$\Delta E = E_f - E_i. \qquad (3.41)$$

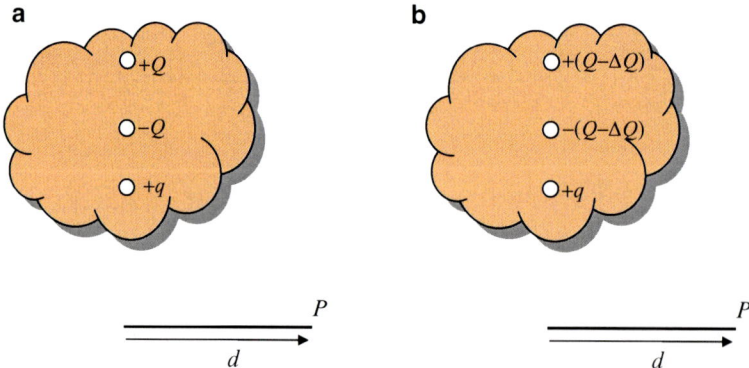

Fig. 3.12 Charges in a thundercloud (**a**) before and (**b**) after a cloud lightning flash. Note that the cloud flash neutralizes a charge of magnitude ΔQ. The parameters used to define the heights of the charge centers are given in Fig. 3.10 (Figure created by author)

3.12 Electric Field Change Due to Ground Flash

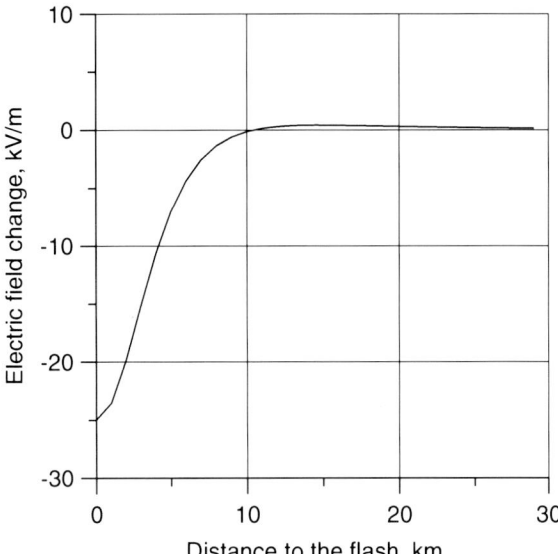

Fig. 3.13 Electric field change at ground level produced by cloud flash depicted as function of distance d to thundercloud (Fig. 3.12). In the calculation, $Q = 100$ C and the charge neutralized by the cloud flash $\Delta Q = 10$ C; $h_2 = 6$ km and $h_3 = 9$ km (Figure created by author)

After substitution from the preceding expressions we obtain

$$\Delta E = -\frac{(Q-Q')}{2\pi\varepsilon_0}\frac{h_2}{\left(h_2^2+d^2\right)^{3/2}} + \frac{(Q-Q')}{2\pi\varepsilon_0}\frac{h_3}{\left(h_3^2+d^2\right)^{3/2}}. \qquad (3.42)$$

Figure 3.13 depicts this field change as a function of the horizontal distance from the thundercloud. In the calculation, it is assumed that $\Delta Q = 10$ C, $h_1 = 4$ km, $h_2 = 6$ km, and $h_3 = 9$ km. Note the change in polarity of the field changes with distance.

3.12 Electric Field Change Due to Ground Flash

Let us consider a tripolar cloud. As before, the heights of the charge centers are h_1, h_2, and h_3. The amount of charge in the respective charge centers before a ground lightning flash is taken to be q, Q, and $-Q$, respectively. In the calculation, we assume that the ground lightning flash leads to the complete neutralization of the positive charge pocket and a transfer of $-\Delta q$ to ground from the negative charge center (Fig. 3.14). The total charge, $-\Delta Q$, removed from the negative charge center is equal to $-\Delta q - q$. Let $Q' = Q - \Delta Q$. Thus, the charge in the negative charge center after the ground flash is equal to $-Q'$. The electric field at ground level at a horizontal distance d from the cloud just before the lightning flash is

$$E_i = -\frac{q}{2\pi\varepsilon_0}\frac{h_1}{\left(h_1^2+d^2\right)^{3/2}} + \frac{Q}{2\pi\varepsilon_0}\frac{h_2}{\left(h_2^2+d^2\right)^{3/2}} - \frac{Q}{2\pi\varepsilon_0}\frac{h_3}{\left(h_3^2+d^2\right)^{3/2}}. \qquad (3.43)$$

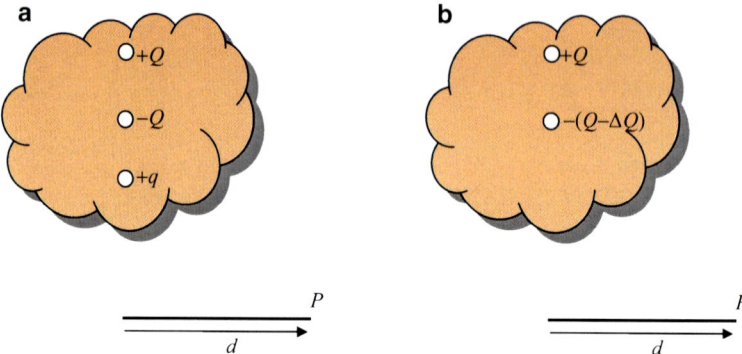

Fig. 3.14 Charges in thundercloud (**a**) before and (**b**) after ground lightning flash. Note that a charge of $-\Delta Q$ is removed from the negative charge center and part of this negative charge is used to neutralize the charge q of positive charge pocket. The rest of the charge $(-\Delta Q + q)$ is transferred to the ground. The parameters used to define the heights of the charge centers are given in Fig. 3.10 (Figure created by author)

After the lightning flash the electric field at the same point is

$$E_f = +\frac{Q'}{2\pi\varepsilon_0} \frac{h_2}{\left(h_2^2 + d^2\right)^{3/2}} - \frac{Q}{2\pi\varepsilon_0} \frac{h_3}{\left(h_3^2 + d^2\right)^{3/2}}. \qquad (3.44)$$

Thus, the field change at the point of observation caused by the ground flash is

$$\Delta E = E_f - E_i. \qquad (3.45)$$

After substitution of the respective parameters we obtain

$$\Delta E = \frac{q}{2\pi\varepsilon_0} \frac{h_1}{\left(h_1^2 + d^2\right)^{3/2}} - \frac{(Q - Q')}{2\pi\varepsilon_0} \frac{h_2}{\left(h_2^2 + d^2\right)^{3/2}}. \qquad (3.46)$$

Figure 3.15 depicts the electric field change generated by a ground flash. In the calculation, it is assumed that $\Delta Q = 10$ C, $h_1 = 4$ km, $h_2 = 6$, and $h_3 = 9$ km. Note that, unlike the field change due to a cloud flash, the polarity of the field change remains the same with distance.

3.13 Electric Field Change Caused by Stepped Leader

For simplicity let us consider a bipolar cloud with negative and positive charge centers at heights h_2 and h_3, with charges $-Q$ and $+Q$, respectively. At time $t = 0$ a stepped leader starts moving down from the negative charge center. We do not specify here how the negative charges are removed from the charge center and

3.13 Electric Field Change Caused by Stepped Leader

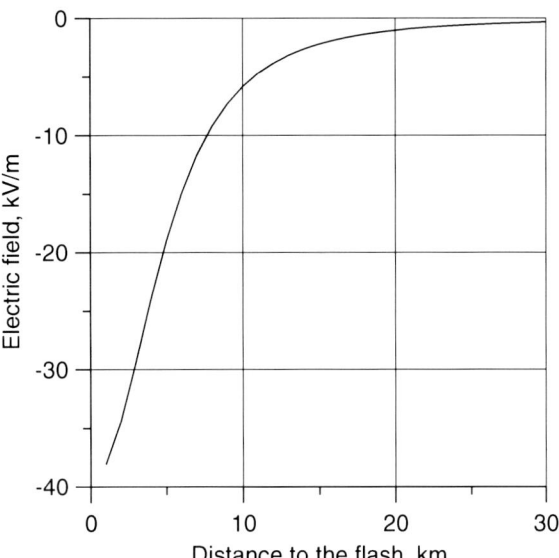

Fig. 3.15 Electric field change at ground level produced by ground flash depicted as function of distance d to thundercloud. In the calculation, it is assumed that $Q = 100$ C and the total charge removed from the negative charge center is -10 C. Part of this charge is used to neutralize the positive charge pocket ($q = 5$ C) and the rest is transported to the ground. In the calculation, $h_1 = 4$ km, $h_2 = 6$ km, and $h_3 = 9$ km (Figure created by author)

transferred to the leader channel (this point will be discussed in a subsequent chapter). Let us assume that the charge on the leader channel is uniform. Let us denote the magnitude of this uniform linear charge density by λ. Other charge distributions could also be assumed, but the procedure for calculation is the same. We also assume that the leader channel moves down at a uniform speed equal to v. Just before the creation of the leader the electric field at a horizontal distance d from the thundercloud is

$$E_i = +\frac{Q}{2\pi\varepsilon_0}\frac{h_2}{\left(h_2^2 + d^2\right)^{3/2}} - \frac{Q}{2\pi\varepsilon_0}\frac{h_3}{\left(h_3^2 + d^2\right)^{3/2}}. \tag{3.47}$$

At time t after the generation of the leader, the physical situation is depicted in Fig. 3.16a. During this time the leader travels a distance of vt. To calculate the electric field caused by the leader, let us divide the leader channel into elementary sections. Consider an element dz located at height z from ground level (Fig. 3.16b). The charge on this element is λdz. The electric field produced by this charge element at a point located at ground level at a horizontal distance d is (note that the charge on the leader channel is negative)

$$dE = \frac{\lambda dz}{2\pi\varepsilon_0}\frac{z}{\left(z^2 + d^2\right)^{3/2}}. \tag{3.48}$$

The total field produced by the leader can be obtained by integrating this from $h_2 - vt$ to h_2. That is,

$$E_l = \int_{h_2-vt}^{h_2} \frac{\lambda dz}{2\pi\varepsilon_0\left(z^2 + d^2\right)^{3/2}}. \tag{3.49}$$

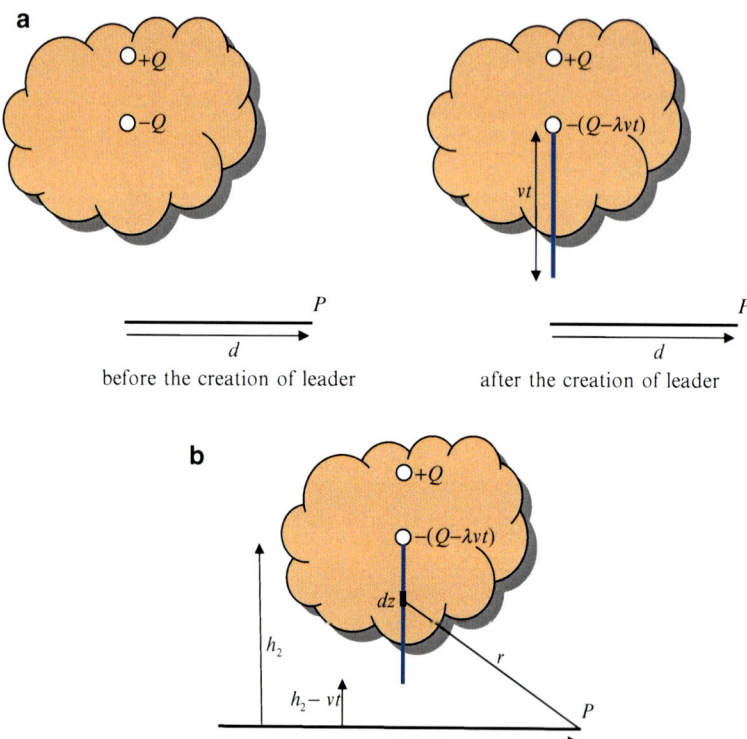

Fig. 3.16 (**a**) Idealized physical situation at time t after creation of leader. The leader initiates from the negative charge center and travels straight to ground at speed v. At time t it had traveled a distance vt. For ease of analysis only the main charge centers are considered in the calculation. (**b**) Geometry relevant to calculation of electric field produced by downward moving stepped leader. In the figure dz is an element of the stepped leader channel located at a distance r from the point of observation P. The length of the leader channel at time t is vt. The tip of the leader channel is located at a height of $h_2 - vt$ (Figure created by author)

This reduces to

$$E_l = \frac{\lambda}{2\pi\varepsilon_0} \left[\frac{1}{\sqrt{(h_2 - vt)^2 + d^2}} - \frac{1}{\sqrt{h_2^2 + d^2}} \right]. \qquad (3.50)$$

The total vertical electric field at the point of observation at time t is

$$E_t = +\frac{(Q - \lambda vt)}{2\pi\varepsilon_0} \frac{h_2}{\left(h_2^2 + d^2\right)^{3/2}} - \frac{Q}{2\pi\varepsilon_0} \frac{h_3}{\left(h_3^2 + d^2\right)^{3/2}} + E_l. \qquad (3.51)$$

3.13 Electric Field Change Caused by Stepped Leader

Thus, the field changes caused by the leader and other charges at time t are

$$\Delta E = E_t - E_i. \qquad (3.52)$$

Substituting for different components we obtain

$$\Delta E = -\frac{\lambda v t}{2\pi\varepsilon_0} \frac{h_2}{\left(h_2^2 + d^2\right)^{3/2}} + \frac{\lambda}{2\pi\varepsilon_0} \left[\frac{1}{\sqrt{(h_2 - vt)^2 + d^2}} - \frac{1}{\sqrt{h_2^2 + d^2}} \right]. \qquad (3.53)$$

This can also be written as

$$\Delta E = -\frac{\lambda v t}{2\pi\varepsilon_0 d^3} \frac{h_2}{\left((h_2/d)^2 + 1\right)^{3/2}} + \frac{\lambda}{2\pi\varepsilon_0 d} \left[\frac{1}{\sqrt{((h_2/d) - (vt/d))^2 + 1}} - \frac{1}{\sqrt{(h_2/d)^2 + 1}} \right]. \qquad (3.54)$$

Figure 3.17 depicts the field change caused by the stepped leader as a function of time at different horizontal distances. In the calculation it is assumed that $\lambda = 0.001$ C/m, $v = 10^6$ m/s, and $h_2 = 6$ km. Note how the polarity of the field change varies with distance.

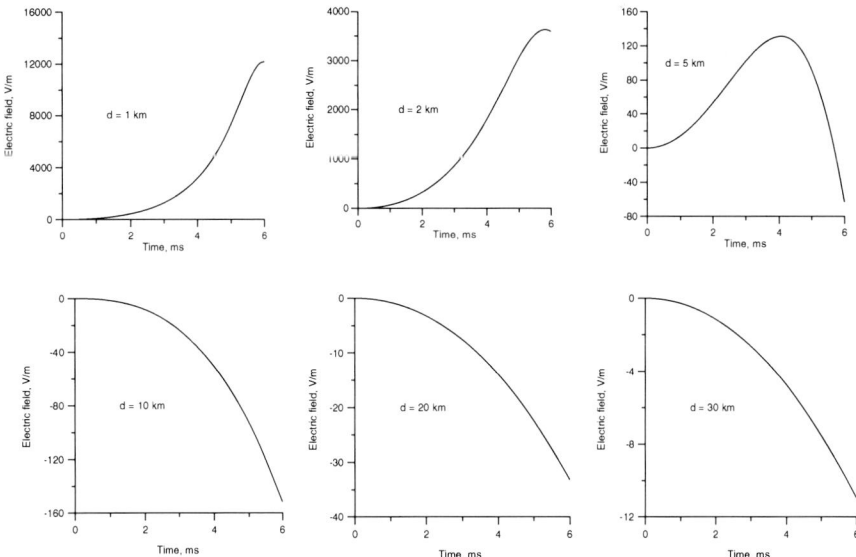

Fig. 3.17 Electric field change at ground level caused by stepped leader as function of time at different horizontal distances (Fig. 3.16b). In the calculation, the speed of propagation of the leader is $v = 10^6$ m/s, and the linear charge density on the leader channel is 0.001 C/m. The height of origin of the leader is 6 km (Figure created by author)

3.14 Electric Field Change Caused by Leader Return Stroke Combination

The electric field at the point of observation at any time during the progress of the leader, i.e., $t < h_2/v$, is given by Eq. 3.51. The electric field at the point of observation when the leader reaches ground, i.e., $t = h_2/v$, is

$$E_{t=h_2/v} = \frac{(Q - \lambda h_2)}{2\pi\varepsilon_0} \frac{h_2}{(h_2^2 + d^2)^{3/2}} - \frac{Q}{2\pi\varepsilon_0} \frac{h_3}{(h_3^2 + d^2)^{3/2}}$$

$$+ \frac{\lambda}{2\pi\varepsilon_0 d} \left[1 - \frac{1}{\sqrt{(h_2/d)^2 + 1}} \right]. \quad (3.55)$$

Thus, the total field change caused by the leader is

$$(\Delta E)_{leader} = \frac{\lambda h_2}{2\pi\varepsilon_0 d^3} \frac{h_2}{((h_2/d)^2 + 1)^{3/2}} + \frac{\lambda}{2\pi\varepsilon_0 d} \left[1 - \frac{1}{\sqrt{(h_2/d)^2 + 1}} \right]. \quad (3.56)$$

The return stroke removes the charge on the leader channel, and therefore, after the return stroke the electric field at the point of observation is

$$E_{t>h_2/v} = \frac{(Q - \lambda h_2)}{2\pi\varepsilon_0} \frac{h_2}{(h_2^2 + d^2)^{3/2}} - \frac{Q}{2\pi\varepsilon_0} \frac{h_3}{(h_3^2 + d^2)^{3/2}}. \quad (3.57)$$

Thus the field change caused by the return stroke (i.e., $E_{t>h_2/v} - E_{t=h_2/v}$) is

$$(\Delta E)_{return} = -\frac{\lambda}{2\pi\varepsilon_0 d} \left[1 - \frac{1}{\sqrt{(h_2/d)^2 + 1}} \right]. \quad (3.58)$$

The total field change caused by the leader return stroke combination is

$$(\Delta E)_{leader} + (\Delta E)_{return} = -\frac{\lambda h_2}{2\pi\varepsilon_0 d^3} \frac{h_2}{((h_2/d)^2 + 1)^{3/2}}. \quad (3.59)$$

This is simply the electric field caused by the removal of charge of magnitude $-\lambda h_2$ from the negative charge center.

3.15 Time-Varying Electromagnetic Fields

Figure 3.18 depicts both the leader field change and the return stroke field change as a function of time at various distances. In the calculation, it is assumed that $\lambda = 0.001$ C/m and $h_2 = 6$ km. Note that in the case of the return stroke only the field change is depicted, assuming that it takes place instantaneously in the time scale of the leader.

3.15 Time-Varying Electromagnetic Fields

In the presence of time-varying currents, electrical charges, electric fields, and magnetic fields, the laws of electricity can be summarized by Maxwell's equations. They are as follows.

Integral form		Point form	
$\oint_l \mathbf{E} \cdot d\mathbf{l} = -\int_s \frac{\partial \mathbf{B}}{\partial t} d\mathbf{s}$	(3.60.1a)	$\text{Curl}\mathbf{E} = -\frac{\partial \mathbf{B}}{\partial t}$	(3.60.1b)
$\oint_l \mathbf{H} \cdot d\mathbf{l} = \mathbf{I} + \int_s \frac{\partial \mathbf{D}}{\partial t} d\mathbf{s}$	(3.60.2a)	$\text{Curl}\mathbf{H} = \mathbf{J} + \frac{\partial \mathbf{D}}{\partial t}$	(3.60.2b)
$\oint_s \mathbf{D} \cdot d\mathbf{s} = \int_v \rho_v$	(3.60.3a)	$\text{Div}\,\mathbf{D} = \rho_v$	(3.60.3b)
$\oint_s \mathbf{B} \cdot d\mathbf{s} = 0$	(3.60.4a)	$\text{Div}\,\mathbf{B} = 0$	(3.60.4b)

The other two equations of importance are

$\mathbf{J} = \sigma \mathbf{E}$	(3.60.5)	$\text{div}\,\mathbf{J} = -\frac{\partial \rho_v}{\partial t}$	(3.60.6)

Equation 3.60.5 defines the relationship between the current density and the electric field through the conductivity σ of the medium. Equation 3.60.6 is the continuity equation based on the fact that electric charges are conserved.

In the preceding equations, $\int_l$ indicates a line integral and $\int_s$ indicates a surface integral. The closed loop around the integral sign, i.e., $\oint$, indicates that the integral is performed around a closed path or over a closed surface. Note also that in isotropic media $E = D/\varepsilon_0$ and $B = \mu_0 H$. The electric and magnetic fields (E and B) can be calculated using time-varying scalar and vector potentials as defined by (Fig. 3.19):

$$\phi = \frac{1}{4\pi\varepsilon_0} \int \frac{\rho_v\left(r, t - \frac{r}{c}\right)}{r} dv, \qquad (3.61)$$

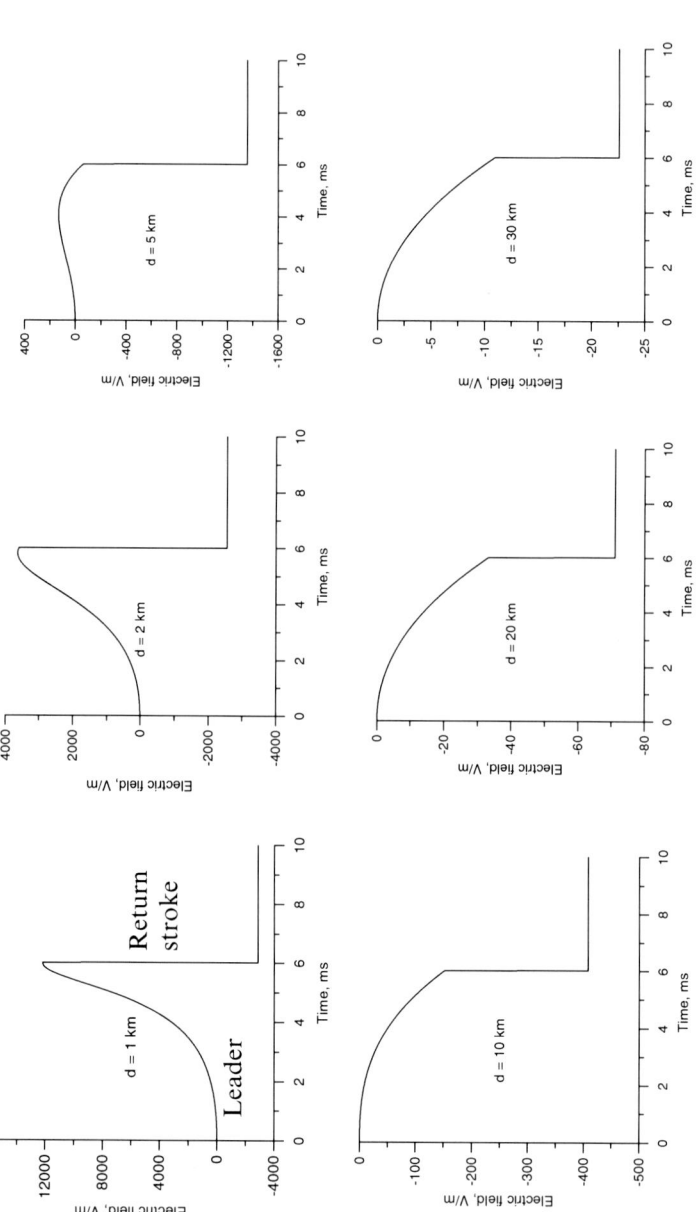

Fig. 3.18 Electric field change caused by leader (slowly varying portion of field change) and return stroke (step in field change) at different distances as function of time. The parameters of the leader are the same as in Fig. 3.17 (Figure created by author)

3.16 Relaxation Time of a Conducting Medium

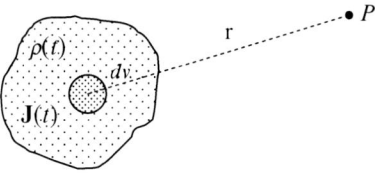

Fig. 3.19 Geometry relevant to definition of scalar and vector potential of time-varying charge and current distributions. In the diagram, $\rho(t)$ and $\mathbf{J}(t)$ are the time-varying charge and current densities; P is the point of observation where the potentials are needed (Figure created by author)

$$\mathbf{A} = \frac{\mu_0}{4\pi} \int \frac{\mathbf{J}\left(r, t - \frac{r}{c}\right)}{r} dv. \tag{3.62}$$

In the preceding equations, ϕ and $\mathbf{A}$ are the time-varying scalar and vector potentials and $\rho(t)$ and $\mathbf{J}(t)$ are the time-varying charge and current densities. The quantity $t - (r/c)$ is called the *retarded time*. The electric and magnetic fields can be calculated from these potentials using the relationships

$$\mathbf{E} = -\text{grad}\phi - \frac{\partial \mathbf{A}(t)}{\partial t}, \tag{3.63}$$

$$\mathbf{B} = \text{curl}(\mathbf{A}). \tag{3.64}$$

These laws must be complemented by the Lorentz force law, which specifies the force on a charge particle q in the presence of electric $\mathbf{E}$ and magnetic fields $\mathbf{B}$ as

$$\mathbf{F} = q\mathbf{E} - q\mathbf{v} \times \mathbf{B}. \tag{3.65}$$

In this equation the first term gives the force on a charged particle due to an electric field, while the second term gives the force of the charged particle caused by a magnetic field. In this equation, $\mathbf{v}$ is the velocity of the charged particle.

3.16 Relaxation Time of a Conducting Medium

Consider a conducting medium or a conductor in an electric field. A conductor contains free electrons and under the influence of the electric field these electrons start moving in the conductor and accumulate at the edges of the conductor. This accumulation of electrons at the edges of the conductor give rise to an electric field that is opposite to that of the applied field. The flow and accumulation of electrons at the edges continue until the two electric fields cancel each other and the electric field inside the conductor is zero. This process of relaxation (or removal) of the electric field takes some time, and this time depends on the conductivity and the dielectric constant of the conducting medium. The same happens if electrical charges are placed inside a conducting medium. These charges create an electric field in the conducting medium and the free electrons move in the conducting medium so as to remove this electric field. The effect is the displacement of the electric charge placed inside the conducting medium to the outer surface of the conductor. Again, this removal of the charge from inside to the outer surface of the conducting medium takes some time, and this time again depends on the conductivity and the dielectric constant of the conducting medium.

Consider an isotropic and homogeneous conductor with a relative dielectric constant ε_r and conductivity σ. Assume that at time equal to zero an excess charge is placed inside the conductor with charge density $\rho_v(r, 0)$. This charge generates an electric field inside the conductor, generating a current that redistributes this charge and displaces it to the surface of the conductor (recall that electric charges may accumulate on the surface of conductors) and causing the electric field inside the conductor to go to zero. Let us evaluate how fast this process takes place.

From the equation of charge conservation we have

$$\mathrm{div}\mathbf{J}(r,t) = -\frac{\partial \rho_v(r,t)}{\partial t}, \tag{3.66}$$

where $\rho_v(r,t)$ is the charge density at any time inside the conductor. Substituting for **J** in the preceding equation from $\mathbf{J} = \sigma \mathbf{E}(r,t)$ we obtain

$$\sigma \mathrm{div}\mathbf{E}(r,t) = -\frac{\partial \rho_v(r,t)}{\partial t}. \tag{3.67}$$

We also know from Gauss's law that

$$\mathrm{div}\mathbf{E}(r,t) = \frac{\rho_v(r,t)}{\varepsilon_o \varepsilon_r}. \tag{3.68}$$

Substituting this into the previous equation we obtain

$$\rho_v(r,t) = -\frac{\varepsilon_o \varepsilon_r}{\sigma} \frac{\partial \rho_v(r,t)}{\partial t}. \tag{3.69}$$

The solution of Eq. 3.69 is

$$\rho_v(r,t) = \rho_v(r,0) e^{-\sigma t/\varepsilon_o \varepsilon_r}. \tag{3.70}$$

The preceding expression for the variation in charge density inside a conductor shows that the charge inside the conductor decreases exponentially in time. The quantity $\varepsilon_o \varepsilon_r / \sigma$ is called the relaxation time of the conductor. In the same way, if we create an electric field inside a conductor, it decreases to zero exponentially with a time constant equal to the relaxation time. Recall that in Sect. 3.4 it was stated that the electric field on the surface of a conductor is perpendicular to the surface of the conductor. If a conductor with finite conductivity is placed in an electric field, it takes the relaxation time for the charges on the conductor to redistribute on the surface in such a way that the electric field becomes perpendicular at every point on the surface. In a perfect conductor the conductivity is infinite, and thus the relaxation time is zero. In this case the redistribution of the charges takes place instantaneously, and all the field components parallel to the conductor vanish instantaneously. This process of relaxation is of importance in understanding, among other processes, the response of the upper atmosphere, which is a conducting medium, to the electric fields generated by lightning flashes.

3.17 Electromagnetic Fields of a Dipole

The electromagnetic fields of a short electric dipole are used frequently in calculating the electromagnetic fields of different processes in a lightning flash. Here we present the electromagnetic fields of a short electric dipole in the frequency domain; the corresponding time domain fields are given in the next section.

Let the dipole length be l, and let it be directed in the positive z-direction with its center at the origin (Fig. 3.20). The current in the dipole is given by

$$I = I_0 e^{j\omega t}. \tag{3.71}$$

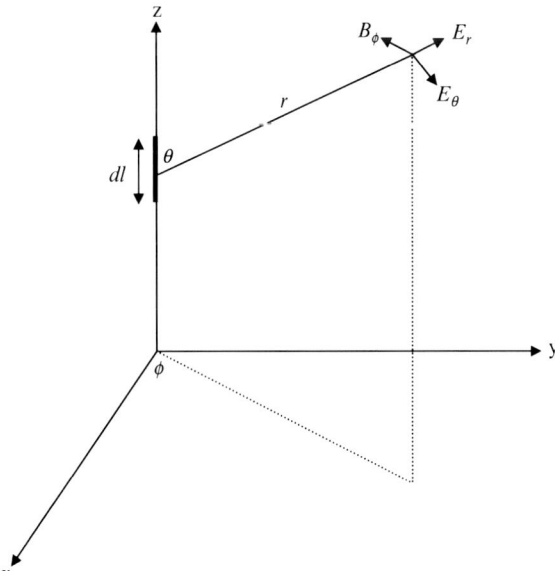

Fig. 3.20 Geometry relevant to calculation of electric and magnetic fields from electric dipole of length dl located at height h from ground plane (Figure created by author)

The electric and magnetic fields at any point in space generated by the short dipole can be calculated using scalar and vector potentials. When $r \gg l$ and $\lambda \gg l$, where r is the distance to the point of observation and λ is the wavelength, the electric and magnetic fields are given by

$$E_r = \frac{I_0 l e^{j\omega t} e^{-j\beta r}}{2\pi\varepsilon_0} \cos\theta \left[\frac{1}{cr^2} + \frac{1}{j\omega r^3}\right], \quad (3.72)$$

$$E_\theta = \frac{I_0 l e^{j\omega t} e^{-j\beta r}}{4\pi\varepsilon_0} \sin\theta \left[\frac{j\omega}{c^2 r} + \frac{1}{cr^2} + \frac{1}{j\omega r^3}\right], \quad (3.73)$$

$$B_\varphi = \frac{\mu_0 I_0 l e^{j\omega t} e^{-j\beta r}}{4\pi} \sin\theta \left[\frac{j\omega}{cr} + \frac{1}{r^2}\right], \quad (3.74)$$

$$\beta = \omega/c. \quad (3.75)$$

The directions of the electric fields in spherical coordinates are indicated in Fig. 3.20. Field components that vary inversely with distance are called *radiation fields*. When the distance to the point of observation is very large, only the radiation fields contribute to the total fields. The other field components attenuate rapidly with distance because they change as $1/r^2$ and $1/r^3$. Thus, the electric and magnetic fields at a point located far from the dipole are given by

$$E_{\theta,\text{rad}} = \frac{I_0 l e^{j\omega t} e^{-j\beta r}}{4\pi\varepsilon_0} \sin\theta \left[\frac{j\omega}{c^2 r}\right], \quad (3.76)$$

$$B_{\varphi,\text{rad}} = \frac{\mu_0 I_0 l e^{j\omega t} e^{-j\beta r}}{4\pi} \sin\theta \left[\frac{j\omega}{cr}\right]. \quad (3.77)$$

Note that the ratio $E_{\theta,\text{rad}}/B_{\varphi,\text{rad}}$ is equal to c, the speed of light in free space (observe that $c^2 = 1/\mu_0\varepsilon_0$).

3.18 Electromagnetic Fields of a Dipole Over a Perfectly Conducting Ground Plane

Let us now consider a dipole located over a perfectly conducting ground plane. The geometry is shown in Fig. 3.21. The electric field at any point over the conducting plane can be calculated by replacing the dipole with an image dipole. The vertical electric field at ground level (note that the horizontal electric field is zero over the surface of a perfectly conducting ground) at a horizontal distance d from the axis of the dipole is then given by

$$E_v = \frac{I_0 l e^{j(\omega t - \beta r)}}{\pi\varepsilon_0} \sin^2\varphi \left[\frac{1}{cr^2} + \frac{1}{j\omega r^3}\right] - \frac{I_0 l e^{j(\omega t - \beta r)}}{2\pi\varepsilon_0} \cos^2\varphi \left[\frac{j\omega}{c^2 r} + \frac{1}{cr^2} + \frac{1}{j\omega r^3}\right]. \quad (3.78)$$

3.18 Electromagnetic Fields of a Dipole Over a Perfectly Conducting Ground Plane

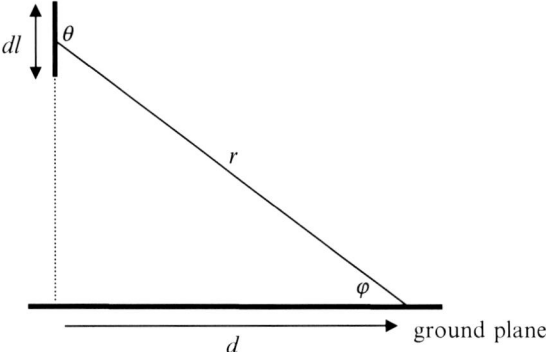

Fig. 3.21 Geometry relevant to calculation of electric fields from electric dipole of length dl located over perfectly conducting ground plane (Figure created by author)

This can be written as

$$E_v = -\frac{I_0 l e^{j(\omega t - \beta r)}}{2\pi\varepsilon_0}\left[\frac{\cos^2\varphi\, j\omega}{c^2 r} + (1 - 3\sin^2\varphi)\frac{1}{cr^2} + \frac{1}{j\omega r^3}(1 - 3\sin^2\varphi)\right]. \quad (3.79)$$

The magnetic field, which is in the azimuthal direction, is given by

$$B_\phi = \frac{\mu_0 I_0 l e^{j(\omega t - \beta r)}}{2\pi}\cos\varphi\left[\frac{j\omega}{cr} + \frac{1}{r^2}\right]. \quad (3.80)$$

When the distance to the point of observation is large, only the radiation fields will contribute to the total field. Thus, the total fields at distances far from the dipole are pure radiation and are given by

$$E_{v,\text{rad}} = -\frac{I_0 l e^{j(\omega t - \beta r)}}{2\pi\varepsilon_0}\left[\frac{\cos^2\varphi\, j\omega}{c^2 r}\right], \quad (3.81)$$

$$B_{\phi,\text{rad}} = \frac{\mu_0 I_0 l e^{j(\omega t - \beta r)}}{2\pi}\cos\varphi\left[\frac{j\omega}{cr}\right]. \quad (3.82)$$

If $d \gg h$, then $r \approx d$ and $\cos\varphi \approx 1$, and when $d/\lambda \approx r/\lambda$, the radiation fields reduce to

$$E_{v,\text{rad}} = -\frac{I_0 l e^{j(\omega t - \beta d)}}{2\pi\varepsilon_0}\left[\frac{j\omega}{c^2 d}\right], \quad (3.83)$$

$$B_{\phi,\text{rad}} = \frac{\mu_0 I_0 l e^{j(\omega t - \beta d)}}{2\pi}\left[\frac{j\omega}{cd}\right]. \quad (3.84)$$

3.19 Electromagnetic Fields of a Current Element in Time Domain Over a Perfectly Conducting Ground Plane

Consider a short current element located over a perfectly conducting ground plane. The current flowing in the current element is given by $I(t)$. The current transports charge from one end of the channel element to the other. The fields generated by the current element are identical to that produced by a short dipole in the time domain. Using Fourier transformation of the equations given earlier, the field components can be written directly as

$$E_v(t) = -\frac{l}{2\pi\varepsilon_0}\left[\frac{\cos^2\varphi}{c^2 r}\frac{dI(t-r/c)}{dt} + (1-3\sin^2\varphi)\frac{I(t-r/c)}{cr^2} + \frac{1}{r^3}(1-3\sin^2\varphi)\int_0^t I(z-r/c)dz\right], \quad (3.85)$$

$$B_\varphi(t) = \frac{\mu_0 dz}{2\pi}\left[\frac{\cos\varphi}{cr}\frac{dI(t-r/c)}{dt} + \frac{\cos\varphi}{r^2}I(t-r/c)\right]. \quad (3.86)$$

When the distance to the point of observation is large, only the radiation fields contribute to the total field. Thus, the total fields at distances far from the dipole are pure radiation fields. If $d \gg h$ then $r \approx d$, $\cos\varphi \approx 1$ and the field components reduce to

$$E_{v,\text{rad}}(t) = -\frac{dz}{2\pi\varepsilon_0}\left[\frac{1}{c^2 d}\frac{dI(t-d/c)}{dt}\right]. \quad (3.87)$$

$$B_{\varphi,\text{rad}}(t) = \frac{\mu_0 dz}{2\pi}\left[\frac{1}{cd}\frac{dI(t-d/c)}{dt}\right]. \quad (3.88)$$

Figure 3.22 depicts field components generated by a small current element at different distances assuming that the current in the short channel is a ramp function given by $I_p t$. In the calculation, the length of the dipole is assumed to be 1 m, $I_p = 30$ kA/s, and for simplicity the dipole is assumed to be located at ground level, i.e., $\cos\varphi = 0$. Note that as the distance to the point of observation from the source increases, the field becomes increasingly more similar to a step function, which is the radiation field associated with the source.

3.20 Electromagnetic Field of a Return Stroke

The electric and magnetic fields generated by a return stroke can be calculated easily by dividing the return stroke channel into a large number of elementary channel sections and treating each section as a short dipole. The geometry relevant to the derivation is given in Fig. 3.23. The total electric field can be calculated by

3.20 Electromagnetic Field of a Return Stroke

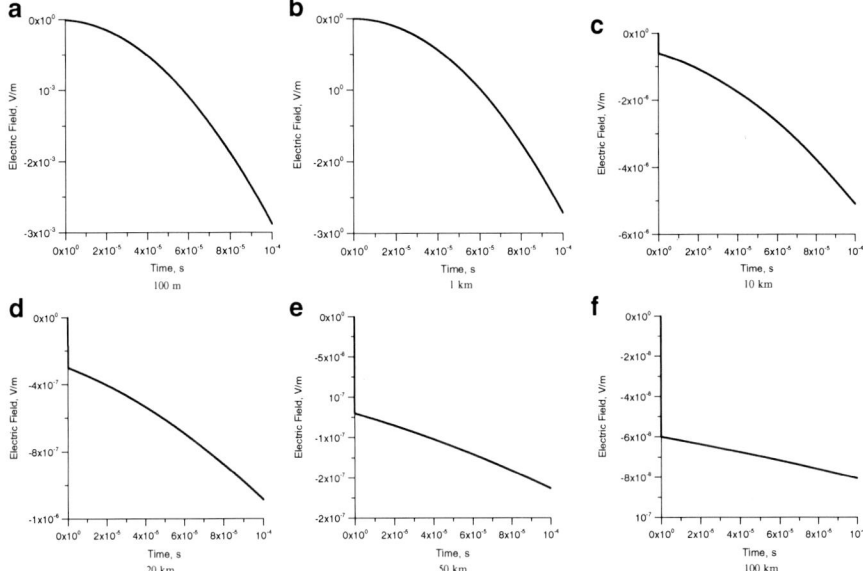

Fig. 3.22 Electric field at surface of perfectly conducting ground generated by small current element located at different distances from point of observation. In the calculation, a current element 1 m in length is assumed to be located at ground level. The current in the current element is a ramp function $I_p t$, where t is the time and $I_p = 30$ kA/s (Figure created by author). Note that in each diagram the electric field starts at zero

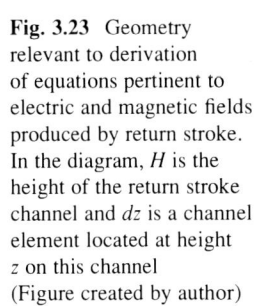

Fig. 3.23 Geometry relevant to derivation of equations pertinent to electric and magnetic fields produced by return stroke. In the diagram, H is the height of the return stroke channel and dz is a channel element located at height z on this channel (Figure created by author)

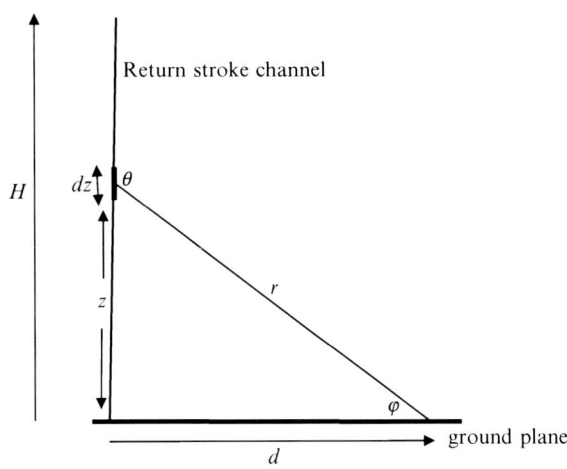

summing the contribution from each dipole. Let us represent the current at height z in the return stroke channel by $I(z, t)$. Then the vertical electric field at a horizontal distance d from the lightning channel is given by

$$E_v(t) = -\int_0^H \frac{dz}{2\pi\varepsilon_0} \left[\frac{\cos^2\varphi}{c^2 r} \frac{dI(t-r/c)}{dt} + \frac{(1-3\sin^2\varphi)}{cr^2} I(t-r/c) + \frac{1}{r^3}(1-3\sin^2\varphi) \int_0^t I(z-r/c)dz \right], \quad (3.89)$$

$$B_\varphi(t) = \int_0^H \frac{\mu_0 dz}{2\pi} \left[\frac{\cos\varphi}{cr} \frac{dI(t-r/c)}{dt} + \frac{\cos\varphi}{r^2} I(t-r/c) \right]. \quad (3.90)$$

In the preceding equations, H is the height of the return stroke channel. When the distance to the point of observation is large, only the radiation fields (i.e., terms varying inversely with distance) remains. When $d \gg H$, it is justified to assume that $\cos\varphi \approx 1$ and $r \approx d$. Under these conditions, the radiation field terms become

$$E_{v,\mathrm{rad}}(t) = -\frac{1}{2\pi\varepsilon_0 c^2 d} \int_0^H \frac{dI(t-r/c)\,dz}{dt}, \quad (3.91)$$

$$B_{\varphi,\mathrm{rad}}(t) = \frac{1}{2\pi\varepsilon_0 c^3 d} \int_0^H \frac{dI(t-r/c)\,dz}{dt}. \quad (3.92)$$

Note that the electric field is directed into the ground (the negative sign). The two components of the radiation field satisfy the condition

$$E_{v,\mathrm{rad}}(t) = c B_{\varphi,\mathrm{rad}}(t). \quad (3.93)$$

This shows that at large distances, where the radiation field is dominant, both the electric and magnetic fields have the same temporal variation, and the ratio of their amplitudes is equal to the speed of light.

References

1. Griffiths DJ (2014) Introduction to electrodynamics. Pearson Education Limited, Harlow
2. Pannofsky WKH, Phillips M (2005) Classical electricity and magnetism. Dover Publications Inc., New York

Chapter 4
Earth's Atmosphere and Its Electrical Characteristics

4.1 Variation of Atmospheric Temperature with Height

Based mainly on the way in which the atmospheric temperature varies with height, Earth's atmosphere is divided into several radial sections. The way in which the air temperature of Earth's atmosphere varies as a function of height is depicted in Fig. 4.1. Observe that as height increases from ground level, initially the temperature decreases with height, reaches a minimum value at a certain height, and then starts to increase. The height where the temperature starts to increase again is called the *tropopause*. The height of the tropopause is not constant around the globe. In tropical regions, its height is approximately 15 km and in midlatitudes it is approximately 10 km. The region below the tropopause is called the *troposphere*. As we will see later, all thunderstorm activity around the globe takes place in the troposphere. With increasing altitude from the tropopause the temperature starts to increase (it could also remain more or less stable over tens of kilometers), and the next inversion point where the temperature starts to decrease with height is called the *stratopause*. The region between the stratopause and the tropopause is called the *stratosphere*. This tendency for the temperature to decrease with height is broken again with increasing altitude at a height known as the *mesopause* (not shown in Fig. 4.1). The region between the mesopause and the stratopause is called the *mesosphere*. As we will see later, all three of these regions of the atmosphere – troposphere, stratosphere, and mesosphere – take part in activities related to lightning flashes. Indeed, it is the troposphere where thunder clouds and lightning flashes manifest themselves, but their effects are felt in both the stratosphere and the mesosphere, and these indirect effects manifest in various forms and are called *upper atmospheric lightning flashes* (a description of these electrical events is given in Chap. 19). But it should be understood that these are not independent events but are always related to the action of lightning flashes taking place in the troposphere. The way in which these indirect effects are manifested depends on the electrical characteristics of the atmosphere and the gas density.

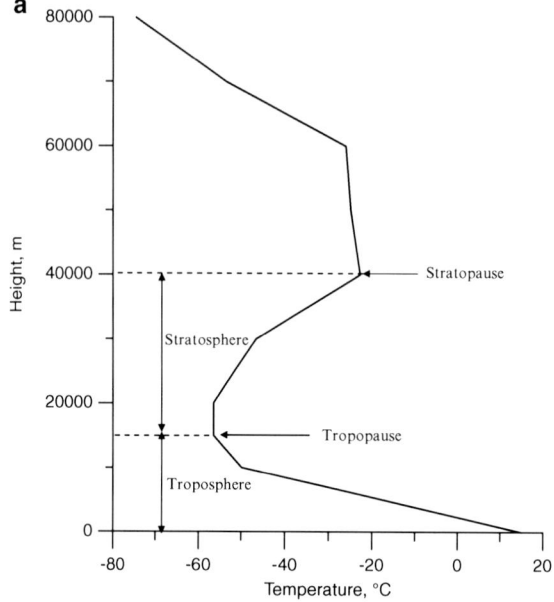

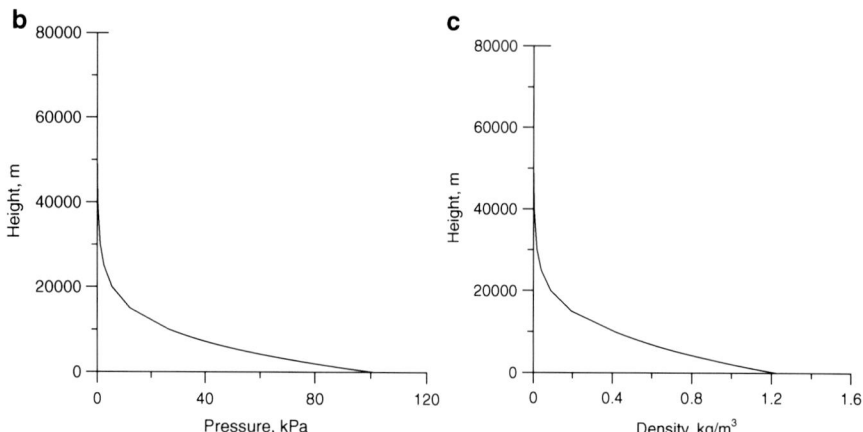

Fig. 4.1 Variation in (**a**) temperature, (**b**) pressure, and (**c**) air density as a function of height of atmosphere (Figures created by author)

4.2 Variation in Air Pressure and Air Density with Height

Since the electric fields necessary to create electrical discharges in the atmosphere and the features of these electrical discharges are closely related to the air density, the features of electrical discharges taking place in the atmosphere vary from one region to another depending on the atmospheric air density. The way in which the air pressure and air density of the atmosphere vary with height is given in Fig. 4.1.

The way in which the atmospheric pressure decreases with height can be represented by the equation

$$p(z) = p(0)e^{-z/\lambda_p}. \tag{4.1}$$

In the preceding equation, $p(z)$ is the pressure at height z, $p(0)$ is the pressure at ground level (101.325 kPa), and λ_p is approximately equal to 8×10^3 m. According to the preceding equation, the pressure decreases by a factor of 0.3 every time the height increases by 8×10^3 m. The density of the atmosphere follows very closely the variation in the pressure with height. It can be represented by the equation

$$\delta(z) = \delta(0)e^{-z/\lambda_d}. \tag{4.2}$$

In the preceding equation, $\delta(z)$ is the density of air at height z, $\delta(0)$ is the air density at ground level (1.208 kg/m³), and λ_d is approximately equal to 7×10^3 m. Let us now consider the electrical characteristics of the atmosphere.

4.3 Electrical Characteristics of Earth's Atmosphere

The electrical characteristics that are of interest in understanding lightning flashes and their effects in the atmosphere are the ion density, electron density, and electrical conductivity of the atmosphere. Ion density consists of both positive ions and negative ions. However, for all practical purposes the atmosphere can be considered neutral. This means that in a given region there is an equal amount of positive and negative charges. That is, the total negative charge in a given volume due to electrons and negative ions is balanced by the positive charges on the positive ions.

4.3.1 Electron Density Profile of Atmosphere

It is important to point out at the beginning that the electron density profile depends on whether the point of observation is on the light side or the dark side of Earth. The reason for this is that on the light side of the atmosphere, energetic rays of the Sun cause the ionization of the atmosphere and increase the electron density. Thus the electron density profile as a function of height may vary over the course of a day. It may also vary with the activity of the Sun, which varies seasonally. Figure 4.2a shows how the atmospheric electron density profile varies during the day and night and how it is related to the activity of the Sun. In this figure, the parameter R is the monthly median solar index, which is a measure of solar activity. As can be seen, increasing solar activity increases the electron density in a given region. Depending on the amount of ionization, the ionosphere can be divided into three layers: D, E, and F (layer F is divided into F1 and F2). These are also shown in

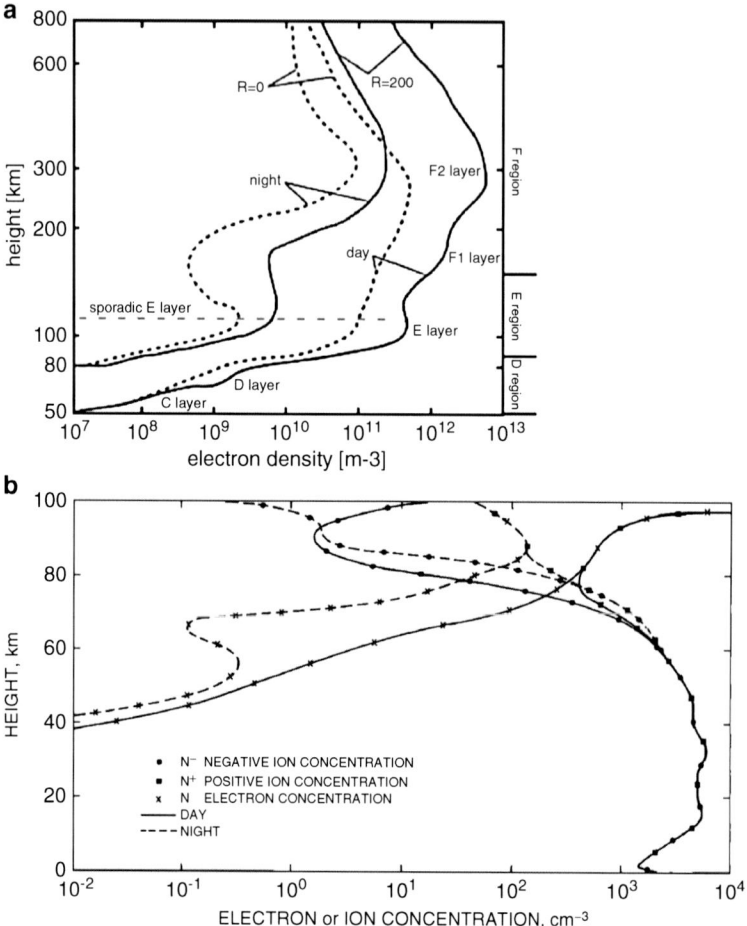

Fig. 4.2 (a) Day and night electron density profile of atmosphere indicating its variation with monthly median solar index R. From reference [1]. (b) Profile of electron and ion concentration in lower ionosphere based on study conducted by Cole and Pierce [2] (Note accumulation of positive ions close to Earth's surface)

Fig. 4.2a. At night the electron densities in different layers become very much depleted. As a consequence, the electron density profile is not fixed, and its variations must be taken into account in any study that requires this parameter as an input. The profiles of electron and ion densities in the lower atmosphere based on another study are shown in Fig. 4.2b. Note the high concentration of positive ions in the vicinity of the Earth's surface. The reason for this is explained later.

One of the topics in lightning research where the electron density profile is needed is in the study of upper atmospheric discharges. The reason for this is the following. The manifestation and properties of electrical discharges in the upper atmosphere depend on the electron density profiles and the information concening the way in which these profiles vary with the time of the day is necessary to

4.3 Electrical Characteristics of Earth's Atmosphere

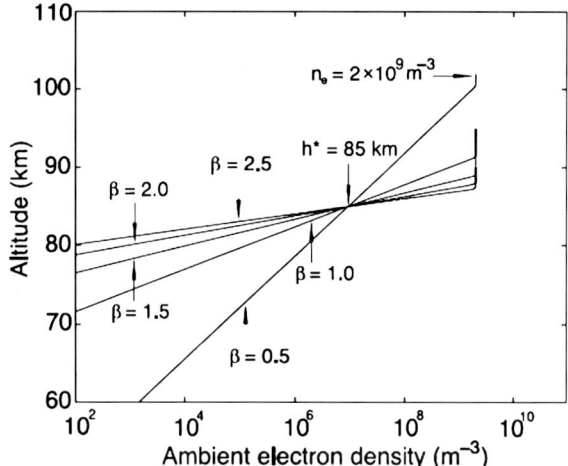

Fig. 4.3 Electron density profiles based on expression derived by Wait and Spies [3] and as used in study by Qin et al. [4]

understand the properties of upper atmospheric discharges taking place at different times of the day. An analytical model of an electron density profile that can be used in such studies was published by Wait and Spies [3]. According to that study, the ambient electron density n_e at any height h is given by

$$n_e = n_{e0} e^{-0.15 h'} e^{(\beta - 0.15)(h - h')} \tag{4.3}$$

where $n_{e0} = 1.43 \times 10^{13}$ m^{-3} and β (km) and h' (km) are parameters that describe the sharpness and reference altitude, respectively. The parameter h' allows the vertical shifting of the electron density profile and β determines how fast the electron density increases with height. This profile was used by Qin et al. [4] in an analysis of lightning-induced upper atmospheric electrical discharges. In their analysis, Qin et al. [4] fixed the value of h' for a given β so that all curves had the same electron density at a height of 85 km. The electron density profiles used by Qin et al. [4] are shown in Fig. 4.3.

4.3.2 Electric Conductivity of Atmosphere

The conductivity of the atmosphere is a measure of the amount of electrical current that can be generated when an electric field is applied to a given part of the atmosphere (Chap. 3). The higher the conductivity, the larger the resulting current due to the application of an electric field. For example, the relationship between the current density J (the current passing through a unit area) under the influence of an electric field E is given by

$$J = \sigma E, \tag{4.4}$$

where σ is the electrical conductivity. The conductivity of the atmosphere is also necessary in evaluating the electrical relaxation time (Chap. 3) of the atmosphere at

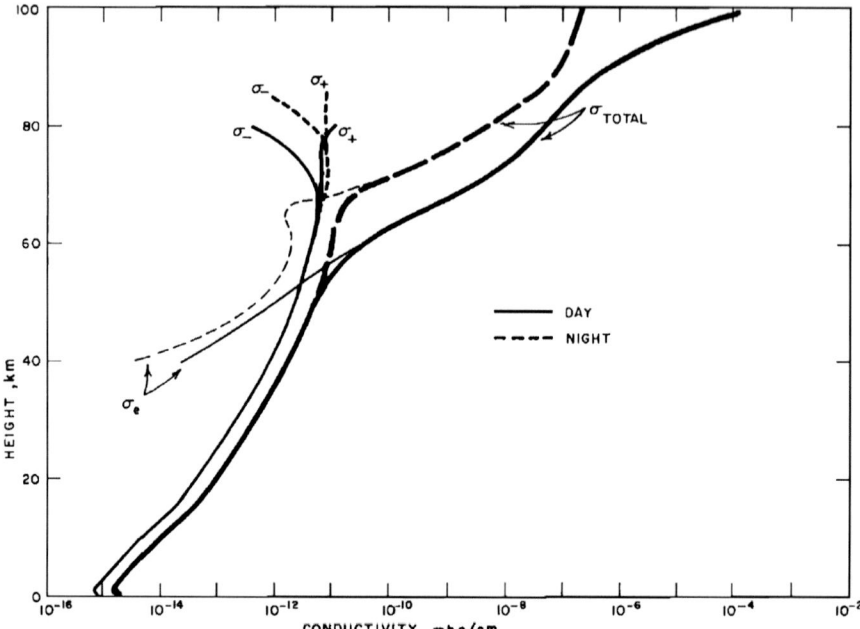

Fig. 4.4 Profile of atmospheric electrical conductivity with height based on study conducted by Cole and Pierce [2]. Suffixes +, −, and e refer respectively to positive ions, negative ions, and electrons

a given time. This parameter determines the rate at which an electric field at a given altitude generated by a lightning flash decays because of the redistribution of the electric charge.

The conductivity of the atmosphere obtained from measurements is shown in Fig. 4.4. The conductivity of the atmosphere is due both to electrons and ions. The conductivity of the lower atmosphere where the electron density is rather low depends mainly on the ion density. On the other hand, the conductivity of the upper atmosphere depends mainly on the electron density. Indeed, the conductivity of the atmosphere above approximately 40 km can be obtained directly from the electron density profile. If the electron density at a height h is equal to $n_e(h)$, then the conductivity of the atmosphere at that height is given by

$$\sigma_e(h) = \frac{q_e^2 n_e(h)}{m_e \nu_{em}(h)}, \tag{4.5}$$

where ν_{em} is the effective collision frequency of electrons and is given by

$$\nu_{em} = \frac{q_e}{\mu_e m_e}. \tag{4.6}$$

4.3 Electrical Characteristics of Earth's Atmosphere

The parameters q_e, m_e, and μ_e in the preceding equations are the electronic charge, electronic mass, and electron mobility, respectively. According to Wait and Spies [3], ν_{em} is given by

$$\nu_{em} = \nu_0 e^{-0.15h}, \qquad (4.7)$$

with $\nu_0 = 10^{13}$ s^{-1}. The conductivity of the atmosphere as a function of height is a parameter that is needed to evaluate the effects of lightning flashes in the upper atmosphere. Lightning flashes interact with the upper atmosphere through electromagnetic fields. In calculating the signature of electromagnetic fields produced by lightning flashes in the upper atmosphere, it is necessary to know the conductivity of the region where the electric field signature is needed. The procedure is as follows. First, the electromagnetic field at the point of interest can be calculated using the field equations developed in Chap. 3 in combination with the return stroke models described in Chap. 10. Once these source signatures at the height of interest are known, the way in which these field signatures are modified by the conductivity of the region must be evaluated. This can be done as follows. Assume that the source electric field generated by the lightning flash at the height of interest is $E_s(t)$. Let the conductivity of the region at the height of interest be σ, and let ε be the relative dielectric constant. Assume for the moment that the electric field produced by a lightning flash at the height of interest is a unit step function. That is, the field rises instantaneously to a unit value and remains at that value as time increases. Because of the relaxation time of the medium, the electric field starts to decay, and the electric field at any time t is then given by

$$e(t) = \exp\left(\frac{-t}{\delta}\right), \qquad (4.8)$$

where

$$\delta = \frac{\varepsilon \varepsilon_0}{\sigma}. \qquad (4.9)$$

Note that δ is the relaxation time of the medium. According to Eq. 4.8, the electric field decays as a function of time. This is in accordance with the theory developed in Sect. 3.16 of Chap. 3. Now, in reality the electric field produced by a lightning flash is not a step but a complicated function of time. Let us denote it by $E_s(t)$. The way in which this field is modified can be understood by dividing the field into a large number of steps and summing the contribution of each step, taking proper care of the time delays between steps using the response function given by Eq. 4.8. Let us denote the resulting electric field by $E(t)$. Then the previously described operation can be mathematically described as

$$E(t) = E_s(0)e(t) + \int_0^t e(t-\tau)\frac{dE_s(\tau)}{dt}d\tau. \qquad (4.10)$$

In general, $E_s(0) = 0$, and only the second term needs to be considered. The preceding operation, where the response of a step function is used to obtain the response of a more complicated function, is known as *Duhamel's theorem*.

4.4 Fair-Weather Electric Field in the Atmosphere

As we will see in the next chapter, a cloud contains electric charges and they extend their influence to Earth by creating electric fields. These electric fields can interact with grounded objects with sharp tips, giving rise to small currents from their tips called *point discharge currents*. These are probably caused by electron avalanche activity very close to the sharp tip where the electric field is enhanced (Chap. 2). Following an experiment by French researchers that demonstrated that there was electricity in clouds (Chap. 1), the French astronomer Lemonnier reported to the Paris Academy of Sciences that he could obtain electrical currents from grounded rods even if there were no thunderclouds nearby. He wrote: "I began to believe that electricity is in air even if clouds are absent." Measurements conducted later indicated that there is an electric field above Earth's surface even under fair-weather conditions and the direction of this electric field is such that it is produced by negative electrical charges on Earth. If this electric field is assumed to be everywhere around Earth at any given time (which is not correct for reasons to be given shortly), one can show that Earth contains an oversupply of approximately -400 kC of charge. With such considerations, it may be concluded that Earth is negatively charged. However, such reasoning may not be correct for the following reason. In the charge estimation given previously, it was assumed that at any given time a fair-weather electric field exists at all points around Earth. This is not true. At any given time, some regions of Earth are covered with thunderstorms. Actually, at any given time, there are approximately 1,500–2,000 thunderstorms taking place in Earth's atmosphere. These regions are called *foul-weather regions*. When we evaluate the total charge on Earth, we must take into account the electric field not only in fair-weather regions but also in foul-weather regions. In those regions, the polarity of the electric field is reversed with respect to the fair-weather one, and its value is much greater than that of the fair-weather region. Thus, estimating a net charge as was done above is incorrect.

The magnitude of the mean electric field that exists in fair-weather conditions is approximately 100 V/m. Observations show that this electric field varies around its mean value by an amount that depends on the time of day (diurnal variation). This amount of variation as a function of time is depicted in Fig. 4.5a. These measurements were carried out over the ocean in an experiment conducted from a ship named the *Carnegie*. The reason for conducting the measurements over the ocean is to remove any disturbances caused by the space charges located close to ground. This curve is known as the *Carnegie curve*. It can be shown that the variation in the fair-weather electric field shown in this curve is closely related to thunderstorm activity around the globe. For example, Fig. 4.5b shows how thunderstorm activity around the globe varies as a function of time. Note that there is a good correlation between thunderstorm activity and the diurnal variation in the fair-weather electric

4.4 Fair-Weather Electric Field in the Atmosphere

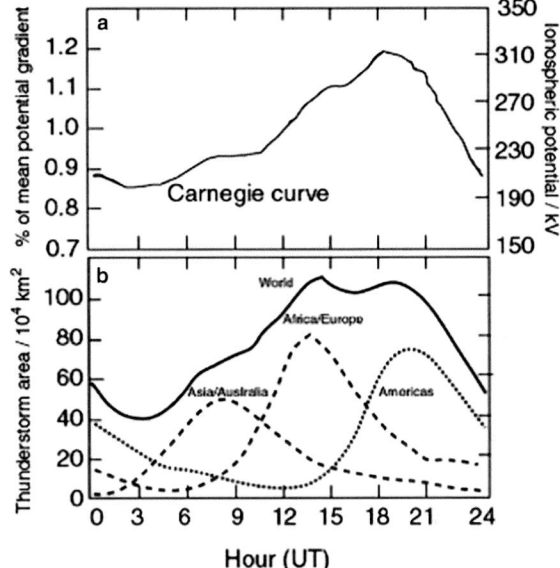

Fig. 4.5 (a) Percentage variation as a function of time of fair-weather electric field around its mean value (*Carnegie curve*). (b) Variation in thunderstorm activity around globe as a function of time. The thunderstorm activity is also given separately for the different continents (Adapted from Whipple and Scrase [5])

field. This shows that the fair-weather electric field is in some way connected to the distribution of global thunderstorms. The classical explanation given for this connection is the following. Consider a thunderstorm located above the Earth. The thunderstorm is bounded by two conducting regions, the surface of the Earth and the ionosphere. The electric field above the thundercloud, which is dominated by the upper positive charge center of the thundercloud, drives positive charges toward the ionosphere. On the other hand, the electric field below the thundercloud is dominated by the electric field generated by the negative change center, and it drives negative charges toward Earth. Moreover, the majority of lightning flashes bring down a negative charge to Earth. Furthermore, the electric field of a thundercloud gives rise to corona currents that generate positive charges that travel toward the cloud either because of the action of the electric field or by the updrafts of the thunderstorm. These corona currents, by transporting positive charge out of the Earth's surface, leave the Earth with a net negative charge. Thus, a thunderstorm acts in a way so as to supply Earth with a net negative charge. This negative charge gives rise to the fair-weather electric field, which completes the current circuit by generating a current that transports negative charges toward the ionosphere or vice versa. This activity is pictorially depicted in Fig. 4.6 and can be represented by a global electrical circuit. In this classical explanation lightning flashes act as a major source that transport negative charge to Earth. It is true that 90 % of the lightning flashes that strike the Earth transport negative charge to Earth while the remaining 10 % transport positive charge. However, the charge transported by a positive ground flash is about ten times larger than that of a negative ground flash. Thus the net charge transported to ground by lightning flashes could be zero. If this is the case then it is the electric fields produced by thunderstorms (and the resulting charge drift) and not the lightning flashes that drives the global electrical circuit.

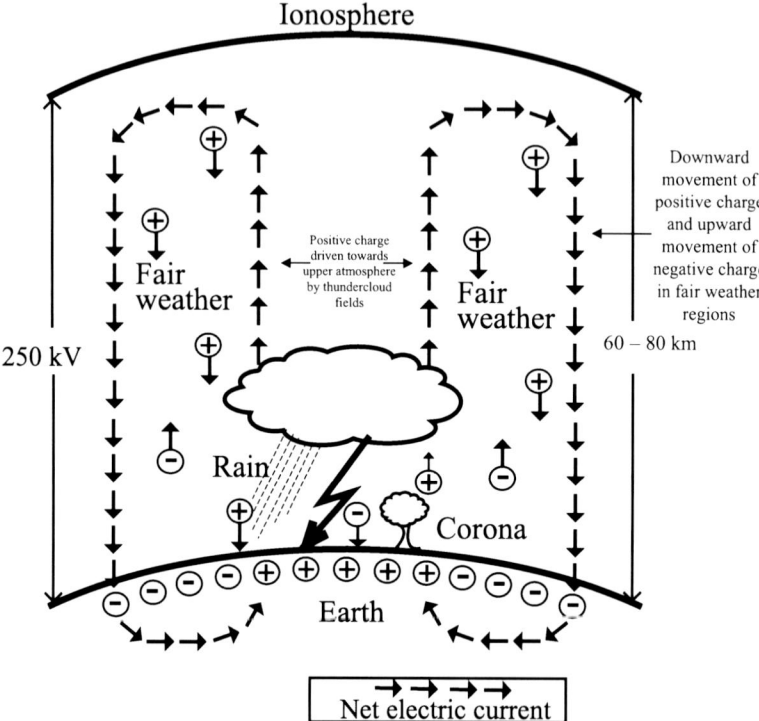

Fig. 4.6 Thunderclouds replenish the net negative charge on Earth by combined action of lightning flashes bringing negative charge to Earth and the corona currents removing positive charge from Earth. Some of the positive charge generated by the corona is captured by rain and brought back to Earth. A positive charge in the upper sections of the cloud creates electric fields that drive positive ions toward the ionosphere. The negative charge gained by the Earth under thunderstorms is drained in fair weather because of the flow of positive charge toward Earth and the movement of negative charge away from Earth (Figure adapted from Uman [6])

4.5 Global Electrical Circuit

The influence of thunderstorms at a global level can be represented by an electrical circuit as depicted in Fig. 4.7. In this electrical circuit, the thunderstorm acts as a generator. By creating an electric field that drives electrons or negative charge downward (or a positive charge toward the upper atmosphere) the thundercloud creates a positive current directed toward the upper atmosphere. At the same time it generates lightning flashes directed toward ground, which brings a mainly negative charge to Earth. This is supplemented by the flow of positive charges generated by the corona toward the cloud. This process leads to a charging of the Earth with negative polarity. This negative charge gives rise to a fair-weather electric field. Since the atmosphere contains both positive and negative ions, in a fair-weather region, negative ions travel upward and positive ions travel toward the ground.

4.5 Global Electrical Circuit

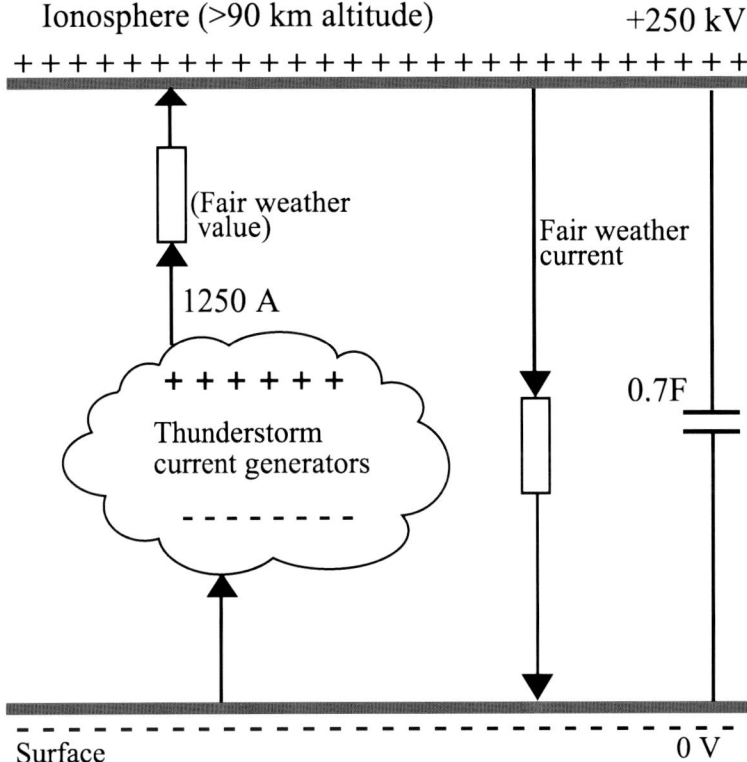

Fig. 4.7 The global electrical activity depicted in Fig. 4.6 can be represented by an *electrical circuit*. The current generators are thunderstorms, which are located predominantly over tropical land masses. The thunderstorms drive a current of approximately 1,250 A toward the ionosphere. The return path is the global fair-weather current flowing between the ionosphere and ground. The resistance between the ionosphere and the ground is approximately 200 Ω, and the capacitance between the ionosphere and the ground is 0.7 F. Note that the decay time constant (product $R \times C$) of the ionosphere–ground spherical capacitor is only approximately 2 min. This means that if global thunderstorm activity were to stop, the fair-weather electric field would decay to zero with a 2 min time constant (Adapted from Kirkby [7])

That means the negative charge brought to ground during thunderstorms leaks away into the ionosphere, completing the circuit. The flow of positive ions onto the Earth's surface under fair-weather conditions is one reason for the accumulation of positive ions close to the Earth's surface, as depicted in Fig. 4.2b.

As mentioned earlier, the total number of thunderclouds active at any given moment may vary from approximately 1,500 to 2,000, and the global positive current flowing above thunderstorms to the ionosphere is approximately 1.25 kA. The current circuit is completed by the positive current flowing from the ionosphere to ground in fair-weather regions. This process can be summarized in a circuit as shown in Fig. 4.7. The load resistance between the ionosphere and the ground through which the fair-weather current is flowing is called the *columnar resistance*.

The columnar resistance in a fair-weather region is approximately 200 Ω, and since the current flowing from the ionosphere to Earth is approximately 1.25 kA, the potential of the ionosphere with respect to ground is approximately 250 kV. The generator in the circuit is the thundercloud. Clearly, the columnar resistance and the capacitance between the ionosphere and Earth is such that if thunderstorm activity ceases immediately, the fair-weather electric field disappears in a matter of minutes. It is also important to recall that the fair-weather electric field at a given location may also vary because of variations in air conductivity. Air conductivity is affected by the magnitude of aerosols present in the air and the amount of ions created by radioactivity in the ground. For example, the generation of fog greatly increases the number of aerosols in the air, and these aerosols get attached to fast moving ions making them bulky and reducing their mobility. This leads to a reduction in conductivity. Furthermore, sunrise and sunset can cause changes in humidity in regions close to the Earth's surface. The hydration of fast ions by the humidity can reduce air conductivity, again affecting the fair-weather electric field. On the other hand, an increase in ionization events taking place in the air can lead to an increase in air conductivity and, hence, a reduction in the fair-weather electric field. For example, when the radioactive cloud caused by the nuclear accident in Chernobyl reached Sweden, a rapid drop in the fair-weather electric field was observed at the meteorological station in Uppsala, Sweden [8]. This drop was caused by the rapid increase in air conductivity caused by the ionization of air by radioactive materials present in the Chernobyl cloud. Actually, it is this rapid reduction in the fair-weather electric field that triggered the alarm in Sweden regarding the accident that happened 1,000 km away.

References

1. The Ionosphere, in the webpage of the Istituto Nazionale di Geofisicae Vulcanologia, Rome. http://roma2.rm.ingv.it/en/research_areas/4/ionosphere, 2014
2. Cole RK, Pierce ET (1965) Electrification in the earth's atmosphere for altitudes between 0 and 100 kilometers. J Geophys Res 70(12):2735–2749
3. Wait JR, Spies KP (1964) Characteristics of the earth-ionosphere wave guide for VLF radio waves, Tech. Note 300. National Bureau of Standards, Boulder
4. Qin J, Celestin S, Pasko V (2011) On the inception of streamers from sprite halo events produced by lightning discharges with positive and negative polarity. J Geophys Res 116: A06304. doi:10.1029/2010JA016366
5. Whipple FJW, Scrase FJ (1936) Point-discharge in the electric field of the earth, Geophys. Mem. VII, No. 68. H. M. Stationery Off., London, pp 1–20
6. Uman MA (1987) The lightning discharge. Academic, London
7. Kirkby J (2007) Cosmic rays and climate. Surv Geophys 28:333–375
8. Israelsson S, Knudsen E (1986) Effects of radioactive fallout from a nuclear power plant accident on electrical parameters. J Geophys Res 91:11909–11910

Chapter 5
Formation of Thunderclouds

5.1 Cumulus Cloud

Almost everyone is familiar with cumulus clouds (Fig. 5.1a), which look like a piece of floating cotton, with a sharp outline and a flat bottom. When a growing cumulus resembles the head of a cauliflower, it is called a *cumulus congestus* (Fig. 5.1b). When conditions are just right, cumulus congestus continues to grow vertically, and the result is *cumulonimbus*, the thundercloud (Fig. 5.1c). It is a giant heat engine that converts the energy of the Sun into the mechanical energy of air currents and the electrical energy of lightning.

5.2 Formation of a Thundercloud

It all starts with the rays of the Sun shining on the face of the Earth. The sunlight warms the ground, and the ground warms the air in contact with it. The first requirement for the creation of a thundercloud is now satisfied. As the air in contact with the ground gets warmer, it expands and becomes lighter than the surrounding air. As a consequence, it starts rising. As we saw previously, the atmospheric pressure decreases with height, and this causes the rising air parcel to expand further. As it expands, it cools down. The second requirement for the creation of a thundercloud is that this rising air parcel should contain water or humidity. The ability of air to hold moisture in it decreases as the air temperature decreases. Consequently, as the rising air parcel cools, the moisture in the air starts condensing. It condenses into very small microscopic water drops, and millions of them are required to form a single rain drop. Visible clouds are made of these microscopic water droplets. Of course, these cloud droplets evaporate as the dry air surrounding a cloud mix with it. This happens mostly at the edges of the cloud and is the reason why clouds have a very sharp contour. Now, as we saw in Chap. 4, the air temperature decreases initially with increasing height.

Fig. 5.1 Three stages of cumulus clouds: (**a**) cumulus, (**b**) cumulus congested, and (**c**) cumulonimbus (Photographs courtesy National Oceanic and Atmospheric Administration (NOAA), USA)

Under normal conditions the rate of decrease in temperature is approximately 6.5 °C for each 1,000 m of ascent. This rate of change in temperature is called the *environment lapse rate* [1]. Remember that our parcel of air expands as it rises, and its temperature decreases. If the air parcel does not contain saturated moisture and if we assume adiabatic conditions (i.e., no transfer of heat from air parcel to surroundings or vice versa), then its temperature drops by 10 °C per 1,000 m of its rise. Since this rate of decrease in temperature applies to an air parcel with unsaturated humidity, it is called the *dry adiabatic lapse rate* [1]. However, the atmospheric temperature does not drop so rapidly and very soon the air parcel becomes colder than the surrounding air, and this stops its upward journey (Fig. 5.2a). But if the rising air parcel contains moisture, heat is released as moisture condenses, and this heat warms up the rising air parcel. This heating due to condensation of water decreases the rate of drop in temperature of the air parcel to approximately 5.5 °C per 1,000 m. This is called the *wet adiabatic lapse rate*. Now, whether or not the conditions necessary for the creation of a thunderstorm exist depends on the wet adiabatic lapse rate. If the temperature in the atmosphere decreases faster than the wet adiabatic lapse rate, then the air parcel continues to rise in the atmosphere, i.e., it remains warmer than the surrounding air, and the final requirement necessary for the creation of a thundercloud is satisfied (Fig. 5.4b). Such an atmosphere is called an *unstable atmosphere*. The ability or potential of the atmosphere to generate thunderclouds is measured by a quantity called convective available potential energy, or CAPE. This parameter gives the measure of the potential energy available for a parcel of air as it moves upward in the atmosphere. It is calculated using the equation [2]

$$\text{CAPE} = \int_{Z_f}^{Z_e} g \frac{T_p(z) - T_e(z)}{T_e(z)} \, dz, \qquad (5.1)$$

where g is the acceleration due to gravity, T_p is the temperature of the air parcel at height z, T_e is the temperature of the environment at height z, Z_f is the height of the atmosphere where the temperature of the environment starts to decrease faster than the moist adiabatic lapse rate of the atmosphere, and Z_e is the height of the

5.2 Formation of a Thundercloud

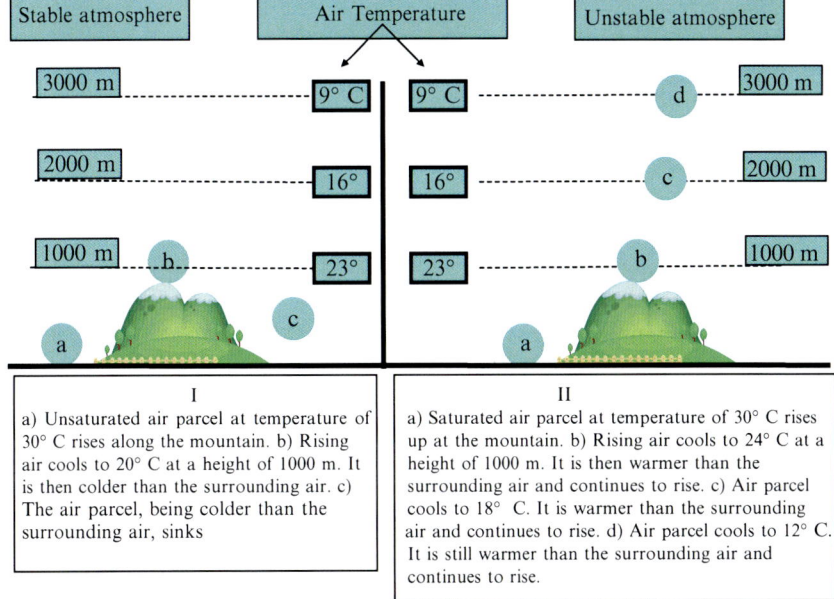

Fig. 5.2 The formation of thunderclouds depends on how fast the air temperature decreases with altitude and on the moisture content of the air at ground level. The atmosphere is unstable when rising air continues to be warmer than the surrounding air. This depends on the temperature of the surrounding air. The rising unsaturated air cools at a rate of 10 °C per 1,000 m. If the air is saturated, the moisture starts condensing as it cools, and the air parcel cools at a rate of 6 °C per 1,000 m. Therefore, the stability of the atmosphere depends on how the atmospheric temperature decreases with altitude and on the moisture content of the air at ground level. Note that in the diagram the environmental lapse rate is 7 °C/km, dry adiabatic lapse rate is 10 °C/km and the wet adiabatic lapse rate is 6 °C/km (Figure created by author)

atmosphere where the temperature of the air parcel becomes identical to that of the environment. The value of CAPE is measured in J kg^{-1} of air. In general, a value of CAPE greater than approximately 1,500 J kg^{-1} is considered to have a higher risk of thunderstorm formation. In summary, three conditions are necessary for the formation of a thundercloud: sunlight, moist air, and an unstable atmosphere.

A rising moist air column in the unstable atmosphere is the first stage of a thundercloud. This stage is called the *cumulus stage*. The air current pushes its way through the atmosphere at speeds as high as 50 m/s. This is the first stage of thundercloud formation (Fig. 5.3a). As the air rises higher and higher, its moisture condenses faster and faster. At the same time, the temperature of the air parcel continues to decrease. When the temperature of the parcel of air drops below 0 °C, the temperature at which water freezes, some of the tiny water droplets freeze and form tiny ice crystals. However, some of these tiny droplets remain as water even at temperatures lower than 0 °C. They are called *supercooled water droplets*. Some of these supercooled water droplets collide with tiny ice crystals and freeze directly on them. In this way, tiny ice crystals can grow in size. Such grown ice particles are called *graupel particles*. As the size of the graupel particles increases, they become heavier

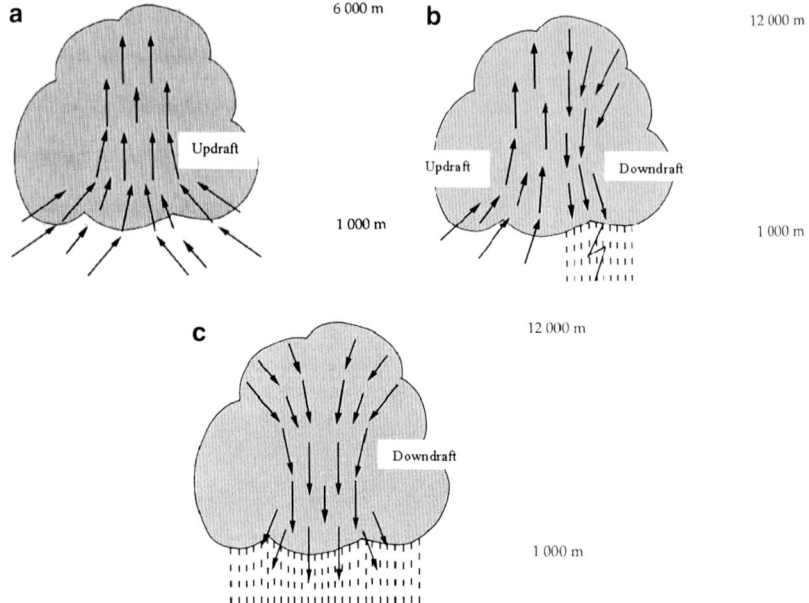

Fig. 5.3 (a) In the cumulus stage of the cloud, updrafts of warm humid air rise from the ground. Rapid updrafts prevent any precipitation from falling from the cloud. (b) As the cloud builds up to a height well above the level at which water freezes (i.e., freezing level), precipitation particles grow larger. At this stage, the cloud contains water droplets, supersaturated water drops, ice crystals, and graupel particles (soft hail). As the graupel particles become heavier, the rising air cannot keep them suspended and they begin to fall. As they fall, they drag the air with them, creating a downdraft. Updrafts continue to feed warm humid air into the cloud. Thunderstorms are very intense during the mature stage, and clouds produce lightning and heavy rain. (c) Falling precipitation causes downdrafts throughout the cloud, choking off the updrafts. With the cloud deprived of its rich supply of warm humid air, cloud droplets no longer form. The rain gradually becomes lighter and the downdrafts weaker. Finally, the cloud dissipates completely (Figure created by author)

and the rising air finds it difficult to push them upward against their weight; they start falling. This signals the mature stage of thundercloud formation. Now, what happens to the parcel of air? Will it go on rising forever? No. As discussed in Chap. 4, the temperature of the atmosphere decreases with height only initially and starts to increase once the tropopause is reached. The temperature inversion at the tropopause limits the growth of thunderclouds. Thus, the activity of thunderclouds is limited to the troposphere. In tropical regions, the tropopause is located approximately 15 km above the ground and in temperate regions approximately 10 km above.

As mentioned earlier, when graupel particles become heavier, the updraft is unable to sustain them, and they start falling. Some graupel particles are carried up and down in air currents for such a long time that they grow by the accretion of supercooled droplets and tiny ice crystals. Sometime they can grow up to a few centimetres in diameter and they are known as hail. The falling graupel particles collide with the tiny ice crystals and supercooled water droplets that are moving up in the updraft. As we will see in the next chapter, it is these collisions that researchers believe is the cause of electric charge generation inside a cloud [3].

When a cloud reaches this stage, it is called a *mature thundercloud* (Fig. 5.3b). It is usually at this stage that a cloud is capable of generating lightning flashes. The cloud top can extend to very large heights during this stage. This stage may last for approximately 15–30 min. At this time the amount of falling graupel particles also increases, and these falling particles start dragging down with them the surrounding air. This gives rise to a downward moving column of air and is called a *downdraft*. This downdraft opposes the movement of air in the updrafts. When the graupel particles reach warmer levels, they melt and give rise to water droplets, which fall to the ground as rain. With increasing precipitation the intensity of the downdrafts increases. This is the final stage of the cloud. In general, lightning activity takes place before strong downdrafts are established. This is why one intense lightning activity is usually observed before strong rain showers. The downdrafts bring cool air from higher altitudes to ground. This, combined with the rain, cools the air at ground level, which interrupts the formation of updrafts. The thundercloud has reached its death bed or the final stage (Fig. 5.3c). With the cessation of updrafts, the fuel necessary for the formation of the cloud is cut off. Deprived of its rich supply of warm humid air, the cloud stops growing. With time, even precipitation decreases. The rest of the cloud vaporizes into the atmosphere and disappears.

However, the preceding description relates to the activity of a single cell of a thundercloud. A thundercloud may have several cells, and when one cell dies down, another one may form. Therefore, lightning activity may continue for some time until all the energy available for updrafts are expended and the hot humid air is replaced by the cold air that comes down in the downdrafts. It is possible that one thundercloud may trigger the formation of another cloud. The cool air that comes down during the final stage of the thundercloud may spread outward and force the surrounding warmer air to rise, generating more thunderclouds. This process may last for hours until all the moist air and the heat are expended inside the thunderclouds.

Thunderstorms that are caused by direct heating of moist air at ground level by sun are called convective thunderstorms. Usually these thunderstorms occur in relatively still air without strong winds. There are cases in which the formation of thunderstorms is aided by the lifting of incoming air along the slopes of a mountain. Thunderstorms formed in this manner are called *orographic thunderstorms* (Fig. 5.4). In addition to such storms, there are others called *frontal thunderstorms*. These occur when large masses of cold air penetrate into a region with warm moist air. The cold incoming air wedges under the warm moist air and provides it with an upward thrust that is necessary for the formation of thunderstorms (Fig. 5.5). This rising warm air creates thunderstorms along the front of the incoming cold air mass. The same thing can happen when warm air flows into a region with cold air. The warm moist air slides up over the cold air mass, triggering thunderstorms. However, along a warm front the rise of warm air above the cold air is more gentle, and consequently the formation of thunderstorms is less probable than in the case of cold fronts. Cold fronts have a more severe boundary between the warm and cold air and are more likely to create thunderstorms.

Fig. 5.4 Thunderstorms are formed when moist hot air is lifted up as it passes across a mountain. These types of thunderstorm are called *orographic thunderstorms* (Diagram courtesy National Oceanic and Atmospheric Administration (NOAA), USA)

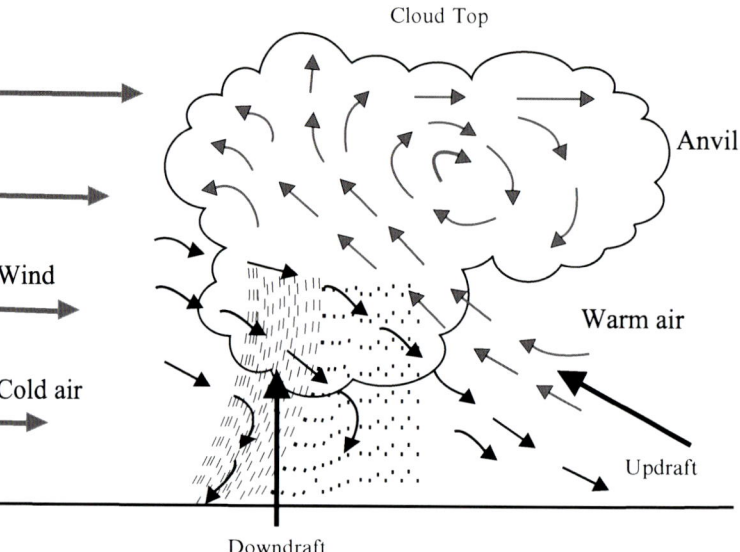

Fig. 5.5 Thunderstorms are formed when cold air masses enter a region with warm air. The cold air wedges under the warm air, resulting in an upward movement of warm air (Figure created by author based on information available in the literature)

A thundercloud is not the only natural element capable of generating lightning. Lightning can be generated in sandstorms, which usually occur in deserts. Lightning can also occur during the eruption of volcanoes. However, the charge generation mechanisms in sandstorms and volcanic plumes are not yet fully understood. Lightning is also produced around the fireballs of nuclear explosions. In this case, the enormous energy released by a nuclear bomb strips off electrons from atoms and in this way produces regions of positive and negative charges. The intense electric fields created by these charges give rise to lightning flashes.

References

1. Ahrens CD (2003) Meteorology today. Brooks/Cole-Thomson Learning, Pacific Grove, West Publishing Company, New York
2. Chandrasekar A (2010) Basics of atmospheric science. PHI Learning Private Limited, New Delhi
3. Jayaratne R (2003) Thundercloud charging mechanisms. In: Cooray V (ed) The lightning flash. IET Publishers, London

Chapter 6
Charge Generation in Thunderclouds and Different Forms of Lightning Flashes

6.1 Charge Generation in Thunderclouds

Many theories have been advanced to explain the generation of electrical charge in thunderclouds [1]. However, one theory has withstood the test of time, and the current consensus is that the mechanism proposed by this theory is the dominant one for charge separation in thunderclouds. This mechanism is based on the collision of graupel and ice crystals in the presence of super-cooled water drops in a cloud.

6.1.1 Ice–Graupel Collision Mechanism

The theory is based on experiments conducted in Japan by Takahashi [2] and in the UK by Jayaratne et al. [3]. In these experiments, a graupel particle, simulated by a riming target, was moved at a certain speed through a cloud of ice crystals and supersaturated water droplets. The charge generated during the collision of ice crystals and graupel particles was measured. Observations revealed that the magnitude and polarity of the charging process depend on the water content of the cloud and the ambient temperature. The results obtained by Takahashi [2] and Jayaratne et al. [3] are shown in Fig. 6.1a, b, respectively. According to Fig. 6.1a, during collisions graupel particles charge positively at higher temperatures, at both low and high water content, and they charge negatively at low temperatures and at intermediate water content. The results obtained by Jayaratne et al. [3] also show that there are two charging regimes, but the temperature and the water content where charges of a particular polarity are achieved are different (Fig. 6.1b). The difference could be due to the different experimental conditions used by the two research groups. Recent experimental data seem to suggest that the experimental conditions used by Takahashi [2] resemble more closely the physical conditions present in clouds [4].

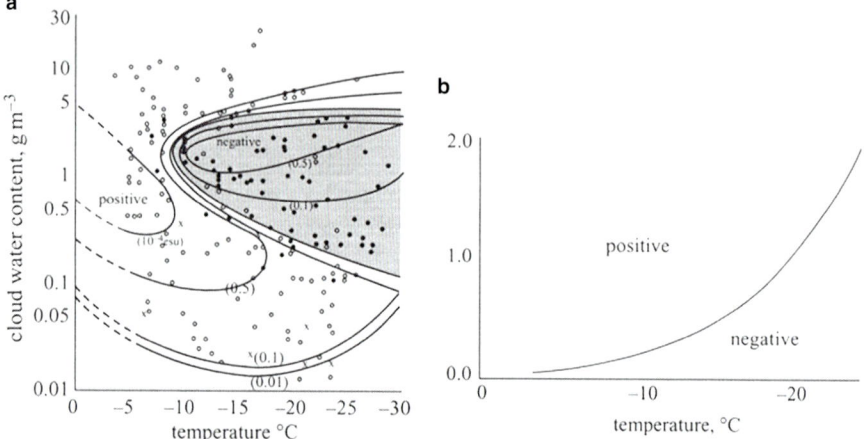

Fig. 6.1 (a) The polarity of the charged acquired by a graupel particle during collision with ice crystals as a function of water content and ambient temperature as determined by an experiment conducted by Takahashi [2] in Japan. (b) Results of a similar experiment conducted in the UK by Jayaratne et al. [3]. Note that whether a graupel particle achieves a charge of negative or positive polarity during collision with an ice crystal is determined by the temperature and the cloud water content

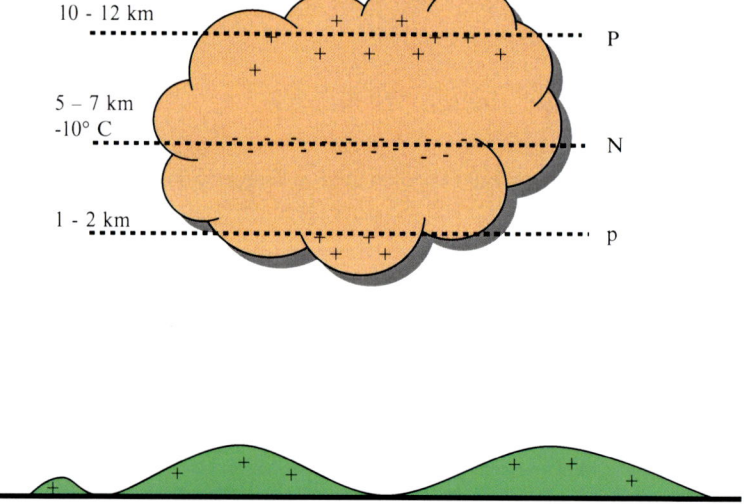

Fig. 6.2 An idealized tripole charge structure of a thundercloud. It contains three charged regions. The main positive charge center is located in the upper part of the cloud (P). The main negative charge center (N) is located below the main positive charge center and it is located in the region of the cloud where the temperature is somewhere between −15 °C to −10 °C. Below the main negative charge center is a positively charged region (p). In reality there could also be a screening layer of negative charge on the upper surface of the cloud formed by the attraction of negative charge to the upper surface of the cloud by the electric field produced by the positive charge (Figure created by author)

Consider a cloud water content of approximately 1 g/m³, which is typical of most clouds. Then, according to Fig. 6.1a, at temperatures below approximately −10 °C, graupel particles receive a negative charge and at higher temperatures a positive charge during collisions. As we will show subsequently, this fact seems to explain the main characteristics of the location of charge centers in thunderclouds.

The exact mechanism through which a charge transfer takes place during graupel–ice collisions is not known at present. One theory, proposed by Baker et al. [5], is based on the temporal growth of the colliding particles. Both graupel particles and ice crystals can grow by accretion of water vapor and because of the riming of the particles by supercooled water droplets that condense on the colliding particles. According to the theory, during the collision of graupel particles and ice crystals, more rapidly growing particles charge positively. This is explained by the following facts. A growing particle transfers negative ions to the surface of the particle. At the ice surface (of both graupel particles and ice crystals) is a liquid layer that is in equilibrium with water vapor, and these negative ions migrate into the liquid layer. The thickness of the water layer increases with increasing surface temperature. As the particle grows by riming and accretion, the latent heat released by the water vapor and supercooled droplets increases the temperature of the surface. As a consequence, the growing particle maintains a thicker water layer on the surface. During collision water is transferred from the thick layer to the thinner one. Since the water contains negative ions, the negative charge is also transferred together with the water. This causes the growing particle to charge positively and the slower growing particle to charge negatively. The growth rate of the particles depends on the liquid water content, and the temperature and, hence, polarity and magnitude of the charge transfer depend on these parameters. As was mentioned earlier, this mechanism can explain some of the field observations on the location of charge centers in clouds. This point is discussed further in the next section.

6.1.2 Consequences of Ice–Graupel Collision Charging Mechanism

6.1.2.1 Tripolar Structure of Cloud Charge

Experimental observations show that a typical cloud contains two main charge centers, one positive and the other negative. An idealized charge structure of a thundercloud is shown in Fig. 6.2. The positive charge center is located above the negative one, and the negative one is very shallow, approximately 1 km in thickness, and located in a region of −15 to −10 °C isotherm. Below the negative charge center is a small positive charge pocket. The ice–graupel collision mechanism can explain this observation satisfactorily. Consider a typical cloud with a liquid water content of approximately 1 g/m³. As the graupel particles fall from greater heights through the clouds, they collide with ice crystals that are being carried upward in

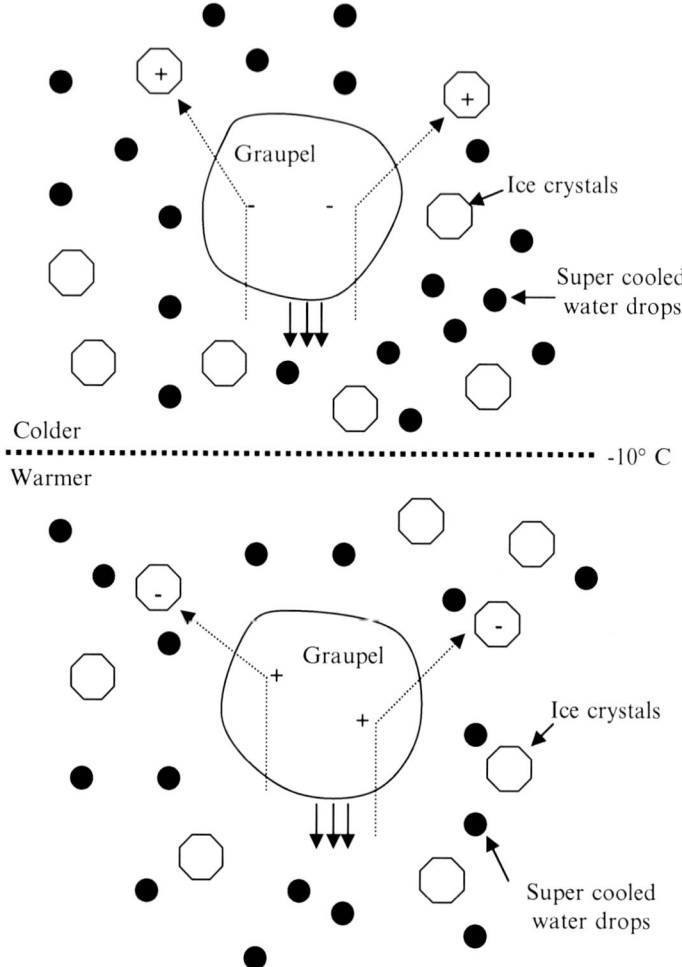

Fig. 6.3 At a temperature below approximately −15 °C to –10 °C graupel particles colliding with ice crystals achieve a negative charge (*the upper diagram*), whereas at temperatures above approximately −15 to –10 °C the graupel particles receive a positive charge (*the lower diagram*) (Figure created by author based on information available in literature)

updrafts. If the temperature is below approximately −15 to −10 °C, the graupel particles charge negatively and the ice crystals positively (upper part of Fig. 6.3). The light positively charged ice crystals travel upward along the updraft, leaving the positive charge at a higher location in comparison with the negatively charged falling graupel particles. As the graupel particles fall further, the temperature increases and the graupel particles start to charge positively (lower part of Fig. 6.3). Thus, there is a region below the height of the isotherms −15 to −10 °C where graupel particles are positively charged. This is the basis of the positive charge pocket located below the negative charge center. For example, the

6.1 Charge Generation in Thunderclouds

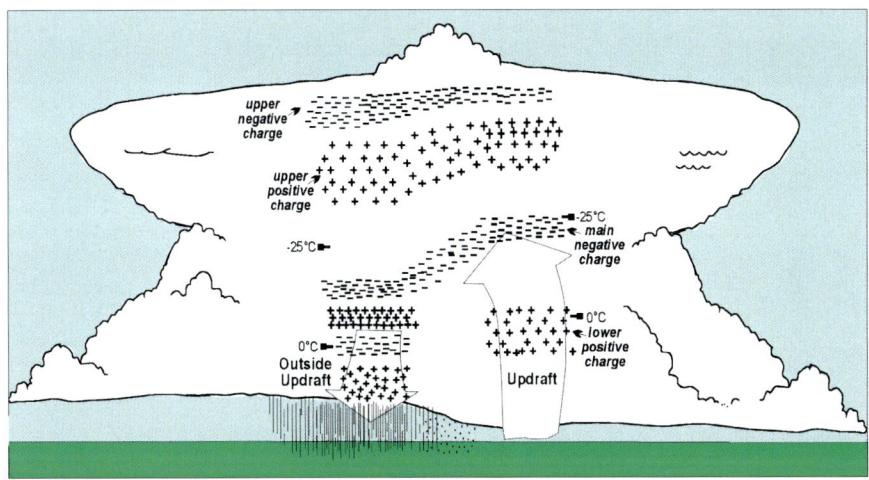

Fig. 6.4 A more detailed charge structure of a thundercloud created using in situ electric field measurements and other meteorological data. The upper negative charge region is probably generated by the flow of negative charges toward the top of the cloud under the influence of the electric field generated by the positive charges (The figure is from Stolzenburg et al. [6]). Note that the detailed charge structure of the cloud is significantly different from the tripolar structure assumed in various analyses (Adapted from Stolzenburg et al. [6])

ice crystals at higher levels of a cloud are positively charged, whereas falling graupel particles at temperatures below approximately -15 to $-10\ °C$ are negatively charged, and graupel particles that fell further in the cloud are positively charged. This creates the observed tripolar structure of the cloud. It also explains why the main negative charge center is located in the region of the -15 and $-10\ °C$ isotherm. Of course, in reality the charge structure is much more complicated than this simple picture suggests [6]. The charge structure of a thunderstorm created using electric field data is shown in Fig. 6.4. Note that even though the basic structure as explained previously is still there, the details of the charge distribution is more complicated. This can be expected because of the large variation in the cloud parameters in different regions of the cloud.

6.1.2.2 Charged Structure of Clouds in Different Geographical Regions

The heights at which different temperatures occur in the atmosphere vary in different geographical regions, and in the same geographical region it may change from one season to another. Since the charging mechanism is controlled by the ambient temperature, the location of charge centers in clouds also changes in different geographical regions. Figure 6.5 shows the location of charge centers in three regions, namely, Florida, New Mexico, and Japan (in winter) [7]. Note that the negative charge center is located in the vicinity of -15 to $-10\ °C$ isotherms in all clouds, but their heights are different because of the difference in the temperature profile.

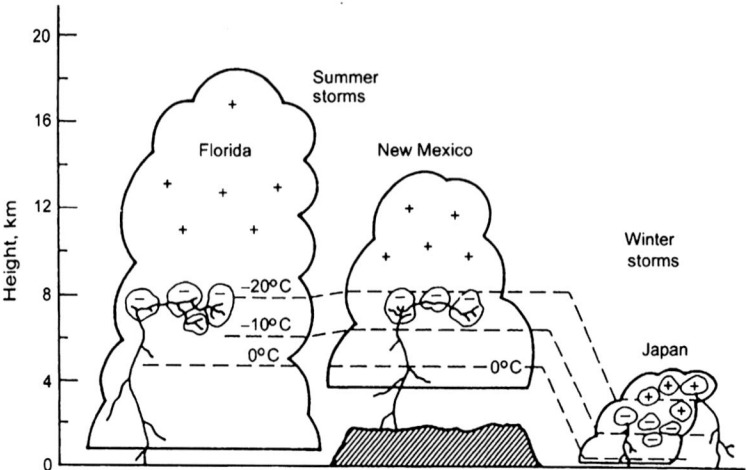

Fig. 6.5 The location of charge centers in thunderclouds is controlled by the ambient temperature, among other parameters. Since the height of a given temperature isotherm depends on the geographical region under consideration, the height of the charge centers in thunderclouds differ in different geographical regions. Figure shows the location of charge centers in summer storms in Florida and in New Mexico and in winter thunderstorms in Japan. Note that in all three regions the negative charge center is located in the vicinity of -15 and $-10\ ^\circ C$ isotherms (Adapted from Krehbiel [7])

6.1.2.3 Different Types of Clouds

What we saw earlier were typical clouds where the water content is approximately 1 g/m³. However, according to experimental data, a variety of cloud charge structures may occur in nature depending on the liquid water content and the temperature in the cloud. The wind shear could also move charge centers with respect to each other. Four types of cloud that can arise due to this variability are shown in Fig. 6.6. The cloud in Fig. 6.6a is the most typical where a negative charge is located below a positive charge. These clouds produce predominantly lightning flashes that bring a negative charge to Earth. However, if there is wind shear, the positive charge center could be sheared away from the negative charge center, making it easier to produce lightning flashes that transport a positive charge to ground (Fig. 6.6b). If the meteorological conditions are such that the depletion of water content by a riming process and rain formation is suppressed, the charging process may take place under conditions of high liquid water content, and in this case it is possible to obtain a cloud with an inverted dipole that generates predominantly positive charges (Fig. 6.6c). Such clouds have been identified in Colorado in experiments conducted within STEPS (Severe Thunderstorm Electrification and Precipitation Study) [8]. When the meteorological conditions are such that less cloud water content is associated with laterally extensive weak updrafts, it is also possible to obtain clouds with an inverted charge dipole. Such meteorological conditions obtain in the stratiform precipitation regions of large mesoscale convective systems (Fig. 6.6d). Since the positive charge center in these clouds is closer to ground, these clouds generate predominantly positive ground flashes.

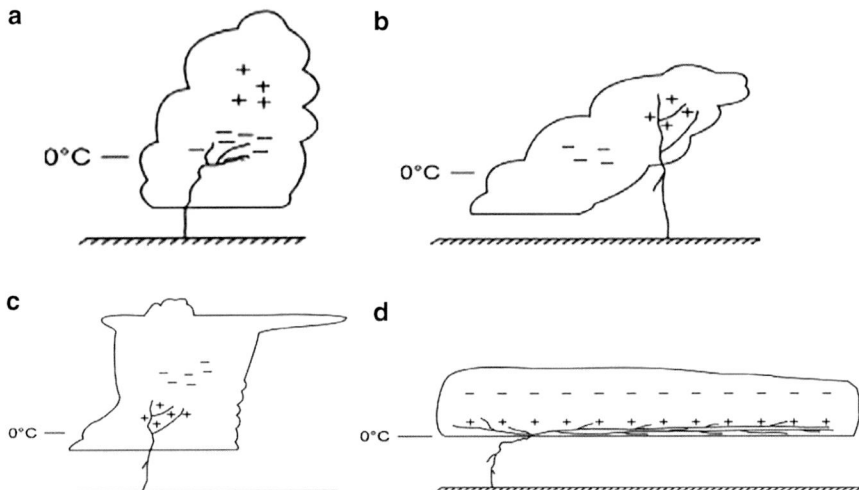

Fig. 6.6 Clouds with different charge structures caused by variation in water content, temperature, and wind shear. See text for explanation. Note that in (**c**) and (**d**) the polarity of the charge centers are inverted. This makes it easier for these clouds to generate positive ground flashes (Adapted from Williams and Yair [8])

6.2 Different Types of Lightning Flash

The basic structure of a lightning flash may be different depending on the two charged regions between which the lightning discharge takes place and the polarity of the charge transported from one region to another. The most common type of lightning flash is the one taking place between the two main charge centers inside a cloud, as depicted in Fig. 6.7a. These lightning flashes are called *intracloud lightning flashes*. Sometimes a lightning flash may take place between opposite charge centers of neighboring clouds. In such cases, they are called *intercloud lightning flashes*. Lightning flashes taking place between the negative charge center and ground (Fig. 6.7b) are called *negative ground lightning flashes*. The majority of lightning flashes that strike ground (90 %) take place between the negative charge center and the ground. A small proportion (10 %) of lightning flashes that strike the ground take place between the positive charge center and the ground (Fig. 6.7c). They are called *positive ground flashes*. It has also been observed that some lightning flashes emerge from a cloud but, instead of reaching the ground, terminate in a region near the base of the cloud (Fig. 6.7d). These types of lightning flash are called *air discharges*. Such lightning flashes might initiate in a manner similar to ground discharges but for some unknown reason could not complete the path to ground. Lightning flashes that strike the ground and those described previously originated inside a cloud. That is, they started in a cloud and ended up on ground. As we will see subsequently, under special circumstances, lightning flashes can also be triggered by tall structures at ground; these have their origin at ground level and

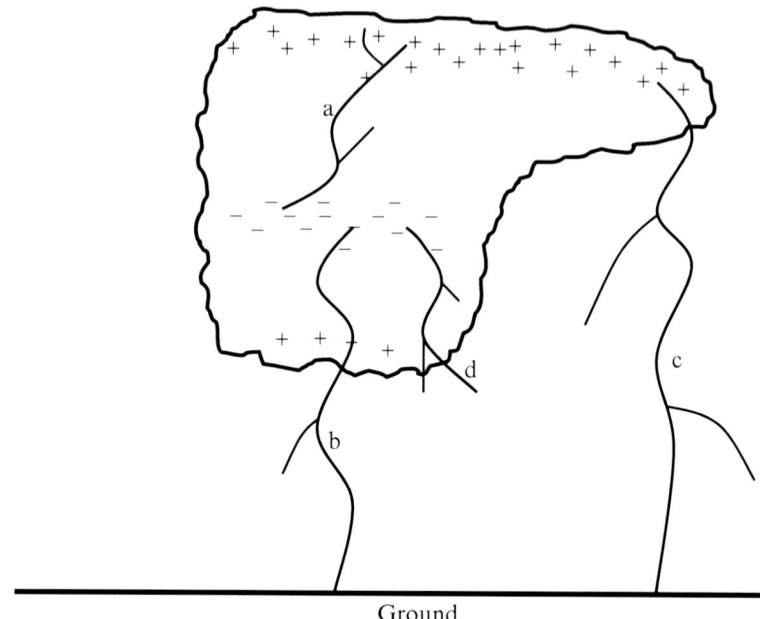

Fig. 6.7 Different types of lightning flash. (**a**) Cloud flash. (**b**) Negative ground flash. (**c**) Positive ground flash. (**d**) Air discharge. The wind shear can shift the upper part of the cloud and, hence, the positive charge center, making it possible for the positive leaders to travel to ground and giving rise to positive ground flashes without the downward moving positive leader being intercepted by the negative charge center and leading to a cloud flash (Figure created by author)

are called *upward initiated lightning flashes*. Depending on whether the upward initiated lightning flash transports negative or positive charge to ground they are called *negative* or *positive upward flashes*, respectively. Sometimes a lightning flash may start as a negative or positive ground flash but at a later stage make a connection to the opposite charge center and transport charge of opposite polarity to ground. These are called *bipolar lightning flashes*.

6.3 Global Distribution of Lightning Flashes

The global distribution of lightning flashes can be determined with satellites that record lightning flashes by using the light generated by them [9]. Since lightning channels are obscured by clouds, observations using satellites cannot distinguish ground flashes from cloud flashes. The latest distribution of lightning flashes obtained from satellite data is shown in Fig. 6.8. According to these observations, approximately 40–100 lightning flashes take place each second around the globe. Observe that most of the lightning flashes take place over land regions rather than over the sea. The reason for this is the less convective intensity in oceanic

6.4 Density of Lightning Flashes Striking Earth

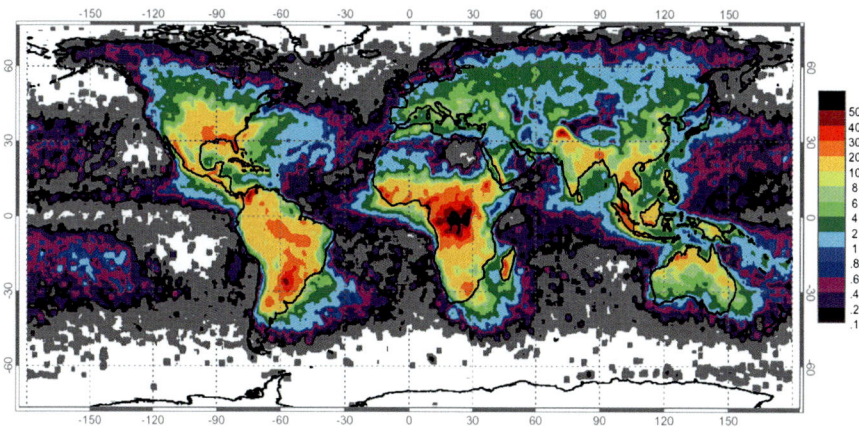

Fig. 6.8 Global distribution of lightning, April 1995–February 2003, from combined observation of NASA OTD (April 1995–March 2000) and LIS (January 1998–February 2003) instruments. The numbers on the color code refer to the annual number of lightning flashes per square kilometer [Courtesy NASA TRMM team]

thunderstorms compared to continental ones. For example, the updraft speeds of oceanic thunderstorms may reach a maximum of approximately 10 m/s, while in continental thunderstorms updraft speed may reach 50 m/s or greater. Updraft speed is closely related to the charging process of thunderstorms, and higher updraft speeds lead to vigorous charging. For this reason, the lightning activity over oceans is an order of magnitude lower than that over land. Furthermore, over land regions, most lightning activity is confined to the tropics.

6.4 Density of Lightning Flashes Striking Earth

In meteorological studies, it is the total number of lightning flashes taking place around the globe that is of interest. However, lightning protection engineers are interested in the number of lightning flashes striking in a given region over a given time interval. Usually, this number is given as the number of lightning flashes striking 1 km^2/year. This number can be obtained by directly locating the lightning flashes that strike in a given region [10]. However, such advanced lightning location systems are not available in many regions of the globe, and scientists have set about to obtain this parameter using what are known as *thunderstorm days*. A thunderstorm day is a day where thunder is heard. Records of thunderstorm days are available from many regions because the parameter is measured routinely by meteorological observations. Figure 6.9 shows the global distribution of annual thunderstorm days. Note again that the number of thunderstorm days is higher in tropical regions and can reach values as high as 150–200 days in certain regions in Asia, Africa and South America. By studying the number of lightning flashes

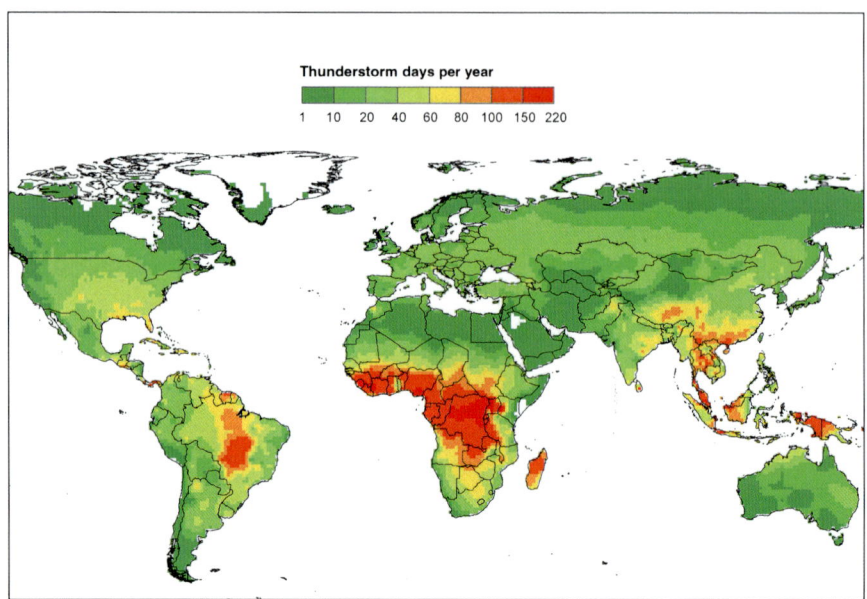

Fig. 6.9 Global distribution of thunderstorm days. (Adapted from [12])

striking in a given region using lightning location systems, scientists have come up with an equation to relate ground flash density to thunderstorm days. This relationship is not a universal one, and several relationships have been proposed for different regions. One equation frequently used by scientists is [11]

$$N_g = 0.04 T_d^{1.25}, \tag{6.1}$$

where N_g is the ground flash density in flashes/km²/year and T_d is the number of thunderstorm days. For example, according to the preceding equation, a region having 100 thunderstorm days/year is expected to receive approximately 13 lightning flashes in each square kilometer per year. Another parameter that is of interest is the annual number of thunderstorm hours. This is the number of hours of thunder activity in a given region in a year. This parameter, T_h, is related to the ground flash density by [13]

$$N_g = 0.054 T_h^{1.1}. \tag{6.2}$$

As we will see in subsequent chapters, the ground flash density is one of the most important parameters in lightning protection.

References

1. Jayaratne R (2003) Thunderstorm electrification mechanisms. In: Cooray V (ed) The lightning Flash. IET Publishers, London, pp 17–44
2. Takahashi T (1978) Riming electrification as a charge generation mechanism in thunderstorms. J Atmos Sci 35:1536–1548
3. Jayaratne ER, Saunders CPR, Hallett J (1983) Laboratory studies of the charging of soft hail during ice crystal interactions. Q J R Meteorol Soc 109:609–630
4. Williams E, Zhang R, Rydock J (1991) Mixed-phase microphysics and cloud electrification. J Atmos Sci 48:2195–2203
5. Baker B, Baker MB, Jayaratne ER, Latham J, Saunders CPR (1987) The influence of diffusional growth rates on the charge transfer accompanying rebounding collisions between ice crystals and soft hailstones. Q J R Meteorol Soc 113:1193–1215
6. Stolzenburg M, Rust WD, Marshall TC (1998) Electrical structure in thunderstorm convective regions. 3. Synthesis. J Geophys Res 103:14097–14108
7. Krehbiel PR (1986) The electrical structure of thunderstorms. In: Krider EP, Roble RG (eds) The earth's electrical environment. National Academy Press, Washington, DC, pp 90–113
8. Williams E, Yair Y (2006) The microphysical and electrical properties of sprite –producing thunderstorms. In: Fullekrug M, Mareev EA, Rycroft MJ (eds) Sprites, elves and intense lightning discharges. Springer, Dordrecht, Netherlands, pp 57–83
9. Christian HJ, Blakeslee RJ, Goodman SJ (1989) The detection of lightning from geostationary orbit. J Geophys Res 94:13329–13337
10. Cummins K, Murphy M (2009) An overview of lightning locating systems: history, techniques, and data uses, with an in-depth look at the U.S. NLDN. IEEE Trans Electromagn Compat 51(3):499–518
11. Anderson RB, Eriksson AJ, Kroninger H, Meal DV, Smith MA (1984) Lightning and thunderstorm parameters. In: Lightning and power systems. IEE conf. publ. no. 236. IEE, London
12. Isaksson L, Wern L (2010) Åska i Sverige 2002-2009. Meteorologi Nr 141 (in Swedish)
13. MacGorman DR, Maier MW, Rust WD (1984) Lightning strike density for the contiguous United States from thunderstorm duration records, NUREG/CR-3759, Office of Nuclear Regulatory Research, US Nuclear Regulatory Commission, Washington, DC

Chapter 7
Mechanism of Lightning Flashes

7.1 Initiation of Lightning Flashes Inside a Cloud

7.1.1 Conditions Necessary for Initiation of Lightning Flashes

As we discussed in Chap. 2, the initiation of an electric discharge requires the electric field in the air to increase beyond a critical electric field, which depends on the air density. At sea level this critical electric field is approximately 3×10^6 V/m. The critical electric field necessary for electrical breakdown decreases with atmospheric density, and at a height of approximately 5 km the value of this field is approximately 1.5×10^6 V/m. It is important to note that these values of the electric fields are applicable in clear air devoid of particles. However, the presence of small particles in air can decrease the background electric field necessary for electrical breakdown due to field enhancement. For example, a spherical particle in a background electric field of strength E gives rise to an electric field that varies as $3E \cos \theta$ on its surface (Fig. 7.1). Thus, the maximum electric field on the surface of the sphere is $3E$. If the particle has a pointed shape, then the field enhancement will be higher. It is important to recognize that to create an electrical breakdown, it is not sufficient for the electric field to reach the critical value at a point. The electric field should increase above the critical value over a critical region so that the electron avalanche process can be initiated. A thundercloud contains a variety of small particles, such as water droplets, ice crystals, and graupel, and their presence will reduce the background electric field necessary for electrical breakdown to a value on the order of 500 kV/m. However, only rarely are such high electric fields observed inside thunderclouds. Measurements conducted inside thunderclouds consistently show typical electric field values of the order of 100–150 kV/m [1]. These values are significantly below the values necessary for electrical breakdown. The question is how the electric fields necessary for an electrical breakdown are

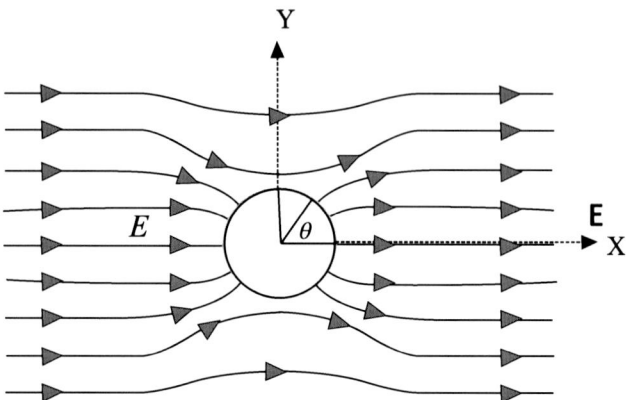

Fig. 7.1 Electric field configuration around a conducting sphere in a uniform electric field. Note the concentration of electric field lines at the two ends of the sphere facing the electric field. This concentration of field lines gives rise to an enhancement of the electric field by a factor of 3 at points where the x-axis (the direction of the electric field) cuts the sphere (Figure created by author)

achieved inside thunderclouds and what the significance is of an overall electric field of approximately 100–150 kV/m in the breakdown process.

7.1.2 Classical Explanation

In the classical explanation, it is assumed that the electric field necessary for the initiation of a discharge is created by the dynamics of the charged particles. This could represent the collision of several charged particles, which momentarily increases the electric field to achieve breakdown. During collisions several particles can act as a single object with an elongated shape, increasing the field enhancement beyond the value necessary for electrical breakdown. Or the field enhancement could be due to the sudden compression of a volume of charged particles resulting from turbulence, thereby increasing the volume charge density and leading to an increase in the electric field. The reason for not observing such high field regions in experiments could be due to the local nature of the high field regions and the transitory nature of the high electric field.

Once electrical breakdown is initiated in a small region, it gives rise to electrical streamer discharges. If the electric field exceeds approximately 250 kV/m (the critical electric field necessary for the propagation of positive streamers in reduced air density in a cloud [2]) over a region of a few meters or so, the discharges may culminate in generating a leader discharge. Once a leader is initiated, it needs approximately 100 kV/m of background electric field in the initial stages of its propagation [3]. As the leader becomes longer, the electric field necessary for its propagation decreases. Thus, a leader may propagate inside a cloud without much hindrance because of the background field of 100–150 kV/m available inside the

7.1 Initiation of Lightning Flashes Inside a Cloud   93

cloud. If the field inside the cloud is below this value, it impedes the propagation of leaders in the cloud, even if they were generated inside high field regions, thereby preventing the initiation of lightning discharges.

7.1.3 Electron Runaway Breakdown Mechanism

The other mechanism proposed for the initiation of lightning flashes is known as the *electron runaway mechanism* [4]. Let us consider this mechanism in detail. A free electron located in a gaseous medium when exposed to an electric field experiences a force equal to $-eE$, where E is the applied electric field and e is the electronic charge. Under the influence of this force and governed by the Lorentz force (Chap. 3) and Newton's second law, the electron continues to accelerate. As the electron accelerates through the gaseous medium, it collides with atoms and molecules, and this causes the electron to lose energy. Thus, this interaction of the electron with atoms and molecules generates a frictional (or drag) force that opposes the force exerted on the electron by the electric field. For a given electron energy, the drag force decreases linearly with decreasing density. Let us denote by F_D the energy lost by an electron moving a unit length because of this frictional force. The unit of this parameter is eV/m. Figure 7.2 shows how this frictional force (in eV/cm) varies as a function of the energy of the electron at standard

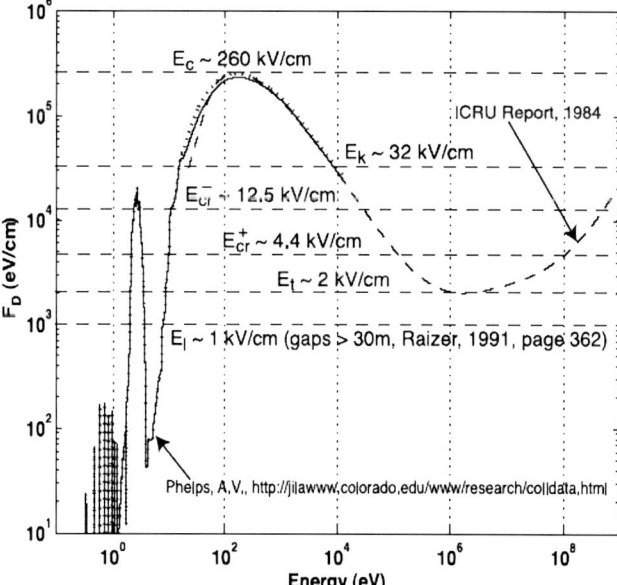

Fig. 7.2 Magnitude of so-called *drag force* as a function of electron energy. The diagram corresponds to normal atmospheric density (Adapted from Moss et al. [5])

atmospheric density. As mentioned earlier, the value of F_D scales inversely with the atmospheric density. For example, the values of F_D corresponding to half the atmospheric density (corresponding to cloud height) can be obtained by dividing the values of F_D given in Fig. 7.2 by 2. Note that the maximum value of this frictional force is (neglecting the maximum around 30 eV), approximately 2.5×10^5 eV/cm, corresponds to an electron energy of approximately 100 eV. After reaching a peak at this electron energy, the frictional force decreases with increasing electron energy and reaches a minimum when the electron energy is approximately 10^6 eV. It is interesting to note that the frictional force can be expressed as an electric field that opposes the acceleration of the electron caused by the applied field. For example, the peak frictional force experienced by an electron having an energy of 100 eV can be translated to an opposing electric field at a magnitude of approximately 250 kV/cm. If the magnitude of the background electric field that accelerates the electron is larger than the opposing electric field, the gain in the energy of the electron per unit length will be larger than the losses and the electron continues to gain energy and becomes a runaway electron. As shown in Fig. 7.2, once the electron exceeds the threshold energy of approximately 100 eV, the external electric field necessary to push the electron to runaway status decreases with increasing energy. For example, a 10^6 eV electron located inside the cloud continues to gain energy and becomes a runaway in a background electric field of approximately 100 kV/m (recall that the atmospheric density at a given cloud height is approximately half the value at ground level and therefore the values given in Fig. 7.2 should be divided by 2). Thus, a relativistic electron having an energy on the order of 10^6 eV generated by a cosmic ray in a background field of approximately 100 kV/m in a cloud continues to gain energy, and during collisions it generates one or more electrons having an energy greater than or equal to 10^6 eV. These electrons also accelerate in this field and generate more electrons. This gives rise to an avalanche of runaway electrons. These runaway electrons not only produce other runaway electrons; they also give rise to a large number of slow moving electrons (thermal electrons). In the runaway breakdown hypothesis, it is assumed that this charge separation and redistribution caused by the runaway electron avalanches changes the electric field in the cloud in such a way that it generates the conditions suitable for electrical breakdown in the cloud.

7.2 Mechanism of Lightning Flash

Once an electrical breakdown is initiated as a result of either of the aforementioned mechanisms, it gives rise to streamer discharges, and these streamer discharges in turn give rise to a leader discharge. Once a leader is created, it propagates in the background electric field, leading to either a cloud flash or a ground flash.

Unfortunately, the cloud flash remains hidden inside the cloud, and its mechanism cannot be studied by direct photography. On the other hand, since part of the ground flash occurs below the cloud, its mechanism can be inferred by studying the physical events of a ground flash that takes place below the cloud base [6].

7.2 Mechanism of Lightning Flash

However, note that the cloud base can be located at a height of 1–2 km, whereas the initiation of a lightning flash may take place at a height of approximately 7 km. Thus, a major portion of the lightning channel of a ground flash remains hidden inside the cloud.

Before describing the mechanism of ground and cloud flashes, let us first describe the mechanism by which a leader discharge, initiated inside a cloud without the aid of conducting bodies that can supply charges to the electrical discharges, propagates in the background electric field that exists inside the cloud.

7.2.1 Bidirectional Propagation of a Leader

As we discussed in Chap. 2, a leader is an electrical discharge that generates a conducting channel in air. In the case of leaders generated from grounded objects, the current necessary for the propagation of the leader is supplied from ground. However, the situation is different in the case of leaders initiated in clouds. Since there are no conducting or grounded objects to supply the charges and currents necessary for the propagation of the leader, the only mode of propagation available for the leader is to propagate as a bidirectional leader [7]. The mechanism is as follows. Once a neutral conducting channel is placed in a background electric field (the electric field inside the cloud), charges of opposite polarity are concentrated at the two ends of the conducting channel. This concentration of electric charge increases the electric field at the ends, and this in turn gives rise to streamer discharges of opposite polarity from the two ends of the conducting channel. As streamer discharges of one polarity propagate forward, the charges of opposite polarity being accumulated in the conductor are utilized by the streamer discharges of opposite polarity moving out from the other end. By concentrating their current into the streamer stem and causing it to conduct, these streamer discharges (Chap. 2) add to the length of the leader channel. Thus, the leader extends from both directions (Fig. 7.3). The total charge on the leader remains zero while positive charges are concentrated on one end and negative charges on the other. This mode of propagation of the leader is called *bidirectional leader propagation*. All lightning leaders that originate in space without the advantage of a conducting grounded

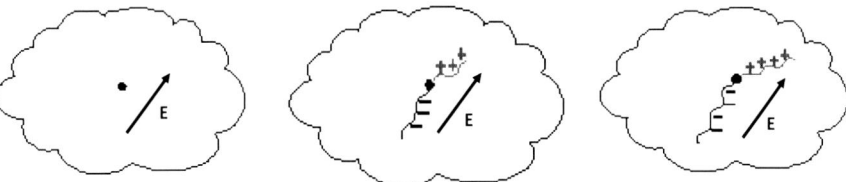

Fig. 7.3 Bidirectional propagation of a leader. Initiation of a leader inside a cloud gives rise to two leader branches, one charged positively and the other negatively, and moving in opposite directions along the field lines of the background electric field. The net charge on the leader channel is zero. In other words, the positively charged leader branch provides the negative charge for the development of the negatively charged branch, and vice versa (Figure created by author)

body propagate in this mode. The initiation of the leader may take place according to the mechanisms describes earlier. Once initiated, the proto leader channel acts as a conductor placed in a uniform electric field, and if the background electric field is strong enough, the bidirectional mode of propagation takes over.

It has been observed experimentally that the positive end of a bidirectional leader propagates more or less in a continuous manner, whereas the negative end moves in intermittent steps. The reason for the intermittent nature of the negative tip of the leader is explained in Chap. 2.

Theoretical studies and inferences made from interferometric studies (Chap. 13) show that from time to time the tip of a positive leader branch of a bidirectional leader could be cut off from the origin of a bidirectional leader, and this shutting off of the current path connecting the positive leader tip to the flash origin can give rise to discharges called *recoil leaders* [8]. In the next section, we consider how a recoil leader is created.

7.2.2 Concept of Recoil Leaders

Though the mechanism of recoil leaders is not clearly understood, some proposals based on experimental observations have been made. The mechanism by which recoil leaders are created is shown in Fig. 7.4. Consider the positive part of a bidirectional leader channel growing upward. As the tip of the positive section of the leader propagates, the conductivity of the channel at a point behind the tip could decrease to such a level that the flow of current to (or from) the tip is cut off

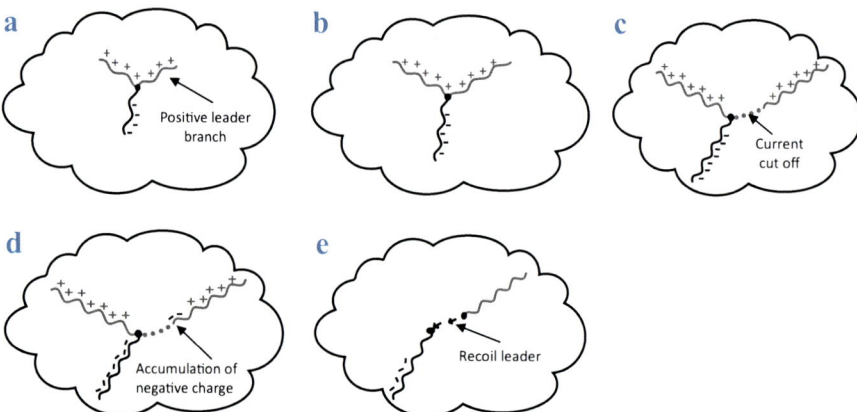

Fig. 7.4 Mechanism through which recoil leaders are created. (**a, b**) A bi-directional leader with two positive branches are created inside the cloud and one of the positive polarity branchs of the leader channel gets cut off (i.e., the channel decays) at the origin. (**c**) As this positive leader branch continues to propagate, negative charges are accumulated at the end of the positive leader channel at the cutoff point. (**d**) When the negative charge accumulated at the channel end reaches a critical value (i.e. the electric field exceeds the breakdown electric field), a negative leader will start from the negatively charged end of the channel and propagate through the cutoff gap, reestablishing the conducting path. (**e**) This negative leader is called a *recoil leader* (Figure created by author)

(as shown in Fig. 7.4). As the positive tip of the leader continues to propagate upward, a negative charge accumulates at the cutoff point. As the negative charge accumulates at the cutoff point, the electric field at the cutoff point increases. When this electric field reaches a critical value, a negative leader that travels toward the point of origin of the bidirectional leader channel is initiated. This discharge is called a recoil leader (Fig. 7.4c). Recoil leaders play a significant role in the development of lightning flashes.

7.2.3 Mechanism of Ground Flashes

The processes taking place during a ground flash can be divided into many parts. Let us consider them individually in order of their occurrence. Most features of flashes that take place below clouds were identified in the first half of the last century using streak cameras containing rotating lenses, i.e., a Boys camera [6]. The mechanisms of events taking place inside clouds were inferred from the electric and magnetic fields generated by lightning flashes. From the latter quarter of the last century on, more details of the mechanism of lightning flashes have been obtained using fast video photography and interferometric techniques [8–10].

Figure 7.5 shows an idealized sketch of a possible streak camera photograph of a ground flash. In a streak camera photograph, one can observe stepped leaders, first return strokes, connecting leaders, dart leaders, subsequent return strokes, continuing currents (not shown in Fig. 7.5), and M-components (not shown in Fig. 7.5). Such photographs provide no clues as to how lightning flashes are initiated inside clouds and how the preliminary breakdown inside clouds leads to a stepped leader. Moreover, the way in which dart leaders and M-components are initiated by recoil streamers and K-changes remains hidden inside clouds. Let us now consider these events one at a time.

7.2.3.1 Preliminary Breakdown

A ground lightning flash is usually initiated at the lower edge of a negative charge center, and there is evidence to support the suggestion that this breakdown process is aided by the electric field enhancement caused by a positive charge pocket located below the negative charge center [11]. Thus, the first breakdown event may take place between the negative charge center and the positive charge pocket (Fig. 7.6). This initial electrical breakdown is called the *preliminary breakdown*, which, as can be inferred from the electromagnetic fields generated by this process (Chap. 9), consists of a multitude of discharges taking place either in the same channel or in multiple channels. This gives rise to a burst of electromagnetic radiation. This is the first event that signals the initiation of a ground flash, and it can be identified easily in the records of remotely measured electric fields. This process cannot be observed directly because it happens inside clouds. However, one might be able to detect the scattered light emitted by this process at the base of clouds.

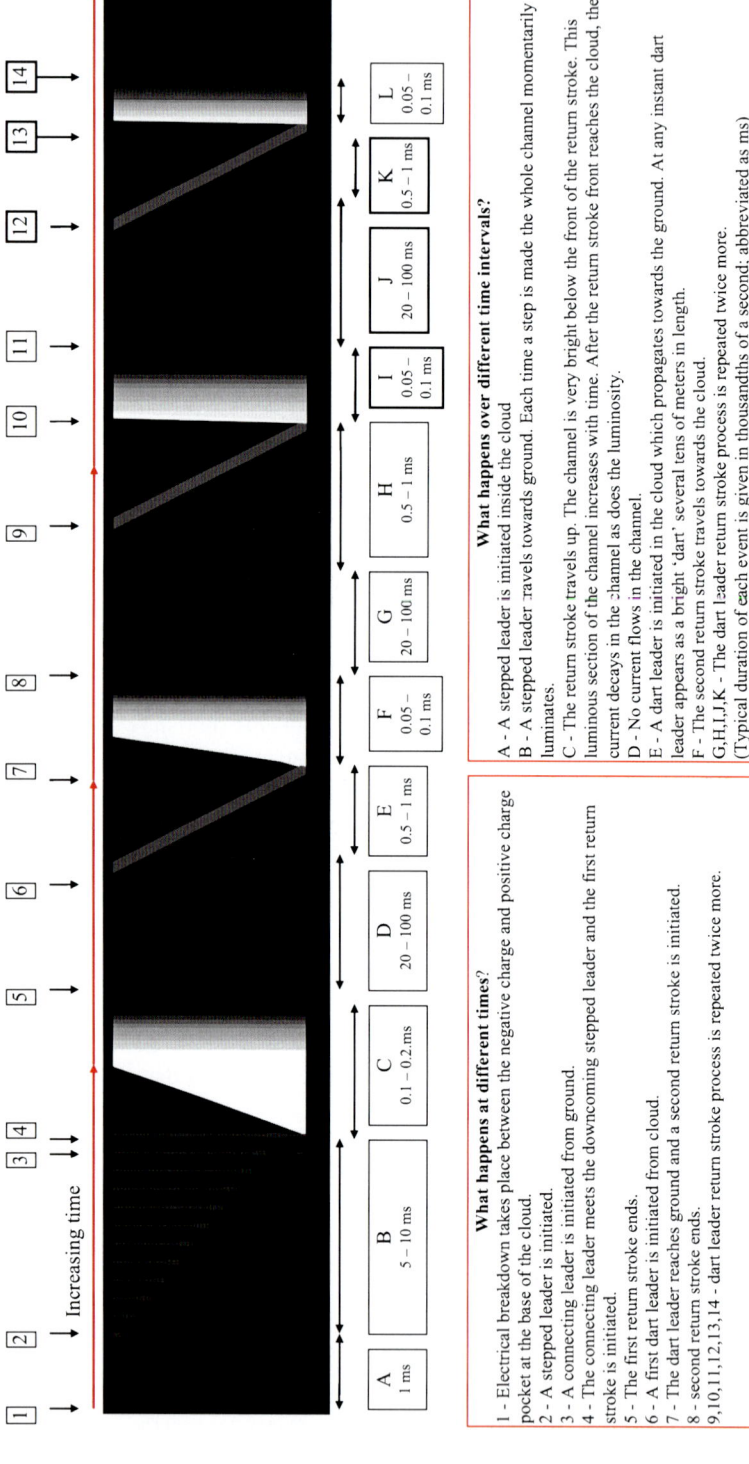

Fig. 7.5 A time-resolved picture of a lightning flash may appear like this on a photographic plate. Time increases to the right (Figure created by author)

7.2 Mechanism of Lightning Flash

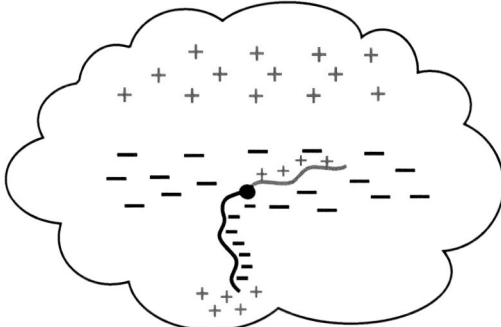

Fig. 7.6 The preliminary breakdown in a ground flash may take place between the negative charge center and the positive charge pocket. This gives rise to a negative leader branch propagating toward the ground and a positive leader branch propagating into the charge center under the influence of the background electric field produced by the negative charge center (Figure created by author)

7.2.3.2 Stepped Leader

The preliminary breakdown gives rise to a bidirectional leader inside the cloud. The positive part of the channel moves (mostly along a horizontal direction) into the negative charge center, and the negative end of the channel starts moving away from the negative charge center and toward ground. Typical for the negative end of a bidirectional leader, it moves in steps toward ground. This negative leader traveling toward the ground in a stepped manner is called a *stepped leader*. Figure 7.7 shows an idealized sketch of a stepped leader moving toward ground.

In a photographic record, the stepped leader appears to propagate toward ground in intermittent steps [6]. The length of each step is approximately 10–100 m, and the time interval between steps is approximately 10–100 μs. During the stepping process the step is illuminated for a few microseconds. The new step usually starts at the end of the previous step (see Chap. 2 for a description of the stepping process).

The stepped leader consists of a hot channel or core that is created during each step, surrounded by the charge associated with streamer discharges that led to the formation of the leader step. Experimental data show that the temperature of the core of a newly created leader channel is approximately 20,000 K [12]. One would expect this temperature to decrease with time, but since the current associated with new leader steps created ahead of this channel section passes through it, the temperature of the section does not decrease below approximately 10,000 K as the stepped leader moves forward. The core of the stepped leader is approximately a few centimeters in diameter. The leader transports a negative charge from the cloud toward ground, and in any given channel section this charge is stored both in the core and in a region around the core called the corona sheath. The charge on the corona sheath is partly due to the charge on the streamer discharges that led to the creation of the leader step and partly due to the corona charges that leak out from the central core, which is at a very high potential (Chap. 2). The diameter of the corona sheath can be a few to a few tens of meters, depending on the amount of

Fig. 7.7 Idealized sketch of a stepped leader moving toward ground. As the negatively charged branch of the leader channel travels toward ground, the positive counterpart of the bidirectional leader moves into the negative charge center, depositing a positive charge in there. Effectively, the process removes the negative charge from the negative charge center and deposits it on the negative branch of the leader channel. The downward moving leader may give rise to several branches as it travels toward ground (Figure created by author)

charge on the leader channel. The larger the charge, the larger the diameter of the corona sheath. A typical stepped leader may have a corona radius of approximately 10 m. The current flowing during the formation of the step can be a few kilo Amperes. There is a continuous current flow along the leader channel from the tip toward the cloud, and this current can be approximately 100 A. The stepped leader stores an electrical charge on it, and a typical stepped leader may contain approximately 0.001 C of charge in each 1 m of the channel close to its tip. This linear charge density may not be uniform on the stepped leader channel. As we will see subsequently, the amount of charge on the stepped leader channel and its distribution along the length of the leader channel is controlled by the potential of the cloud.

With respect to stepped leaders, two speeds are of interest: the speed of formation of the steps and the average downward speed of the stepped leader. Experimental data show that the speed of step formation can be up to approximately 10^8 m/s [13]. Since there is a pause during steps, the average downward speed of the stepped leader is approximately 3×10^5 m/s [10].

On its way toward ground a stepped leader may give rise to several branches. These branches, being negative discharges, also exhibit a stepping behavior.

7.2.3.3 Connecting Leader

As the stepped leader approaches ground, the electric field at ground level increases steadily. When the stepped leader reaches a height of approximately a few hundred meters or less from ground, the electric field at the tip of grounded structures increases to such a level that leader discharges are initiated from them (Fig. 7.8). These discharges, called *connecting leaders*, travel toward the downward moving stepped leader. The length of the connecting leader can be several tens of meters,

7.2 Mechanism of Lightning Flash

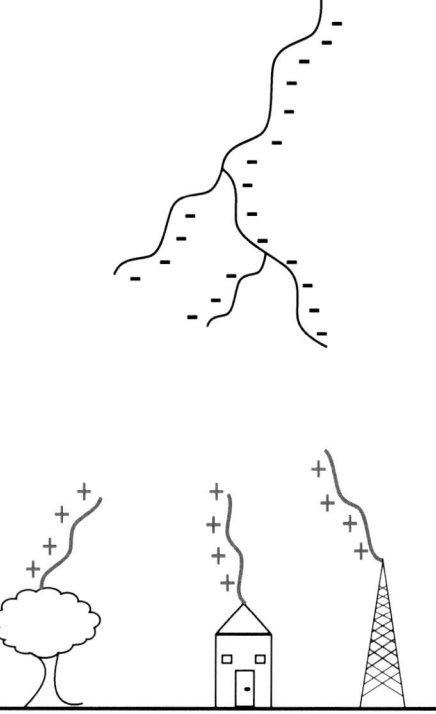

Fig. 7.8 Under the influence of the electric field of the downward moving negative stepped leader, structures at ground level launch positive connecting leaders that attempt to connect with the downward moving stepped leader (Figure created by author)

and its final length depends on several parameters. For a given stepped leader the connecting leader grows longer when the height of the structure from which it issued increases. The reason is that the taller the structure, the higher the electric field at its tip produced by the stepped leader. This is due to the field enhancement caused by the accumulation of positive charge on the structure. Thus, a connecting leader is initiated from a taller structure when the stepped leader is at a greater height from the structure than in the case of a shorter structure. Consequently, a taller structure gives rise to long connecting leaders. For a given structure, the length of the connecting leader increases with increasing charge on the stepped leader. The reason is that a stepped leader with a higher charge can create conditions suitable for the initiation of a connecting leader from the structure when the stepped leader is at a higher height. Thus, connecting leaders generated by either tall structures or stepped leaders with a large charge are longer than connecting leaders generated by either short structures or stepped leaders transporting a small amount of charge.

A connecting leader caused by the electric field of a stepped leader carries a positive charge on it. The average speed of a connecting leader is approximately 10^5–10^6 m/s [14].

Many structures at ground level can issue connecting leaders as the stepped leader approaches ground. The length of the connecting leader may depend on (a) the geometry of the structure, (b) the lateral distance of the stepped leader from the structure, or (c) the ability of the structure to supply the current necessary for the propagation of the connecting leader. Connecting leaders can also be issued from different points of the same structure. Connecting leaders issued from different structures or from different points of the same structure compete with each other to make a connection to the stepped leader.

7.2.3.4 Striking Distance

As the stepped leader approaches ground, the connecting leaders issued from different structures or from different points on the same structure accelerate toward it to make the final connection. However, very soon one of the connecting leaders takes the lead and starts approaching the downward moving stepped leader. As the distance between the tips of the two leaders (i.e., the leading connecting leader and the stepped leader) decreases, the electric field between them increases. When the separation between them reaches a critical distance, the average electric field between the tips of two leaders increases beyond the electric field necessary for streamer propagation (Chap. 2). The propagation of streamers in the intervening region leads to imminent electrical breakdown between the two leader tips, and a connection between them becomes imminent. This condition is called the *final jump condition*, and the distance between the tip of the downward moving stepped leader and the point of origin of the connecting leader at the instant of the final jump condition is called the *striking distance* (Fig. 7.9). Note that the striking distance should increase with increasing charge on the stepped leader channel and should also increase with the height of the structure. The reason for this is that both these conditions promote long connecting leaders (Chaps. 14 and 18).

It is important to note that other definitions have been given to the striking distance. For example, it was defined as the height of the tip of the stepped leader when a connecting leader is generated [15]. But since many connecting leaders are issued from grounded structures, this definition cannot be used to define the striking distance ambiguously.

The preceding sections describe the action of a connecting leader that successfully bridges the gap between a structure (or ground) and a stepped leader. The moment the connection is made between the successful connecting leader and the stepped leader, the resulting return stroke (see the next section) reduces the background electric field significantly, and the other connecting leaders stop their advance because of a lack of electric field to support their propagation.

In the preceding discussion, we tacitly avoided the description of the movement of the connecting leader and the stepped leader at times close to the final jump. For example, what is the direction of propagation of the leaders and to what extent are they influenced by the presence of the other? It is correct to say that both leaders move in the general direction of the highest electric field. Some scientists assume that both the stepped leader and the connecting leader move toward each other at times close to the final jump, whereas others assume that it is the connecting leader that moves toward the stepped leader while the path of the latter is little influenced by the connecting leader. This remains an open question at present in the lightning research community.

7.2 Mechanism of Lightning Flash

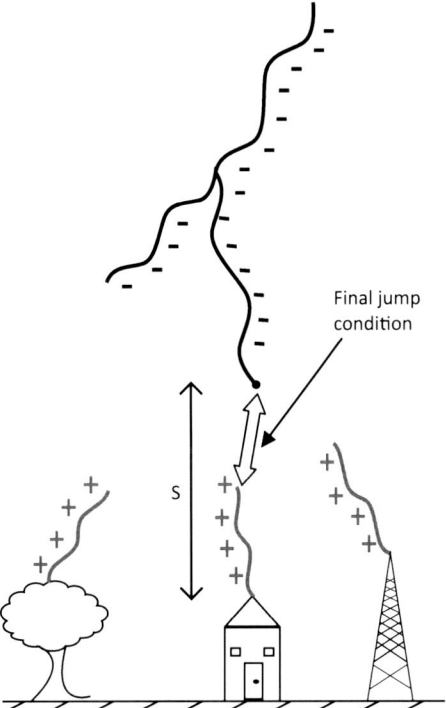

Fig. 7.9 Definition of striking distance. As the connecting leader generated by a grounded structure approaches the downward moving stepped leader, the electric field in the gap between the tips of two leaders continues to increase. When the electric field in the gap reaches a value of approximately 500 kV/m, streamer discharges start moving into the gap, making the final contact imminent. This situation is called a *final jump condition*. The distance to the tip of the stepped leader from the tip of the structure that initiated the connecting leader is the striking distance. In the figure the striking distance is denoted by S (Figure created by author)

7.2.3.5 Return Stroke

As explained previously, a connecting leader may successfully bridge the gap between ground and the downward moving stepped leader. The object that initiated the successful connecting leader is the one that will be struck by lightning. Once the connection is made between the stepped leader and ground, a wave of near-ground potential travels along the leader channel toward the cloud, and the associated luminosity event that travels upward at a speed close to that of light in free space is called a *return stroke*. Let us see what happens during this process. The discussion given below is pertinent to a negative return stroke but the physical process is identical but of opposite polarity in positive return strokes. The stepped leader channel is at a potential very close to the cloud potential. For a typical lightning flash this could be approximately 50 MV (Chap. 14). To maintain this potential, it must carry a negative charge along the channel. Now, the ground is at zero potential, and the tip of the connecting leader is also at a potential close to that of the ground. When a connection is made between the connecting leader and the stepped leader, the potential of the first element on the stepped leader that is being "touched" by the connecting leader (i.e., the tip of the stepped leader) changes from the cloud potential to ground potential. As a result, the charge located at that channel element that was necessary to maintain that element at the cloud potential is released, and this charge flows to ground along the connecting leader. Consequently, the potential at the next element on the leader channel that was adjacent to the grounded element also changes and the charge located on it is also released. In this manner a wave of ground potential travels upward and releases the charge bound on the leader channel by the cloud potential. The released charge travels to ground along the stepped leader channel. This flow of charge toward ground generates a large current in the channel, causing it to glow brilliantly as it is being heated. As the charges located at higher and higher levels along the stepped leader join in the dash toward ground, this high current region and the accompanying luminosity move up along the channel. This process of sweeping the electric charge stored on a stepped leader channel to ground is called a *return stroke* (Fig. 7.10).

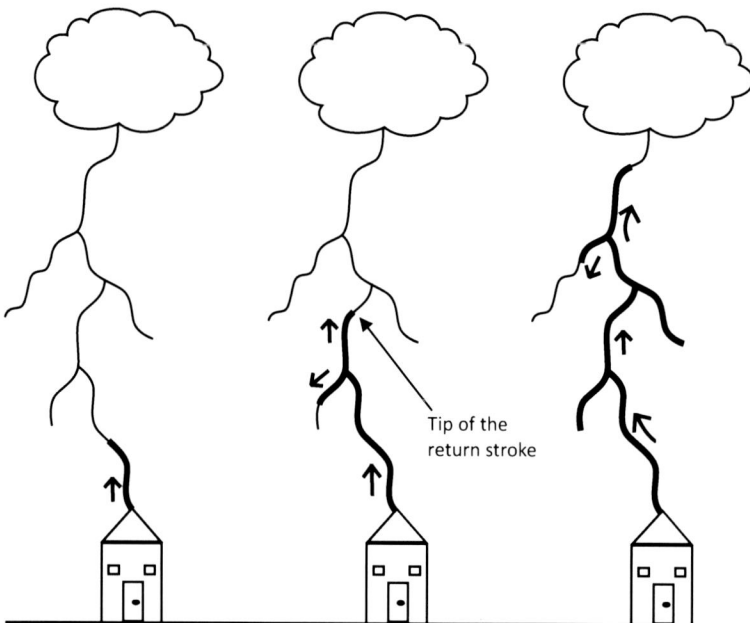

Fig. 7.10 When the downward moving stepped leader meets a connecting leader or a grounded structure (in the figure, the connecting leader is not shown for clarity), a potential discontinuity propagates upward along the stepped leader channel. This upward moving potential discontinuity is called the *return stroke*. The return stroke also moves along the branches of the stepped leader channel. In the case of negative return strokes, electrons move downward along the return stroke channel and the direction of positive current flow associated with the return strokes is as shown in the diagram. An alternative way to visualize the return stroke is to assume that it moves positive charge from ground and deposit it on the leader channel neutralizing the negative charge of the leader (Figure created by author)

7.2 Mechanism of Lightning Flash

In a return stroke initiated by a stepped leader (containing a negative charge), electrons move downward as the return stroke travels upward, much like a tube of sand that has just been opened at the bottom: the action moves up while the sand particles fall down. The speed of upward growth of a return stroke is approximately 10^8 m/s [16]. The rapid flow of electric charge out of the channel and into the striking point generates a large current. The average peak current can be as high as 30,000 A, with individual strikes reaching 80,000 A or more [17]. The total charge brought to ground during a return stroke is approximately 5 C (more details on the properties of return-stroke currents can be found in Chap. 8).

The rapid generation of heat during a return stroke raises the temperature of the air in the discharge channel to approximately 30,000 K in a few microseconds [18]. This temperature is approximately six times higher than that of the surface of the Sun. This almost instantaneous heating of the air in the return-stroke channel causes it to expand with such force that it generates a shock wave in the air. We experience the remnants of this shock wave as thunder.

Usually, a stepped leader consists of several branches, and only one of the branches of a stepped leader reaches the ground first. What happens to the other branches that were making their way through the air toward ground? When the upward moving return-stroke front reaches a branch point in the stepped leader channel, it forms two fronts, one moving along the original direction (i.e., along the main channel) and the other along the branch (Fig. 7.10). When the return stroke moves along a branch (which may still be trying to reach ground), the charges in it stop their abortive attempt to create a path to ground and follow the hot channel of the return stroke to ground. Because the branch is deprived of its electric charge, it cannot develop any further. This current flow from the branch into the main channel gives rise to a sudden increase in the current flow in the main channel. The increase in current flow gives rise to an increase in brightness of the main channel. The increase in the brightness of the main channel that happens when the return-stroke front encounters a branch is called a *branch component*.

The thin bright channel that can be observed by the naked eye during a ground lightning flash is a manifestation of a return stroke. The eye is not fast enough to resolve either the propagation of the return stroke or the short time interval between the passage of the stepped leader and the return stroke. To the naked eye it appears that the whole channel becomes bright simultaneously.

In the previous section it was mentioned that a leader channel consists of a thin core surrounded by a region consisting of defunct streamer discharge channels carrying a charge i.e., a corona sheath. A return stroke travels along the central core of a leader channel. It is this channel whose potential is transferred from cloud potential to ground potential during a return stroke. This change in potential of the central core generates a high electric field on its surface, giving rise to positive streamers. These positive streamers carrying a positive charge move into the corona sheath and neutralize the negative charge on it (Fig. 7.11). As the positive streamers originating from the central core move into the corona sheath, they leave behind a negative charge on the central core. Once released into the central core, the negative charge flows to ground along the return-stroke channel. The net effect is the transfer of negative charge from the corona sheath along the return-stroke channel to ground.

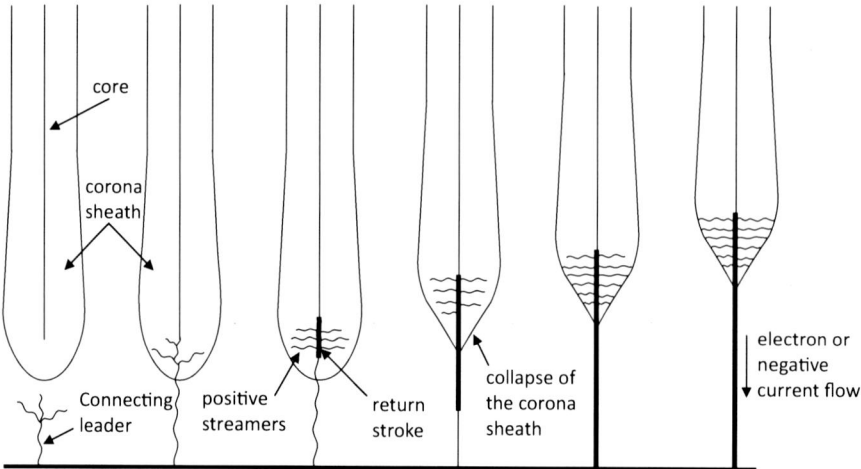

Fig. 7.11 Detailed mechanism of return strokes. A return stroke is initiated when the connecting leader, after penetrating the corona sheath of the stepped leader, meets the hot core of the stepped leader. Initially, the return stroke propagates in two directions, one front toward ground and the other front toward the cloud. As the return stroke passes through the stepped leader channel, it extends positive streamers and neutralizes the negative charge on the corona sheath (i.e., collapse of corona sheath). This process of neutralization proceeds upward, while the electrons released by the neutralization process generate a (negative) current that travels downward (i.e., the positive current travels upward) (Figure created by the author)

One interesting question is whether the return-stroke channel ends up being neutral or charged at the end of the return-stroke process. Actually, whether the channel is charged or not depends on the background electric field present at the moment of the leader propagation. Since the return-stroke channel is near zero potential and is located in the background electric field of the cloud, electric charges are induced in this channel. More charges are induced at higher sections and fewer at lower sections. Thus, in principle, the return-stroke channel ends up being positively charged following the return stroke. The positive charge density is zero close to the ground and increases with increasing height. In other words, the return stroke not only removes negative charge from the stepped leader channel but also deposits a positive charge on it.

It is important to note that positive charges do not move a significant distance during the return stroke. Thus, all charge transfer processes, i.e., removal of negative charge from the leader channel and charging of the return-stroke channel with a positive charge, take place by the removal of electrons from the leader channel.

The active lifetime of a return stroke is approximately a few hundred microseconds. After this time the channel ceases to glow, but it remains hot at a few thousand degrees for several tens of milliseconds. Sometimes, this is the end of the lightning flash, but in many cases events take a more dramatic turn.

7.2.3.6 Recoil Leaders

The arrival of the first return-stroke front at the point of origin of a lightning flash inside the cloud leads to a change in potential in the vicinity of this point. This change in potential may initiate a positive discharge that travels away from the end of the return-stroke channel into the charge center. Recall that as the negative stepped leader was traveling downward toward ground, its positive counterpart was traveling into the negative charge center. From time to time, recoil leaders travel along branches of this positive leader channel toward the flash origin. On occasion, these discharges may die out before they make contact with the cloud end of the return-stroke channel. Such events are called *K changes*. If these discharges make contact with the previous return-stroke channel, which is still hot and therefore a preferable path for an electrical discharge to ground, it will lead to an electrical discharge that travels toward ground. These electrical discharges are called *dart leaders*.

7.2.3.7 Dart Leaders and Dart Stepped Leaders

If the return-stroke channel happens to be in a *partially conducting stage with no current flow* during the arrival of a recoil leader to the flash origin, it may initiate a dart leader that travels along the still hot but defunct return-stroke channel toward ground (Fig. 7.12). Similarly to a stepped leader, it transports a negative charge toward the ground, but unlike a negative stepped leader, its tip travels continuously instead of making steps. The reason for the absence of steps in the dart leader is that the hot, defunct return-stroke channel makes it much easier to create an electrical breakdown at the tip of the dart leader, and the physical process that is needed for a leader to travel in virgin air becomes unnecessary. In a time-resolved photograph, a dart leader appears as a dart that travels toward ground. It has this appearance because most of the electrical activity that gives rise to the generation of light takes place over a few meters of the tip of the dart leader. The length of this bright dart is approximately 10–70 m. Recall that there is a channel connecting this dart to the cloud, and, though not visible, a current flows along this channel as the dart moves forward. Usually, the majority of dart leaders do not create any branches. However, occasionally, a dart leader immediately following the first return stroke may give rise to a few branches. The current associated with the dart leader is approximately 1 kA, and its speed of propagation is approximately 10^7 m/s [19]. As mentioned previously, the dart leader carries a negative charge, and the charge density at the ground end of a typical dart leader is approximately 0.0001–0.00015 C/m [20] (see also Chap. 14).

In some instances, as the dart leader progresses toward ground, it may encounter, especially close to ground, a defunct return-stroke channel section whose ionization has decayed to such an extent that it cannot support the continuous propagation of the dart leader. In this case, the dart leader may start to propagate toward ground as a stepped leader. Such a leader is called a *dart-stepped leader*. Sometimes the lower part of the channel has decayed to such an extent that the dart leader stops *before actually reaching the* ground. These are termed *attempted leaders*.

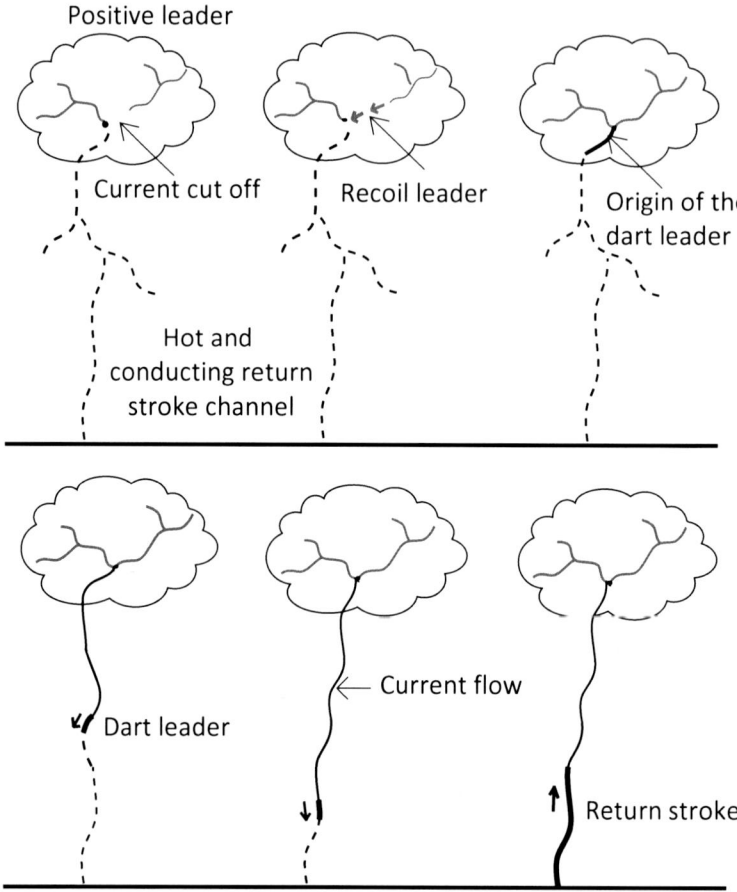

Fig. 7.12 A dart leader originates when a recoil leader generated by a positive branch of a leader meets the hot and conductive (but no current flow) return-stroke channel at the point of origin of the lightning flash. The encounter gives rise to a dart leader that propagates toward ground. Only the region close to the head of the dart leader (a few meters) is bright, even though a current is flowing along the whole channel behind the head of the dart leader. Because of this, the dart leader appears as a bright dart (hence the name) that travels from cloud to ground. The dart leader charges the channel with a negative charge. When the dart leader reaches the ground, a return stroke is initiated (Figure created by author)

7.2.3.8 Subsequent Return Strokes

When the dart leader reaches the ground, another return stroke, called a *subsequent return stroke*, is initiated. Recall that the potential of a dart leader is much greater than the ground potential (though not as high as the potential of the stepped leader because the removal of the charge during the first return stroke reduces the could potential), and the subsequent return stroke is a discharge that transports the ground potential along the channel. The mechanism of charge transfer is identical to that of the first return stroke. However, its speed of upward propagation can be somewhat higher than that of the first return stroke. Since the charge transported by a dart leader is less than that of a stepped leader, the current generated by subsequent return strokes are usually less than that of the first strokes. Typically, the peak current of a subsequent return stroke is approximately 12 kA (see Chap. 8 for more details on subsequent return-stroke currents). When the subsequent return-stroke front reaches the cloud, activity similar to that which occurred after the arrival of the first return stroke at the cloud's end may take place. This activity may cause another dart leader to travel along the subsequent return-stroke channel, leading to another subsequent stroke.

In the literature on lightning, the electrical activity in a cloud that takes place between return strokes are collectively known as *junction processes* or *J processes*. Note also that branch components occur mainly in the first return strokes and occasionally in the first subsequent stroke (i.e., the one immediately following the first return stroke). This is because, in general, dart leaders do not give rise to branches. A typical ground flash may last for approximately 0.2–0.3 s, with a mean number of strokes of between 3 and 4.

7.2.3.9 Continuing Currents

In certain instances when the return-stroke front reaches the flash origin, the electrical activity in the cloud may continue to feed the return-stroke channel with a low-level current for a long duration. These currents are called *continuing currents*. Continuing currents may last anywhere from 10 ms to hundreds of milliseconds, and the current flowing along the channel may have magnitudes in the range of several tens to hundreds of amperes [21, 22]. In many cases, continuing currents are initiated by subsequent strokes, and on average the percentage of lightning flashes that support continuing currents is approximately 30–50 % [23].

7.2.3.10 M-Components

If the return-stroke channel happens to be carrying a continuing current at the time of arrival of a recoil leader to the flash origin, the encounter results in a discharge that travels toward the ground along the channel supporting the continuing current (Fig. 7.13). These discharges are called *M-components*. When M-components reach the ground, no return strokes are initiated, but recent analyses of the electric fields generated by M-components show that the current wave associated with them may reflect from the ground [24].

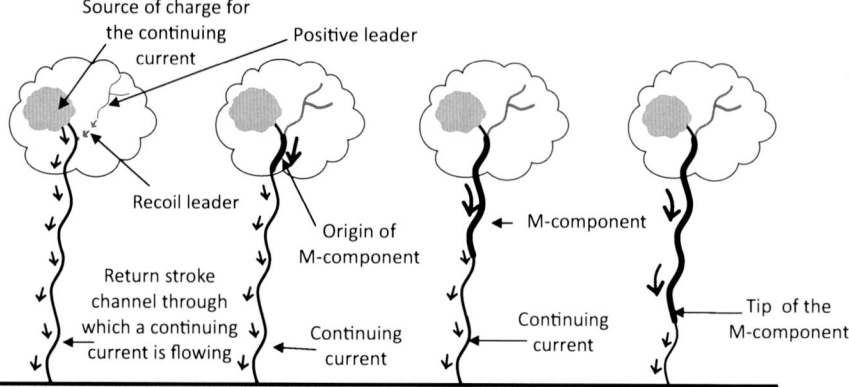

Fig. 7.13 M-components are created in return-stroke channels through which a continuing current is flowing. A recoil leader generated by a branch of a positive leader with a current cutoff makes a connection with the return-stroke channel through which a continuing current flows. The interaction gives rise to an enhancement of the current flow in the channel, and this disturbance (or the current pulse) travels toward ground along the return-stroke channel. This is called an *M-component*. Research has shown that when an M-component reaches the ground, the current pulse associated with the M-component is reflected at ground end with a reflection coefficient equal to approximately one (Figure created by author)

7.2.4 Mechanism of Cloud Flashes

Cloud flashes normally occur between the main negative and upper positive charge of a cloud. Most of the information available today on the mechanism of cloud flashes is based on electric field measurements. More recently, scientists have made important discoveries utilizing very high-frequency (VHF) radio imaging techniques [8, 9, 25]. The following description of a cloud flash is based on these observations (Fig. 7.14).

A cloud flash commences with a movement of negative discharges from the negative charge center toward the positive one in a more or less vertical direction. The vertical channel develops within the first 10–20 ms from the beginning of the flash. This channel is a few kilometers in length and developed at a speed of approximately 1.5×10^5 m/s. Even after the formation of the vertical channel, it is possible to detect an increase in the electrostatic field, which is indicative of a negative charge transfer to the upper levels along the vertical channel.

The main activity following the formation of a vertical channel is the horizontal extension of channels in the upper level (i.e., channels in the positive charge center). These horizontal extensions of the upper-level channels are correlated to brief electrical breakdowns at the lower levels, followed by discharges propagating from the lower level to the upper level along the vertical channel. Thus, upper-level breakdown events are probably initiated by electric field changes caused by the transfer of charges from lower levels. For approximately 20–140 ms of a cloud flash, repeated breakdowns occur between the lower and upper levels along the

7.2 Mechanism of Lightning Flash

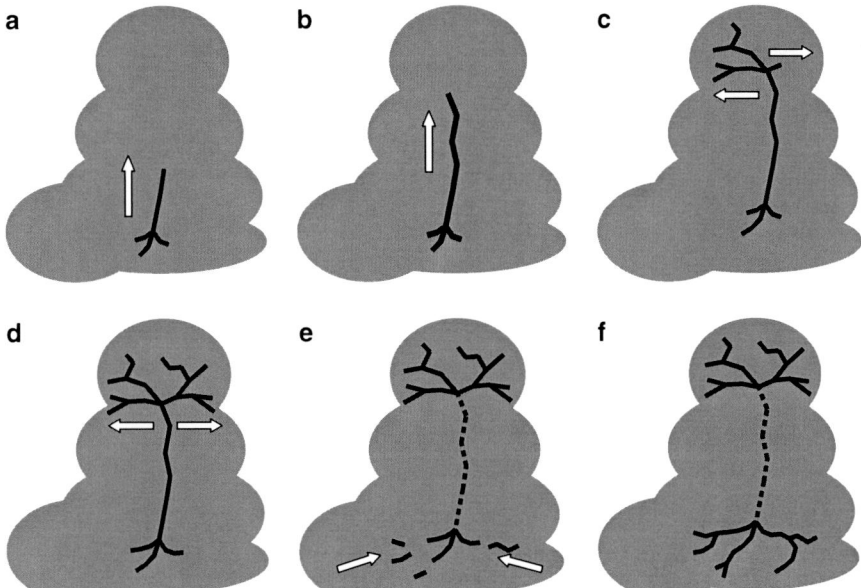

Fig. 7.14 Mechanism of a cloud flash. The cloud flash commences with a movement of negative discharges from the negative charge center toward the positive one in a more or less vertical direction. This is the initial stage (**a, b**). This stage is followed by an active stage in which horizontal extension of the upper-level channels takes place while charge is being transported from the lower level to the upper level along the vertical channel (**c, d**). In the latter part of this active stage, significant extensions of the lower-level channels take place, but the extensions take place retrogressively (**e**). In the final stage, the conductivity of the vertical channel decreases and the upper-level channels are cut off from the low-level channels (**f**). The *arrows* indicate the general direction of the discharge development (Figure created by author)

vertical channel. These discharges transport negative charges to the upper levels. Breakdown events of this type can be categorized as *K changes*. In general, the vertical channels through which these discharges propagate generate no radiation in the radio frequency range, indicating that they (i.e., the vertical channels) are conducting. This is so because, in general, conducting channels do not generate radio frequencies as discharges propagate along them. Occasionally, however, a discharge makes the vertical channel visible at radio frequencies, and then propagation can be observed at a speed of approximately $(5-7) \times 10^6$ m/s, which is typical of K changes. This active stage of discharge may continue for up to approximately 200 ms.

In the latter part of this active stage (140–200 ms), significant extensions of the lower-level channels (i.e., channels in the negative charge center) take place, but they occur retrogressively. That is, successive discharges, or K changes, often start just beyond the outer extremities of the existing channels and then move into and along these channels, thereby extending them further. These K changes transport negative charges from successively longer distances to the origin of the flash, and sometimes even to the upper level of the cloud flash, as inferred from radio frequency emissions from the vertical channel. Sometimes, these K changes give rise to discharges that start at the origin of the flash and move away from it toward the origin of the K changes. Such discharges can be interpreted as positive recoil events that transport positive charges away from the flash origin and toward the point of initiation of the K change.

In the final part of the discharge, the vertical channel and the upper-level channels are cut off from the lower-level channels. This is probably caused by a decrease in the conductivity of the vertical channel.

7.2.5 Mechanism of Upward Initiated Lightning Flashes

The basic features of an upward initiated lightning flash are shown in Fig. 7.15. Consider a tall structure similar to that of a telecommunications mast. Consider a situation when this tower is located under a growing thundercloud. As the charging process inside the cloud intensifies, the electric field strength below the cloud increases. Because of the geometry of the tower, this background electric field strength increases by several tenfold or even several hundredfold at the tip of the tower. If this electric field exceeds a critical value over a critical distance, it will give rise to a positive leader (similar to a connecting leader) that travels toward the cloud (see Chap. 2 for details concerning the initiation of positive leaders). When it reaches the cloud, it makes a connection with the charge centers, and a current similar to a continuing current starts to flow along this channel to ground. Once this continuing current ceases, dart leaders may travel along the channel, initiating return strokes. The only difference between this lightning flash and a normal lightning flash is the absence of a stepped leader and the first return stroke and the presence of a continuing current at the beginning of the flash. The processes taking place during the dart leader–return stroke sequences are identical to those of normal lightning flashes.

7.2.6 Mechanism of Positive Ground Flashes and Their Main Difference with Negative Ground Flashes

Positive ground flashes transport positive charges from cloud to ground. A positive flash usually originates at a point close to the positive charge center. As in the case of a negative ground flash, the leader is bidirectional, but in this case the negative leader end travels toward the positive charge center while the positive leader end travels toward ground. Usually, positive ground flashes originate from positive charge centers displaced laterally from the negative charge center as a result of wind shear [26]. This makes it possible for a positive leader to propagate unhindered by the negative charge center, which otherwise would have interacted with the positive leader to generate a cloud flash. As is typical for leaders carrying a positive charge,

7.2 Mechanism of Lightning Flash   113

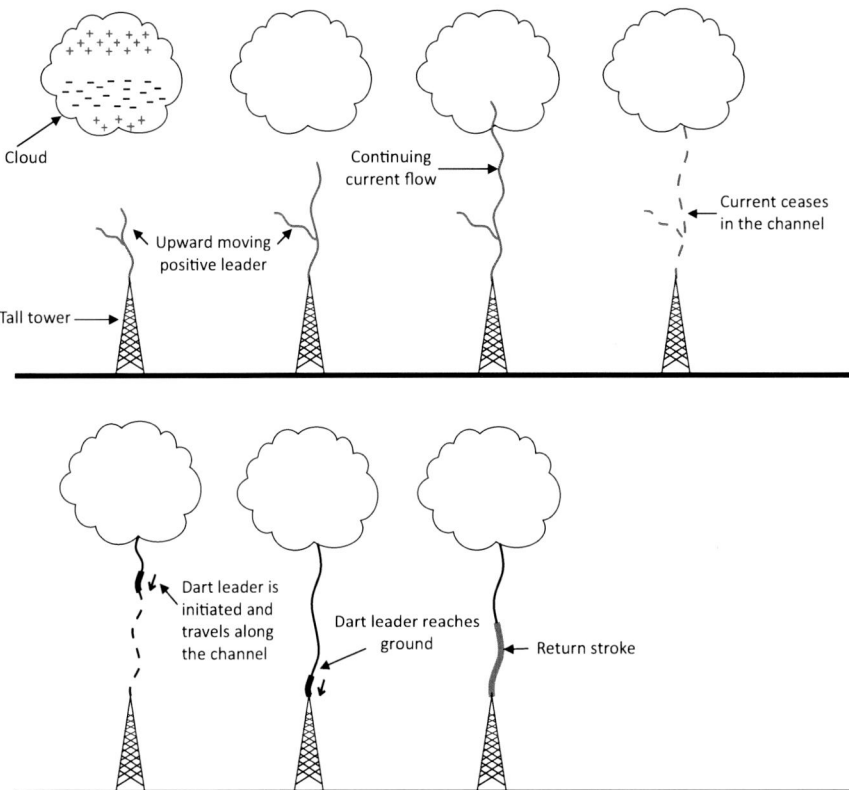

Fig. 7.15 Mechanism of upward initiated lightning flashes. Under the influence of the electric field created by a thundercloud, a leader (the polarity depends on the electric field, but let us assume it to be a positive leader) is generated from a tall structure. It travels upward (sometimes creating branches that are directed upward) toward the cloud. Once it reaches the cloud, a continuing current may start to flow along the channel to ground. After several to many tens of milliseconds the continuing current ceases. At this stage, because of the action of a recoil leader, a dart leader is created in the channel. This dart leader travels toward ground, and when it reaches the ground, a return stroke is initiated. This dart leader–return stroke process may be repeated several times in the channel (Figure created by author)

the positive leader travels more or less continuously toward ground, but the channel luminosity may change from time to time during its propagation. For this reason the term *stepped leader* is reserved for the first leaders associated with negative ground flashes. A positive leader may generate large amounts of very faint branches as it travels toward ground, and from time to time bright recoil leaders travel along these branches toward the main channel [27]. The luminosity variation observed during the propagation of a positive leader may be a result of this action of recoil leaders (Sect. 7.2.2). As the leader approaches the ground, a negative connecting leader is issued from a grounded structure, and this leader may show stepping behavior typical of negative leaders. When a negative connecting leader meets the downward moving positive leader, a return stroke is issued, and the mechanism of the return stroke is identical to that of a negative return stroke.

Experimental data show that positive lightning flashes seldom support subsequent return strokes. Most flashes contain only one return stroke. Positive return strokes may harbor very large peak currents (on the order of 200 kA) and the duration of positive currents could be much longer (several milliseconds) than those of negative return strokes (Chap. 8). The reason for the single-stroke nature of positive lightning flashes could be that the cloud-probing leader of positive flashes is a negative one, and, unlike positive leaders, negative leaders are not hindered by current cutoff (which leads to recoil leaders). Thus the current continues to flow during the return stroke until the charge reservoir is depleted.

7.3 Nomenclature of Ground Lightning Flashes

Lightning ground flashes can be divided into two types based on the point of initiation. If the point of initiation of a lightning ground flash is in a cloud, then it is categorized as a downward flash. If it initiates by a structure at ground level, it is called an upward lightning flash. Depending on the polarity of the charge brought to ground, both types can be further subdivided into positive and negative polarities. These categories are illustrated in Fig. 7.16. Typical parameters of negative ground flashes are summarized in Table 7.1.

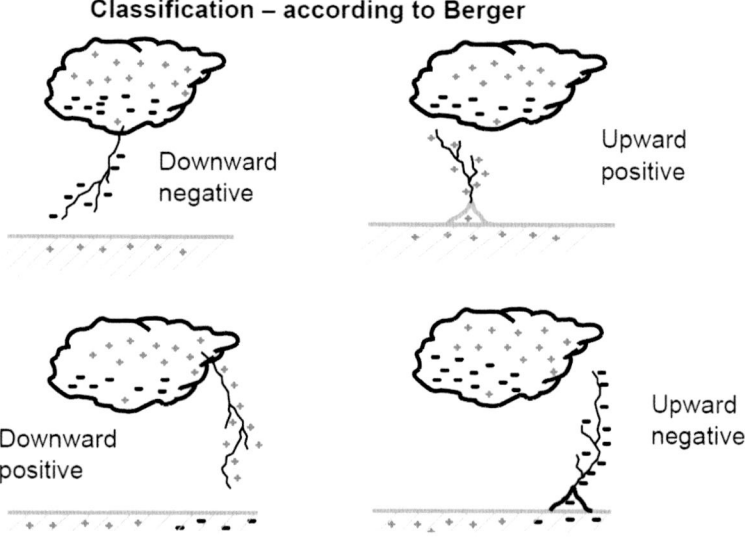

Fig. 7.16 Four categories of lightning ground flashes (Adapted from Berger [28]). Whether the lightning flash is termed up or down is determined by the direction of the leader; the polarity is determined by the charge on the leader channel

Table 7.1 Typical parameters of lightning flashes

Parameter	Typical value or range
Duration of lightning flash	200–300 ms
Number of return strokes per flash	3–4
Time interval between strokes	40–60 ms
Percentage of flashes with single strokes	20 %
Speed of stepped leaders	3×10^5 m/s
Length of steps of stepped leaders	10–100 m
Time interval between steps of stepped leaders	10–100 µs
The charge per unit length on the stepped leader (close to ground end)	0.001 C/m
Peak current in first return stroke	30 kA
Rise time of current of first return strokes	5 µs
Rate of change of current of first return strokes	10–20 kA/µs
Charge associated with first return stroke	5 C
Speed of dart leaders	10^7 m/s
Length of dart of dart leader	10–70 m
Charge on dart leader	1 C
Peak current of subsequent strokes	12 kA
Rise time of current of subsequent strokes	0.5 µs
Rate of change of current of subsequent strokes	50–100 kA/µs
Charge associated with subsequent strokes	1 C
Percentage of flashes with continuing currents	30–50 %
Duration of continuing currents	100 ms
Amplitude of continuing currents	100–200 A
Peak current of M-components	100–200 A
Rise time of M-component current	400 µs
Duration of M-component current	2 ms

References

1. Marshall TC, McCarthy MP, Rust WD (1995) Electric field magnitudes and lightning initiation in thunderstorms. J Geophys Res 100:7097–7103
2. Bazelyan EM, Raizer YP (1997) Spark discharge. CRC Press, New York
3. Les Renardiéres Group (1974) Research on long air gap discharges – 1973 results. Electra 35:47–155
4. Gurevich AV, Medvedev YV, Zybin KP (2004) New type discharge generated in thunderclouds by joint action of runaway breakdown and extensive atmospheric shower. Phys Lett A 329:348–361
5. Moss GD, Pasko VP, Liu N, Veronis G (2006) Monte Carlo model for analysis of thermal runaway electrons in streamer tips in transient luminous events and streamer zones of lightning leaders. J Geophys Res 111:A02307
6. Schonland BFJ (1956) The lightning discharge. Handb Phys 22:576–628
7. Kasemir HW (1960) A contribution to the electrostatic theory of a lightning discharge. J Geophys Res 65:1873–1878
8. Shao XM (1993) The development and structure of lightning discharges observed by VHF radio interferometer. PhD thesis, New Mexico Institute of Mining and Technology, Socorro

9. Shao XM, Krehbiel PR, Thomas RJ, Rison W (1995) Radio interferometric observations of cloud-to-ground lightning phenomena in Florida. J Geophys Res 100:2749–2783
10. Saba MMF, Ballarotti MG, Pinto O (2006) Negative cloud-to-ground lightning properties from high-speed video observations. J Geophys Res 111:D03101
11. Murphy MJ, Krider EP (1996) Lightning charge analyses in small Convection and Precipitation Electrification (CaPE) experiment storms. J Geophys Res 101(D23):29615–29626
12. Orville RE (1968) Spectrum of the stepped leader. J Geophys Res 73:6999–7008
13. Wang D, Takagi N, Watanabe T, Rakov VA, Uman MA (1999) Observed leader and return stroke propagation characteristics in the bottom 400 m of rocket triggered lightning channel. J Geophys Res 104:14369–14376
14. Yokoyama S, Miyski K, Suzuki T, Kanao S (1990) Winter lightning in Japan sea coast-development of measuring system on progressive features of lightning discharge. Trans IEEE (Pow Deliv) 5(3):1418
15. Golde RH (1973) Lightning protection. Edward Arnold, London
16. Idone VP, Orville R (1982) Lightning return stroke velocities in the thunderstorm research program. J Geophys Res 87:4903–4916
17. Berger K, Anderson RB, Kröninger H (1975) Parameters of lightning flashes. Electra 40:101–119
18. Orville RE (1968) A high speed time resolved spectroscopic study of the lightning return stroke, Parts 1, 2, 3. J Atmos Sci 25:827–856
19. Idone VP, Orvuille RE, Hubert P, Barret L, Eybert-Berard A (1984) Correlated observations of three triggered lightning flashes. J Geophys Res 89:1385–1394
20. Crawford DE, Rakov VA, Uman MA, Schnetzer GH, Rambo KJ, Stapleton MV, Fisher RJ (2001) The close lightning electromagnetic environment: dart leader electric field change versus distance. J Geophys Res 106:14909–14917
21. Shindo T, Uman MA (1989) Continuing current in negative cloud-to-ground lightning. J Geophys Res 94(D4):5189–5198
22. Thottappillil R, Goldberg JD, Rakov VA, Uman MA, Fisher RJ, Schnetzer GH (1995) Properties of M-components from currents measured at triggered lightning channel base. J Geophys Res 100(D12):25711–25720
23. Rakov V, Uman M (1990) Long continuing current in negative lightning ground flashes. J Geophys Res 95:5455–5470
24. Rakov VA, Thottappillil R, Uman MA, Barker P (1995) Mechanism of the lightning M component. J Geophys Res 100:25701–25710
25. Shao XM, Krehbiel PR (1996) The spatial and temporal development of intracloud lightning. J Geophys Res 101:26641–26668
26. Takeuti T, Nakano M, Brook M, Raymond DJ, Krehbiel P (1978) The anomalous winter thunderstorms of the Hokuriku Coast. J Geophys Res 83:2385–2394
27. Saba MF, Cummins KL, Warner TA, Krider EP, Campos LZS, Ballarotti MG, Pinto O Jr, Fleenor SA (2008) Positive leader characteristics from high-speed video observations. Geophys Res Lett 35:L07802. doi:10.1029/2007GL033000
28. Berger K (1977) The earth flash. In: Golde RH (ed) Lightning. Academic, San Diego, pp 119–190

Chapter 8
Electric Currents in Lightning Flashes

The length of a lightning channel can be tens of kilometers long, and at each point on this channel there is a current flow that changes with time. Unfortunately, it is possible to directly measure only the current at the strike point of a ground lightning flash. When discussing the features of a lightning current, we are actually referring to the features of the lightning current at the strike point. In special cases the current can be measured at elevated strike points by recording the currents in lightning flashes striking airplanes in flight. It is important to note that the features of currents flowing at other points of a lightning channel could be different from that measured at ground level. The farther away the point of interest from the strike point, the larger the possible difference between the two currents. However, before giving the features of lightning currents at strike points, let us understand how the currents in lightning strikes can be measured.

8.1 Measurement of Lightning Currents at Strike Point

There are two ways to measure directly the currents at strike points in lightning flashes: using instrumented towers or triggered lightning experiments. The knowledge we have today concerning lightning currents is obtained from these techniques.

8.1.1 *Measurement of Lightning Currents Using Instrumented Towers*

The number of lightning strikes to a structure increases as the height of the structure increases. The reason for this is as follows. The large enhancement of the background electric field at the extremity of a tall structure makes it easier for the structure to generate connecting leaders in comparison to other, shorter, structures in the vicinity. Because of this, as a stepped leader approaches ground, connecting leaders are initiated first by tall structures, and these leaders have an advantage over connecting leaders generated by shorter structures in intercepting the downward moving stepped leaders. This advantage of tall structures over other grounded objects in their vicinity increases the number of strikes to tall structures. However, in tall structures, a considerable portion of lightning strikes are caused by upward initiated lightning flashes. Indeed, the percentage of upward initiated lightning flashes increases with increasing structure height. For example, the number of lightning strikes (both upward and downward) N per year as a function of structure height H can be described by the Eq. [1]

$$N = 24 \times 10^{-6} H^{2.05} N_g, \tag{8.1}$$

where N_g is the ground flash density (in flashes per square kilometer per year) in the region. It is important to point out that it is not just the physical height of the structure that determines the number of lightning strikes to it. A tower located on a hill has an advantage because the effect of the hill serves to increase the electric field at the tip of the structure. For example, a short tower, say approximately 60 m tall on a 500-m-high hill, may be more efficient than a 100-m-tall structure over ground. Thus, what goes into the preceding equation as H is not just the physical height but the effective height as felt by the electric field.

Since taller structures receive a large number of lightning strikes, these towers could be equipped with current-measuring devices to record lightning strikes. There are two ways to measure the current at the strike point of a lightning strike. The first method is to let the current pass through a known resistor and measure the voltage of the current that passes across the resistor during lightning strikes (Fig. 8.1a).

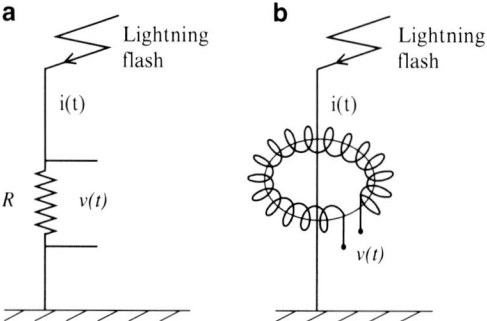

Fig. 8.1 (a) Shunt method for measuring the current in a lightning flash. In this method, the voltage appears across a resistor [$v(t)$ in the diagram] as the lightning current flows through it is measured, and from this the signature of the current is obtained. (b) Rogowski coil method for measuring the current in a lightning flash. The magnetic field forms closed loops around the current-carrying conductor and induces a voltage in the coil that is placed around the conductor. The voltage is proportional to the time derivative of the magnetic field, which in turn is proportional to the time derivative of the current. Thus, integrating the voltage makes it possible to obtain the current signature (Figure created by author)

8.1 Measurement of Lightning Currents at Strike Point 119

Since the voltage that develops across the resistor is equal to the product of the current and the resistance of the resistor, i.e., $I(t)R$, the current can be calculated from the measured voltage. Of course, the resistor should be able to withstand a lightning current, the voltage that develops across the resistor should not create a spark in the air across the resistor, and the voltage-measuring device should be able to record the fast features of the current accurately. Furthermore, the resistor and the measuring system should be arranged in such a way to counteract any voltages caused by time varying magnetic fields. This method is called the *shunt method*. The other way to measure a lightning current is to record the magnetic field generated by the current as it passes along the tower (Fig. 8.1b). As the current passes through the tower, it generates a time-varying magnetic field around it which is proportional to the current passing through the tower. This time-varying magnetic field can be measured by measuring the voltage induced in a coil placed around the tower. This method is called the *Rogowski coil method*. The voltage induced in the loop or the coil is proportional to the rate of change of the magnetic field, which is indeed proportional to the rate of change of the current passing along the tower. Thus, by integrating the measured voltage one can obtain a signal that is proportional to the current flowing along the tower. One drawback of this method is that a voltage is induced in the coil only when the magnetic field and, hence, the current change. The slower the variation in the magnetic field, the lower the voltage induced. For this reason this method has a limit on the measurement of slowly varying currents. Of course, if the current does not change, i.e., dc current, it cannot be detected at all using this method.

8.1.2 Measurements of Lightning Currents Using Triggered Lightning

In the triggered lightning procedure an artificial tall structure is created almost instantaneously below a thundercloud which is mature enough to contain charges large enough to generate lightning flashes. The technique is as follows (Fig. 8.2). Start with a long and thin wire one end of which is connected to the intended strike point of a lightning flash. The wire is wound around a spool and the spool is attached to a rocket. When a mature thundercloud is overhead, the rocket is fired toward the cloud. As the rocket moves upward, the wire unwinds from the spool but remains connected both to the strike point and to the rocket. As the rocket moves upward a vertical conductor that increases in height with time is created below the thundercloud. As the rocket travels upward, the electric field at its tip continues to increase because the vertical length of the wire increases), and when the rocket reaches a certain height, the electric field at the tip of the rocket increases to a value large enough to launch an upward moving leader from its tip. Under the influence of the electric field produced by the cloud, this leader travels toward the charge center in the cloud.

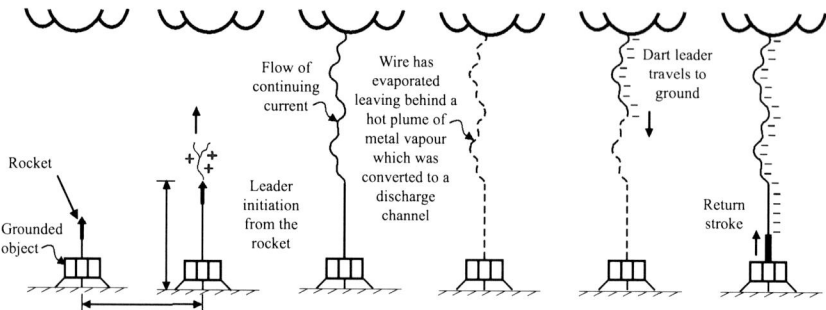

Fig. 8.2 Technique to trigger lightning flashes. A rocket with a trailing wire is shot toward a cloud. As the rocket moves upward, the electric field at the tip of the rocket increases. When it reaches a critical value, a positive leader is initiated from the rocket. This positive leader travels toward the cloud. When the leader reaches the charge center, a long continuing current is initiated in the channel. Because of the heat generated by the current, the wire explodes, leaving behind a channel filled with metal vapor. The high conductivity of the metal vapor supports the formation of a discharge along the wire residue, and the current continues to flow to ground along the channel. This may continue for several tens of milliseconds to several hundred milliseconds. After the cessation of the continuing current, dart leaders may travel along the hot channel, giving rise to return strokes (Adapted from Rakov [2])

When the leader reaches the charge center, a long continuing current is initiated in the channel. Because of the heat generated by the current, the wire explodes, leaving behind a channel filled with metal vapor. The high conductivity of the metal vapor supports the formation of a discharge along the wire residue, and the current continues to flow to ground along the channel. This may continue for several tens of milliseconds to several hundred milliseconds. After the cessation of the continuing current, dart leaders may travel along the hot channel, giving rise to return strokes. The physical processes taking place during the triggered lightning flashes are very similar to those taking place in upward initiated lightning flashes. Since the ground end of the leader channel is connected to the object under study or the current-measuring equipment or both, the current signatures propagating along the lightning channel can be measured, and their effects on the object under consideration, if any, can be studied. Currents can be measured using either a shunt or a Rogowski coil, as in the case of instrumented towers.

8.2 General Features of Currents Measured in Triggered Lightning and Towers

The lightning currents measured in instrumented towers can be separated into two categories: currents in upward initiated lightning flashes and those in lightning flashes that were initiated in a cloud and struck the towers. In the case of triggered lightning, all lightning flashes belong to the upward initiated category. As mentioned above the currents in triggered lightning flashes are similar to those in upward initiated flashes from tall towers [2].

8.2 General Features of Currents Measured in Triggered Lightning and Towers

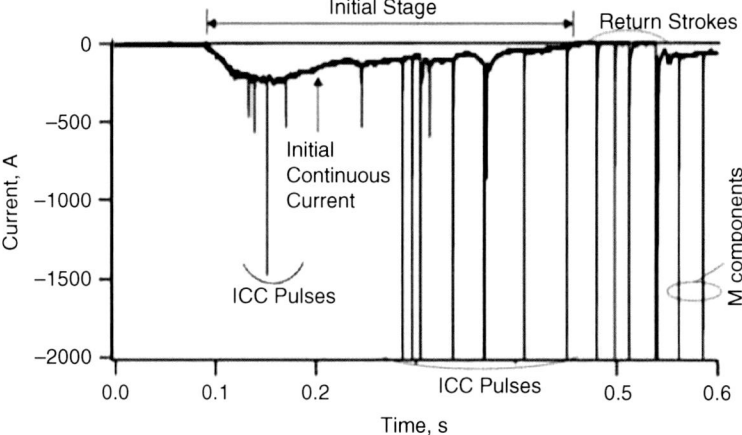

Fig. 8.3 Current waveform generated by upward initiated lightning flash at Peissenberg Tower in Germany. Note that the current waveform contains an initial continuing current stage followed by four return strokes, the last one with a continuing current supporting two M-components (Adapted from Miki et al. [3])

8.2.1 General Features of Upward Initiated Lightning Flashes

Figures 8.3 (from Ref. [3]), 8.4 (from Ref. [4]), and 8.5 (from Ref. [5]) show examples of currents measured in upward initiated lightning flashes in tall towers and triggered lightning. Figure 8.3 is an example of the current in an upward initiated lightning flash that brought a negative charge down to ground, and Fig. 8.4 is an example of a current in an upward initiated lightning flash that brought a positive charge down to ground. Figure 8.5 is an example of a lightning current measured in a triggered lightning flash that brought a negative charge to Ground. Note that in Figs. 8.3 and 8.4, negative charges flowing to ground are assumed to represent a negative current, whereas in Fig. 8.5 negative charges flowing to ground represent a positive current.

Let us consider the signature of the currents depicted in Figs. 8.3 and 8.5. Both these currents transported a negative charge to Earth, but the former is an upward initiated flash from a tower while the latter is measured from a triggered lightning flash. Observe that the general features of the currents are similar. The currents are initiated with a slowly varying current component on which rapidly varying current impulses are superimposed. This stage is called that of *initial continuing currents*. When the initial continuing current ceases, dart leaders travel along the channel, generating return strokes. Some of these return strokes can generate long continuing currents, and during these continuing current phases M-components may be observed. Note that dart leaders occur only after the continuing current goes to zero. As described in the previous chapter, the reason for this lies in the origin of dart leaders. If a recoil leader connects to a channel containing a continuing current, it generates an M-component,

Fig. 8.4 (a) Total current waveform of positive upward initiated lightning flash at Gaisburg Tower in Austria. Note that there were no return strokes in the flash (Adapted from Zhou et al. [4]). (b) First 5 ms of flash shown in **a** with pronounced pulses during rising part of wave shape (Adapted from Zhou et al. [4])

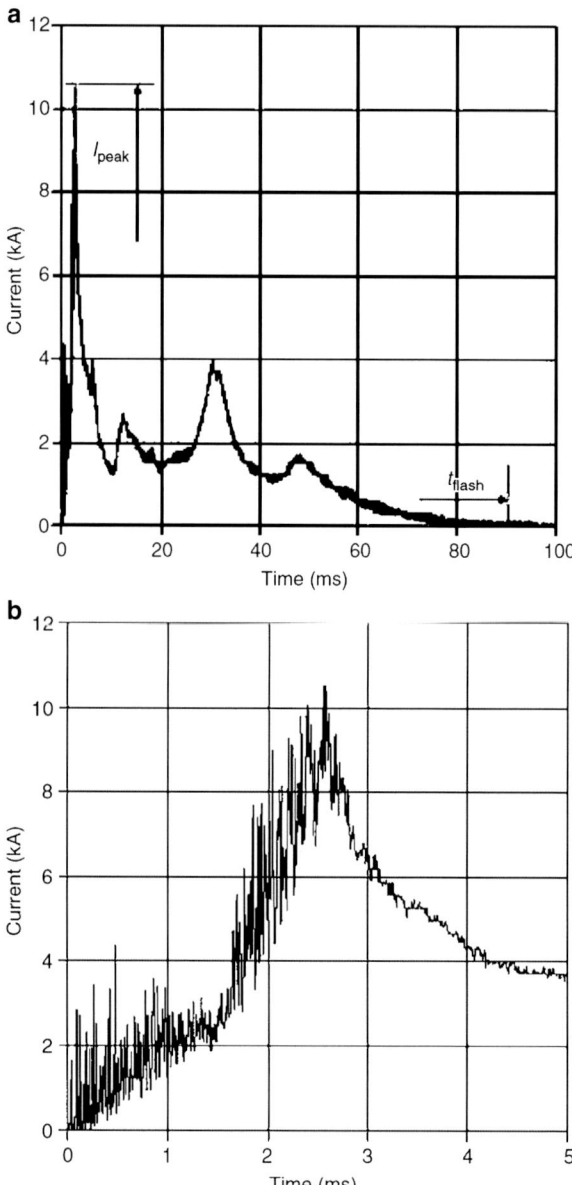

instead of a dart leader. Thus, dart leaders arise in the absence of continuing currents along the channel. Sometimes the currents in both upward initiated lightning flashes and triggered lightning flashes may terminate at the end of the continuing current phase without return strokes. Figure 8.4 shows an example of an upward initiated lightning flash that transported a positive charge to Earth – a positive upward initiated lightning flash. Note that the initial portion of the current contains bursts

8.2 General Features of Currents Measured in Triggered Lightning and Towers   123

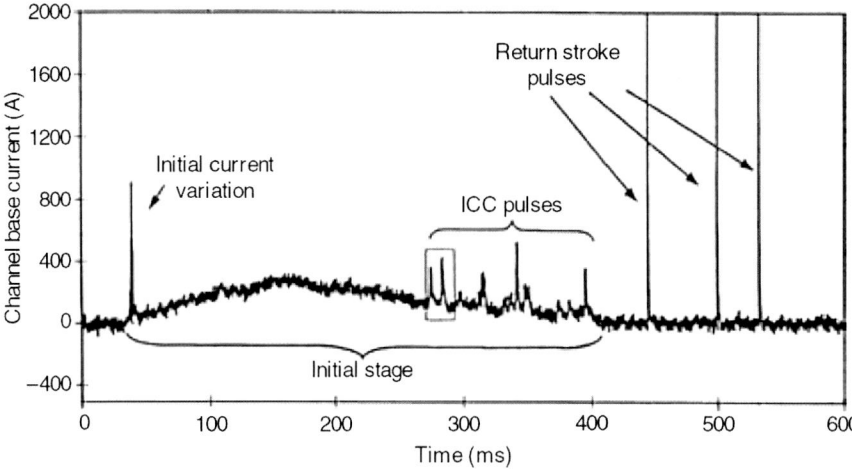

Fig. 8.5 Example of overall current record of triggered lightning flash at Camp Blanding, Florida. The initial stage consists of a continuing current. The current pulses that occur in the continuing current phase are marked as ICC pulses. They could very well be M-components. Once the current ceases, dart leader–return stroke sequences take place. Note that in the drawing, the sign convention for a positive current is opposite to that of Fig. 8.3 (Adapted from Wang et al. [5])

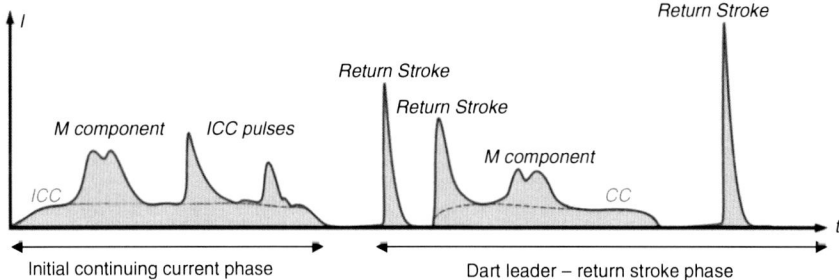

Fig. 8.6 General features of upward initiated lightning flashes. Note that some of the return strokes can give rise to continuing currents, and M-components may occur during the continuing current. The current pulses that occur in the initial continuing current phase are called ICC pulses, but they could very well be M- components. Note that in the drawing, the sign convention for a positive current is opposite to that of Fig. 8.3 (Adapted from Diendorfer [6])

of current pulses. This initial portion of the current is shown in an expanded time scale in Fig. 8.4b. These current pulses are probably generated by the stepping process associated with upward initiated negative leader. Note also that in the example shown in Fig. 8.4, there are no return strokes. Observe also that the basic features of negative ground flashes that transport negative charges to ground initiated by tall towers and triggered lightning are very similar and their current signatures can be described pictorially by the curve depicted in Fig. 8.6, which was adapted from Ref. [6].

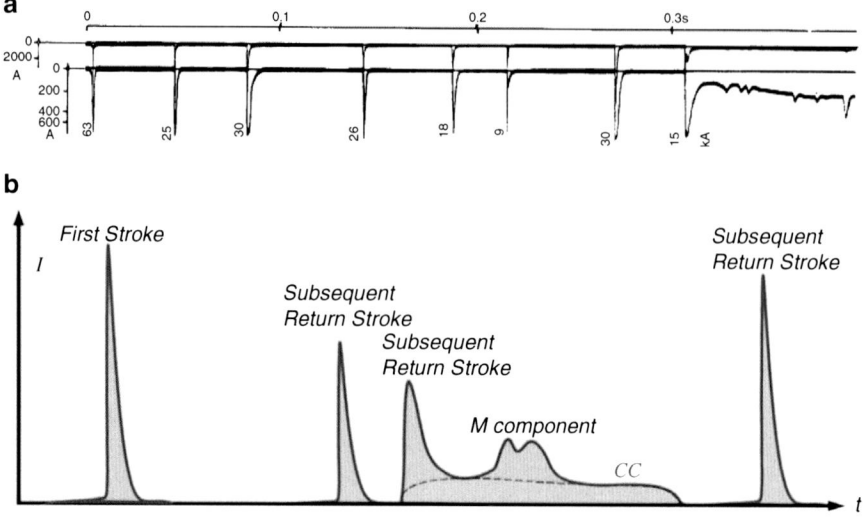

Fig. 8.7 (**a**) Current (shown in two amplitude scales) measured at base of downward negative ground flash containing eight return strokes. The number associated with each peak gives the peak current. Observe that the last return stroke gave rise to a continuing current. Note that the return stroke current appears abruptly without the initial continuing current stage that is present in upward initiated lightning flashes (Adapted from Berger [7]). (**b**) General features of natural downward lightning flashes. Note that in the drawing, the sign convention for a positive current is opposite to that of Fig. 8.3 (Adapted from Diendorfer [6])

8.2.2 General Features of Downward Lightning Flashes

Figure 8.7a shows an example of a current waveform measured by Berger [7] at the base of the channel in a negative downward lightning flash that struck the instrumented tower located at Mount San Salvatore in Switzerland. The typical features of such current waveforms associated with downward lightning flashes can be pictorially depicted by the sketch shown in Fig. 8.7b. Note that the current starts immediately with the first return stroke. The reason for this is that the current is initiated when a connection is made between the connecting leader and the downward moving stepped leader. The first return stroke is followed by several dart leader return stroke sequences. The features of the current waveform taking place after the first return stroke are identical to events that take place in an upward lightning flash after the cessation of the initial continuing current.

8.3 Specific Features of Lightning Currents

From the point of view of lightning protection, the most important parameters of lightning currents are the peak amplitude of the first and subsequent return strokes, the peak current derivative, the total charge transported by the flashes, and the

8.3 Specific Features of Lightning Currents

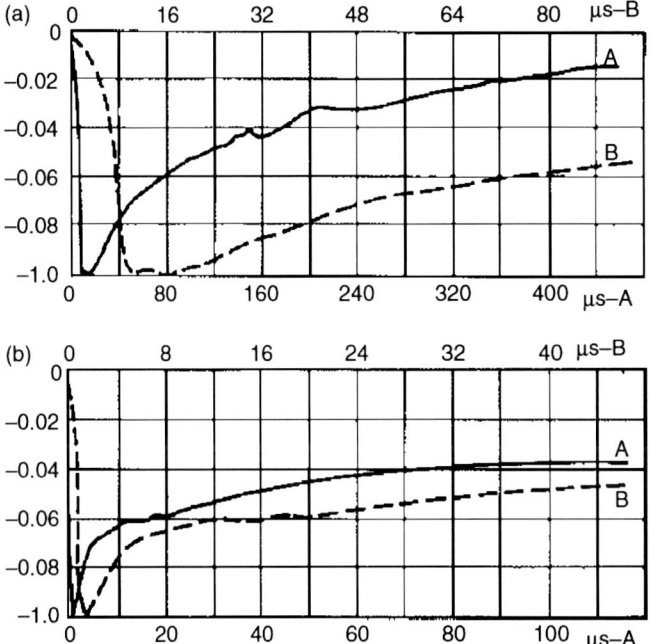

Fig. 8.8 Typical normalized (to unit amplitude) negative return stroke current wave shapes: (**a**) first return stroke, (**b**) subsequent return strokes. Note that the waveforms are shown in two time scales (Adapted from Berger [7])

action integral. In the case of first return strokes, recall that the information is based on downward flashes to tall towers.

8.3.1 Features of First and Subsequent Return-Stroke Currents

The general shape of first and subsequent return stroke currents are shown in Fig. 8.8. These shapes are based on the lightning currents measured by Berger [7] at Mount San Salvatore. Note that the currents rise to their peak values in approximately a few microseconds and then continue to decay. The currents last several hundred microseconds. The important current parameters that are of interest in lightning protection are the distribution of peak currents, the distribution of peak current derivatives, the distribution of the charge, and the distribution of the action integral. These parameters, as measured by Berger and his coworkers, are tabulated in Table 8.1. In this table, the 5, 50, and 95 % values of the distributions of different parameters are given. For example, the 5 % value of the negative first return stroke peak current is 80 kA. That means 5 % of the negative first return stroke currents exceed 80 kA. The other percentages are defined accordingly. The impulse charge is the charge brought to ground over the first 2 ms of the return stroke.

Table 8.1 Parameters of downward lightning ground flashes

Parameter and process	Unit	N	Percentage of cases exceeding tabulated value		
			95 %	50 %	5 %
Peak current	kA				
Negative first strokes		101	14	30	80
Negative subsequent strokes		135	4.6	12	30
Positive first strokes		20	4.6	35	250
Impulse charge	C				
Negative first strokes		90	1.1	4.5	20
Negative subsequent strokes		117	0.22	0.95	4.0
Positive strokes		25	2.0	16	150
Maximum current derivative	kA/µs				
Negative first strokes		92	5.5	12	32
Negative subsequent strokes		122	12	40	120
Positive first strokes		21	0.2	2.4	32
Action integral	A^2s				
Negative first strokes		91	6.0×10^3	5.5×10^4	5.5×10^5
Negative subsequent strokes		88	5.5×10^2	5.5×10^3	5.2×10^4
Positive first strokes		26	2.5×10^4	6.5×10^5	1.5×10^7
Total charge	C				
Negative first strokes			1.1	5.2	24
Negative subsequent strokes			0.2	1.4	11
Negative flashes			1.3	7.5	40
Positive flashes			20	80	350

Adapted from Berger [7]
N number of observations

8.3.1.1 Peak Current Distribution

In lightning protection, the safety level of a lightning protection system is designed according to different protection levels [8]. These protection levels are based on the peak current of first return strokes. For example, a structure protected at level I allows only lightning flashes with first return stroke peak currents of less than 3 kA to strike the structure. The peak currents associated with levels II and III are 5 kA and 10 kA, respectively (see Chap. 17). Thus, it is necessary to know the distribution of peak return stroke currents in a given region to make decisions concerning the lightning protection system. Moreover, as we will see subsequently, the striking distance of stepped leaders are described in terms of first return stroke peak current. Therefore, without knowing the distribution of peak currents in the first return strokes, it is impossible to evaluate the number of lightning flashes striking a given structure.

Figure 8.9 shows the distribution of peak currents in negative first, positive first, and negative subsequent return strokes. Observe that the geometric mean value of

8.3 Specific Features of Lightning Currents

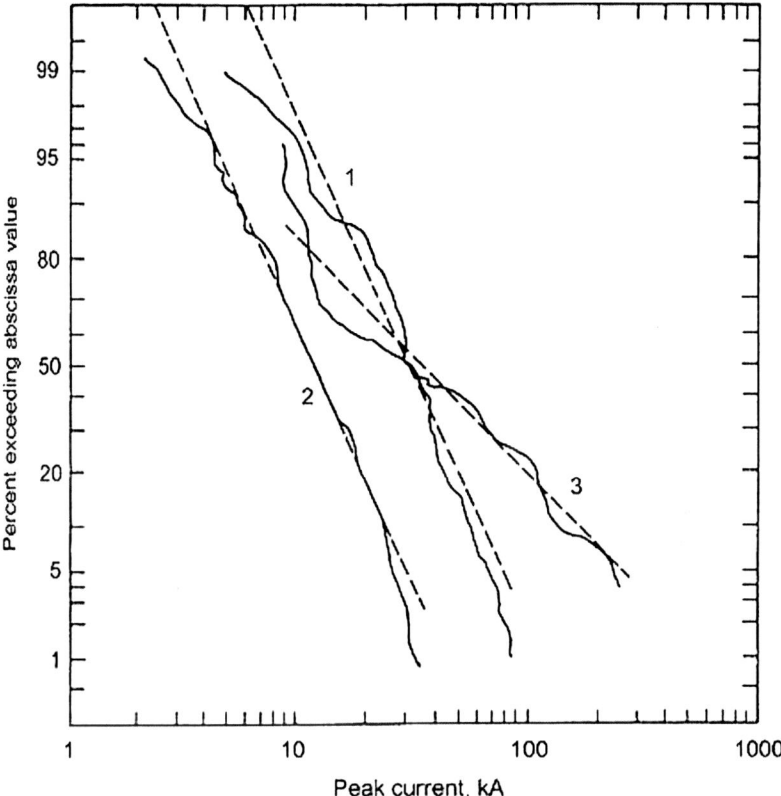

Fig. 8.9 Cumulative statistical distributions of return stroke current peak (*solid line curves*) and their log-normal approximations (*slanted dashed lines*) for (1) negative first strokes, (2) negative subsequent strokes, and (3) positive strokes (Adapted from Berger [7])

the peak current is approximately 30 kA for first and 12 kA for subsequent return strokes. It is important to point out here that the distribution of return stroke currents may change from one geographical region to another, and this variation may be approximately 20 %.

8.3.1.2 Peak Current Derivative

The magnetic field generated by a lightning flash in the vicinity of a strike point is proportional to the current flowing to ground at the strike point. If the lightning flash injects its current into a conductor, the resulting magnetic field in the vicinity of the conductor is also proportional to the current in the lightning flash. The voltage produced by this magnetic field in a conducting loop as a result of magnetic induction is proportional to the rate of change of the magnetic field, and this in turn depends on the rate of change of the lightning current. Thus, the peak voltage induced in the conducting loop depends on the peak current derivative. In the case of lightning

Table 8.2 Peak current derivative measured in return strokes of negative rocket-triggered lightning

Location/Year	Number of observations	Geometric mean (kA/μs)	Arithmetic mean
Kennedy Space Center, Florida, 1985–1991 [10]	134	118	–
Saint-Privat d'Allier, France, 1986, 1990–1991 [10]	47	43	–
Camp Blanding, Florida [11]	15	80	73
Camp Blanding, Florida [9]	64	117	97

flashes, the most interesting components with respect to magnetic induction are the return strokes because they generate the largest currents with the highest peak current derivatives. Note that in Table 8.1, the geometric mean values of the peak current derivatives (i.e., 50 % value) of negative first and subsequent return strokes are 14 and 40 kA/μs. Since the time resolution of the recording system used by Berger [7] is somewhat limited in comparison to modern ones, researchers believe that the geometric mean value of the subsequent return stroke current derivatives as obtained by Berger should be regarded as a possible lower limit of the geometric mean value. Fortunately, subsequent return strokes are similar in natural and triggered (or upward initiated) lightning flashes. Thus, the peak current derivatives of subsequent strokes can be obtained from triggered lightning or from upward initiated lightning flashes. Such values as obtained in several triggered lightning experiments are tabulated in Table 8.2, which was adapted from Ref. [9]. According to these studies, the geometric mean value of a negative subsequent return stroke peak current derivative lies in a range of 50–100 kA/μs.

8.3.1.3 Distribution of Charge Brought to Ground

When a lightning flash get attached to a metal surface, the lightning current flowing in the air (i.e., in the discharge channel) must pass onto the metal. This transition of the current flow from air to metal takes place over a very thin layer called a *cathode layer*. Across this layer the voltage drop is approximately 15–20 V. Let us denote this by V_c. Thus, the heat energy that is transferred to the metal across the cathode layer is given by

$$W = QV_c, \tag{8.2}$$

where Q is the charge transported across the cathode layer. To evaluate the heat generated during a return stroke, Q in the preceding equation should be replaced by the charge dissipated by the return stroke. To know the total energy dissipated by a flash, Q in the preceding equation must be replaced by the total charge dissipated by the flash. Once the energy dissipated is known, it can be used, for example, to calculate how much material will be melted by a lightning strike (Chap. 15). This is important in dimensioning the radius of lightning terminals and in evaluating

8.3 Specific Features of Lightning Currents

the thickness of a metal surface that can withstand a lightning strike without melt-through.

The 5, 50, and 95 % values of the measured charge of first return strokes, subsequent return strokes (i.e. impulse charge), and of the whole flash are presented in Table 8.1. Note that on average a negative first return stroke dissipates approximately 5 C and a negative lightning flash dissipates approximately 7.5 C. Positive lightning flashes dissipate nearly ten times more charge than negative lightning flashes. Recently, on the basis of tower measurements, some upward initiated negative lightning flashes dissipating more than 300 C were identified.

8.3.1.4 Action Integral of Return-Stroke Current

Let us consider a situation where a lightning flash strikes an object having a resistance. Let us denote the resistance of the conductor by R. Then as the lightning current passes through the object, the amount of energy dissipated is given by

$$W = R \int_0^{t_d} i^2(t)dt. \tag{8.3}$$

The integral $\int_0^{t_d} i^2(t)dt$ is known as the *action integral*. The upper limit of this integral, t_d, is the duration of the event under consideration (i.e. the duration of the return stroke etc.). If this integral is known, then the energy dissipated in the object, and hence the increase in temperature of the object, under study can be calculated. Once this energy is known, the increase in temperature of the object under study can be estimated. Assume that the object under study is a lightning conductor. If the radius of the conductor is too small, then the amount of energy released may raise the temperature of the conductor to such a value that it explodes. Thus, in dimensioning lightning terminals (i.e., those units in the lightning protection system that intercept the lightning flash; see also Chap. 17), it is necessary to know the value of energy dissipated in it by lightning flashes. This can be estimated using the preceding equation.

The action integral also determines the force between two conductors sharing a lightning current. Assume that a lightning flash will strike two conductors placed close to each other and the current is equally shared between them. The magnetic forces will try to bring the two conductors together, and the mechanical impulse, I, that both conductors experience is (see Chap. 3)

$$I = K \int_0^{t_d} i^2(t)dt, \tag{8.4}$$

where K is a constant. This information is necessary to secure fast the conductors that carry lightning current so that they are secured mechanically as the lightning current flows through them.

It is the return strokes that generate the largest action integral in various processes associated with lightning flashes. Thus, they constitute an important parameter in engineering studies. In this case, the upper limit of integration of the action integral t_d is the duration of the return strokes. Table 8.1 presents 5, 50, and 95 % levels of the action integral of negative first return strokes, positive first return strokes, and subsequent strokes. Note that the geometric mean of the action integral is approximately 5×10^4 A^{2}s for negative first return strokes.

8.3.2 Features of M-Component Current

As described in the previous chapter, M-components occur during the continuing current phase of return strokes. They can be easily identified in current records obtained from either triggered lightning or tower measurements. For example, M-components superimposed on the continuing current of a return stroke in a triggered lightning flash are shown in Fig. 8.10, which was adapted from Ref. [12]. An expanded view of the M-component current is shown in Fig. 8.11. The M-component currents measured at the channel base typically have hundred amperes or so of current and a rise time of several hundred microseconds.

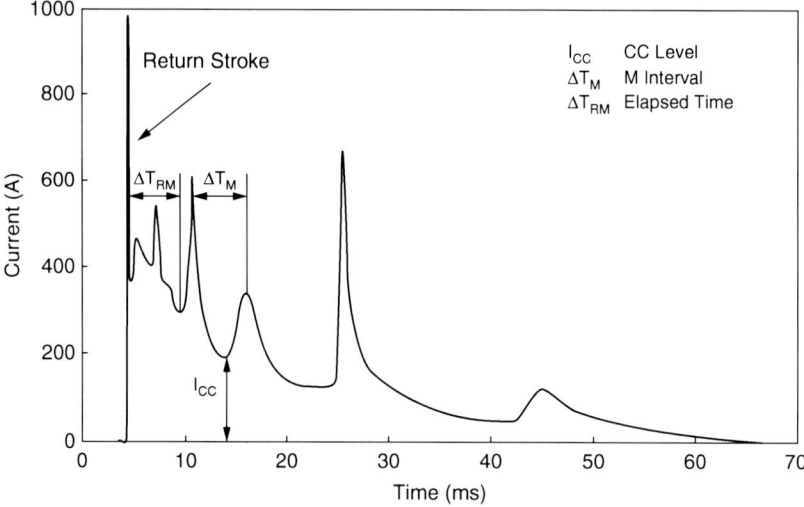

Fig. 8.10 Current record showing one return stroke (saturated at approximately 1 kA level) followed by several M-components in a flash triggered at Kennedy Space Center in 1990. Shown are the definitions of the continuing current level I_{cc}, M interval ΔT_M, and elapsed time ΔT_{RM}. Note that in the drawing, the sign convention for a positive current is opposite to that of Fig. 8.3 (Adapted from Thottappilli et al. [12])

8.3 Specific Features of Lightning Currents

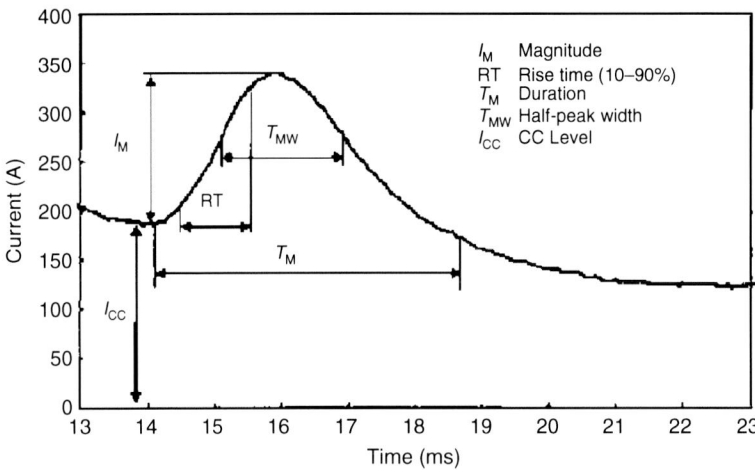

Fig. 8.11 Expanded portion of record in Fig. 8.10 illustrating current associated with M-component. I_M is the peak current, RT the 10–90 rise time, T_M the duration, and T_{MW} the half-peak width (Adapted from Thottappilli et al. [12])

Table 8.3 Statistics of M-components; parameters defined in Fig. 8.11

Parameter	Sample	Geometrical Mean	Standard deviation $\log_{10}(x)$	Cases exceeding tabulated value		
				95 %	50 %	5 %
Magnitude	124	117	0.50	20	121	757
Rise time (µs)	124	422	0.42	102	415	1,765
Duration (ms)	114	2.1	0.37	0.6	2.0	7.6
Half-peak width (µs)	113	816	0.41	192	800	3,580
Charge (mC)	104	129	0.32	33	131	377
Continuing current level (A)	140	177	0.45	34	183	991
M-interval (ms)	107	4.9	0.47	0.8	4.9	23
Elapsed time (ms)	158	9.1	0.73	0.9	7.7	156

Adapted from Thottappilli et al. [12]

The parameters (identified in Fig. 8.11) of M-components are presented in Table 8.3. In this table are given the statistics of the rise time (10–90 %), peak value, duration, and half-width of M-component currents, together with the charge transferred by M-components and the amplitude of the continuing current during which the M-components had occurred.

8.3.3 Features of Continuing Currents

Even though the currents associated with return strokes usually go down to zero in several hundred microseconds, in some return strokes (in approximately 30–50 % of negative lightning flashes) the current amplitude decreases to approximately 100 A or so within this time but then, instead of going down to zero, may maintain this current amplitude for a few milliseconds to a few hundred milliseconds. Such currents are known in the literature as *continuing currents*.

A continuing current can be divided into [13] long continuing currents (longer than 40 ms), short continuing currents (between 10 and 40 ms), and very short continuing currents (1–10 ms). However, it is important to mention that this division, while helpful in bookkeeping, has no physical basis. That is, the physical process that gives rise to continuing currents irrespective of their duration could be the same.

According to measurements taken by Saba et al. [14] in Brazil, of 233 negative ground flashes containing 608 strokes, 50 % of the strokes supported continuing currents longer than approximately 1 ms, and 35.6 % of the strokes were followed by short or long continuing currents.

References

1. Eriksson AJ (1987) The incidence of lightning strikes to power lines. IEEE Trans Pow Deliv 2(3):859–870
2. Rakov V (2010) Rocket triggered lightning and new insights. In: Cooray V (ed) Lightning protection. IET Publishers, London
3. Miki M, Rakov VA, Shindo T, Diendorfer G, Mair M, Heidler F, Zischank W, Uman MA, Thottappillil R, Wang D (2005) Initial stage in lightning initiated from tall objects and in rocket-triggered lightning. J Geophys Res 110:D02109. doi:10.1029/2003JD004474
4. Zhou H, Diendorfer G, Thottappillil R, Pichler H, Mair M (2012) Characteristics of upward positive lightning flashes initiated from the Gaisberg Tower. J Geophys Res. doi:10.1029/2011JD016903
5. Wang D, Rakov VA, Miki M, Uman MA, Fernandez MI, Rambo KJ, Schnetzer GH et al (1999) Characteristics of the initial stage of negative rocket-triggered lightning. J Geophys Res 104:4213–4222
6. Diendorfer G (2010) LLS performance validation using lightning to towers. In 21st international lightning detection conference, Florida, USA, 2010
7. Berger K (1977) The earth flash. In: Golde RH (ed) Lightning. Academic, San Diego, pp 119–190
8. IEC 62305 (2006) International standard on protection against lightning
9. Schoene J, Uman MA, Rakov VA, Kodali V, Rambo KJ, Schnetzer GH (2003) Statistical characteristics of the electric and magnetic fields and their time derivatives 15 m and 30 m from triggered lightning. J Geophys Res 108:4192. doi:10.1029/2002JD002698
10. Depasse P (1994) Statistics on artificially triggered lightning. J Geophys Res 99:18515–18522
11. Uman MA, Rakov VA, Schnetzer GH, Rambo KJ, Crawford DE, Fisher RJ (2000) Time derivative of the electric field 10, 14 and 30 m from triggered lightning flashes. J Geophys Res 105:15577–15595

12. Thottappilli R, Goldberg JD, Rakov VA, Uman MA, Fisher RJ, Schnetzer GH (1995) Properties of M-components from currents measured at triggered lightning channel base. J Geophys Res 100:25711–25720
13. Shindo T, Uman MA (1989) Continuing current in negative cloud-to-ground lightning. J Geophys Res 94(D4):5189–5198
14. Saba MF, Ballarotti MG, Pinto O Jr (2006) Negative cloud-to-ground lightning properties from high-speed video observations. J Geophys Res 111:D03101

Chapter 9
Electromagnetic Fields of Lightning Flashes

9.1 Introduction

Electromagnetic fields of lightning flashes are of interest both in lightning protection studies and in understanding the physics of lightning flashes. In lightning protection knowledge of the characteristics of electromagnetic fields generated by lightning flashes is necessary in estimating the voltages and currents induced in structures by lightning flashes. The features of lightning-generated electromagnetic fields are of interest in lightning physics because they can be used as a vehicle to probe the physical mechanism behind lightning flashes. They can also be used to estimate the signature of currents in lightning flashes. In this chapter the characteristics of electromagnetic fields of lightning flashes are described. First, let us consider the basic principles of the procedures used to measure both the electric and magnetic fields. The two most common techniques used to measure electric fields are the vertical antenna and the field mill [1]. In the case of magnetic fields a loop antenna is used in the measurements.

9.2 Measurement of Electric Fields

9.2.1 Principle of the Vertical Antenna

Consider a small metal sphere (in principle, this can be of any shape) located at height h above the ground plane. If there is a background electric field directed towards the ground, the sphere will be raised to a potential of $E \times h$. Because of the influence of the electric field, charges will be displaced in the sphere (Fig. 9.1a). Now, assume that this sphere is connected to ground by a very thin wire. In this case, a positive charge moves from sphere to ground, leaving behind a negative

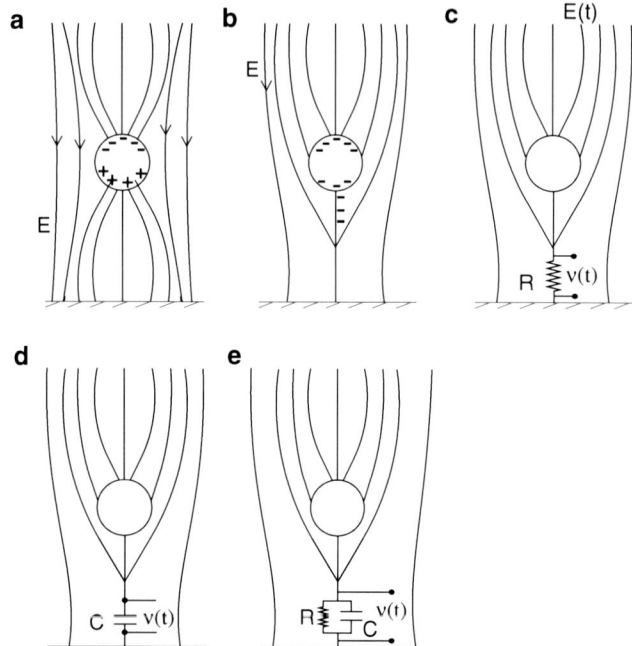

Fig. 9.1 (a) In an isolated sphere in an electric field, electric charges are displaced. The net charge on the sphere is zero. (b) If the sphere is connected to ground by a thin wire, then a positive charge (of course depending on the direction of the electric field) flows to ground, leaving a net negative charge on the sphere. If the electric field changes, then the negative charge on the sphere changes, and this creates a current flow to ground along the wire. (c) If the sphere is connected to ground across a resistor, then as the current flows to ground, a voltage is created across the resistor. This voltage is proportional to the time derivative of the electric field. (d) If the sphere is connected to ground through a capacitor, then the voltage generated across the capacitor is proportional to the time-varying electric field itself. (e) If the wire is connected to ground through a parallel combination of R and C, then the voltage that appears across the combination would depend on the product RC. If RC is much larger than the time during which changes in the electric field take place, then the voltage is proportional to the electric field itself. On the other hand, if RC is much shorter than the time during which changes in the electric field take place, then the voltage is proportional to the time derivative of the electric field. The output is a combination of electric field and the electric field time derivative if the aforementioned conditions are not satisfied (Figure created by author)

charge on the sphere (Fig. 9.1b). The net charge on the sphere is proportional to the voltage $E \times h$. Thus, the charge on the sphere can be written as

$$Q = E \times h \times C_s, \tag{9.1}$$

where C_s is the capacitance of the sphere. The capacitance of the sphere depends on the geometrical parameters of the sphere and its height above the ground plane. Now, as the electric field changes, the charge on the sphere changes according to the equation

$$Q(t) = E(t) \times h \times C_s. \tag{9.2}$$

9.2 Measurement of Electric Fields

For ease of analysis, assume that the electric field changes sinusoidally. That is,

$$E(t) = E_0 \sin(\omega t), \tag{9.3}$$

where ω is the angular frequency. Assume that there is a small resistor between the wire and the ground (Fig. 9.1c). The voltage across the resistor is given by the current times the resistance of the resistor. Now, as the electric field changes, the charge on the sphere varies as given by Eq. (9.2), and the current flowing across the resistor $I(t)$ (neglecting the effect of the capacitance C_s) is given by the rate of change of the charge on the sphere. That is,

$$I(t) = \frac{dQ(t)}{dt} = h \times C_s \times \frac{dE(t)}{dt}. \tag{9.4}$$

Thus, the voltage across the resistor is proportional to the derivative of the electric field. Now assume that instead of a resistor there is a capacitor C connected at the end of the wire (Fig. 9.1d). Then the voltage drop across the capacitance is equal to the voltage of the sphere, and this is given by $E \times h$. Thus, the voltage across the capacitor is proportional to the electric field. Now assume that we have connected both a resistance and a capacitance to the circuit (Fig. 9.1e). Then the voltage across the capacitor–resistor combination depends on the relative impedances of the resistor and the capacitor. The impedance of the resistor for the current flow is simply its resistance. With the capacitor the impedance is equal to $\frac{1}{\omega C}$. If the resistance is such that $R \ll \frac{1}{\omega C}$, then the current flows along the resistor and the voltage across the combination is proportional to the derivative of the electric field. On the other hand, if $R \gg \frac{1}{\omega C}$, then the current flows along the capacitance and the voltage across the combination is proportional to the electric field. Thus, selecting the value of the RC product makes it possible to measure either the electric field or the electric field derivative. However, this choice depends on the frequency of the signal. On the other hand, lightning produces not a sinusoidal signal but a signal that varies in time in a complicated manner. It can be shown, however, that when the signal is not sinusoidal, the parameter $1/\omega$ roughly translates as a time interval, Δt (i.e. $\Delta t \sim 1/\Delta \omega$). Thus, if $RC \ll \Delta t$, for any variation in the electric field that takes place so slowly that any variation over a time interval Δt can be neglected, the output voltage is proportional to the derivative of the electric field. On the other hand, if $RC \gg \Delta t$, for any variation in the electric field taking place over a time scale of Δt, the voltage is proportional to the electric field itself. For example, the electric field duration of a return stroke is approximately 100 μs. Thus, RC should be much larger than 100 μs if the voltage output of the antenna is to give the electric field over this time interval faithfully. The duration of the electric field from a complete flash can be approximately 0.5 s. In this case, RC should be larger than approximately 10 s to obtain a faithful representation of the total electric field.

In our analysis, we have neglected the effect of the capacitance C_s and the effect of the wire connecting the antenna probe to the ground. In reality, the capacitance C_s must be added to C, and instead of C, one must use $(C+C_s)$ in the foregoing analysis. Moreover, the presence of a wire changes the potential of the antenna probe, and this can be accounted for by changing the potential from $E \times h$ to a value $E \times h_{\text{eff}}$, where h_{eff} is called the *effective height* of the antenna. The thinner the conductor connecting the probe to ground, the closer h_{eff} is to h.

9.2.2 Principle of Electric Field Mill

Consider a square plate (the shape of the plate is immaterial in reality) located in flush with a ground plane above which there is an electric field (Fig. 9.2). Let us call this plate the probe. The probe is connected to ground by a thin conducting wire. The charge induced on the plate due to the electric field is $\varepsilon_0 A \times E$, where A is the area of the plate. Now, consider another plate of the same dimensions that is grounded. Let us call this the screening plate. Initially, the screening plate is located as in Fig. 9.2a, and the probe is completely exposed to the electric field. Now assume that the screening plate starts moving at speed v and with time it will screen the probe from the electric field (Fig. 9.2b). The final position of the screening plate is as shown in Fig. 9.2c. Let us study the charge and the current flow along the thin wire as the probe is gradually screened from the electric field by the screening plate. Let us assume that the exposed area of the probe to the electric field is $a(t)$. Then the charge on the plate at any given time is

$$q(t) = \varepsilon_0 E a(t). \tag{9.5}$$

Thus, the current flowing along the conductor to ground is

$$i(t) = \varepsilon_0 E \frac{da(t)}{dt}. \tag{9.6}$$

Now, as the screening plate moves at speed v, the exposed area of the probe varies as shown in Fig. 9.3. Initially, it is equal to A and gradually decreases to zero. Assume that the screening plate reverses its motion at this time. As the screening plate starts moving in the opposite direction, the area of the probe exposed to the electric field increases again and reaches the value A when the screening plate has returned to its original position. Now, this variation of the exposed area of the probe to the electric field with time gives rise to a change in the current flowing along the wire. This varying current is given by Eq. 9.6, and the result is depicted in Fig. 9.4. As shown by Eq. 9.6, the amplitude of the current pulse is proportional to the electric field. Thus, this technique makes it possible to measure the electric field in air. Now, what happens if the electric field changes? Let t_0 be the time for the

9.2 Measurement of Electric Fields

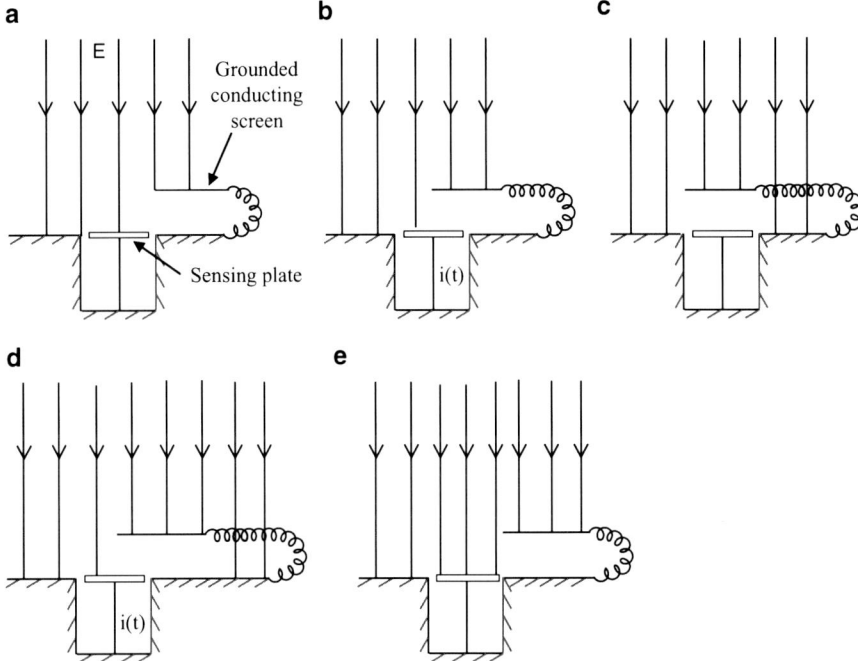

Fig. 9.2 Action of an electric field mill. (**a**) Consider a uniform electric field. The sensing plate of the field mill connected to ground is fully exposed to the electric field, and a net negative charge is induced on the plate. The magnitude of the charge induced on the plate depends on the magnitude of the background electric field. (**b**) As the grounded conducting screen moves above it, the sensing plate is screened from part of the electric field, and as a consequence part of the induced charge flows to ground, creating a current flow to ground along the conductor. (**c**) The flow of current ceases when the screening plate completely covers the sensing plate. (**d**) If the screening plate now starts to move backward, more and more area of the sensing plate becomes exposed to the electric field and the induced charge on the sensing plate increases. This gives rise to a current that is opposite in polarity to the current that flowed when the screening plate was moving forward. (**e**) The full charge on the sensing plate is restored when the sensing plate is fully opened. Thus, if the screen is forced to move back and forth, a current will flow from the sensing plate to ground, and from this current the background electric field can be estimated. See the text to understand what happens if the electric field changes at the same time as the screening plate moves back and forth (Figure created by author)

screening plate to move a complete cycle. If the electric field does not change over a time interval of t_0, then by measuring how the peaks of the current pulses change with time, we can understand how the field changes with time. But if the field changes rapidly so that a significant change takes place over time interval t_0, then the aforementioned device would not be able to resolve how the field changes over time. But allowing the screening plate move faster and faster, i.e., making t_0 shorter and shorter, would make it possible to resolve electric fields that are changing faster. The previously described device is called a *field mill*. Different methods have been created to perform this action of screening and nonscreening of the probing

Fig. 9.3 The area of the plate changes as the grounded screen moves forward and backward. Note that at time $t=0$ the exposed area of the sensing plate is equal to A and A is the area of the sensing plate (Figure created by author)

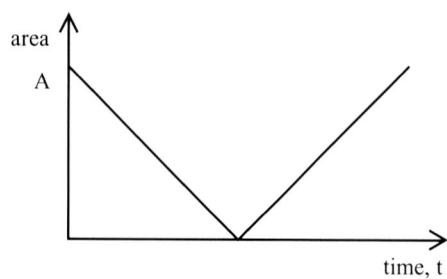

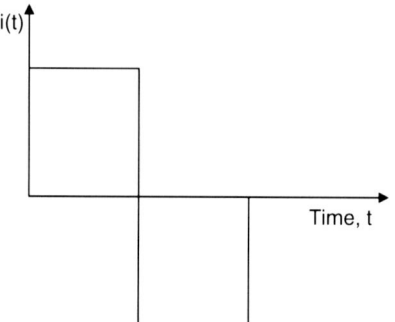

Fig. 9.4 Variation in current flowing to ground from sensing plate as area of plate varies as in Fig. 9.3. Note that the current jumps instantaneously to a finite value if the screening plate starts to move instantaneously. In the diagram it is assumed that the background electric field remains the same as the screen moves forward and backward (Figure created by author)

plate from an electric field to be measured. For example, field mills are arranged so that a screening plate is allowed to rotate very rapidly above the probing plate so that its area changes very rapidly. Modern field mills can resolve electric fields in the range of milliseconds.

9.2.3 Measurement of Magnetic Fields

Assume that there is a magnetic field that changes with time in space. If a conducting loop with a small gap is placed in this magnetic field, then a voltage is induced in the loop and, by Faraday's law, this voltage is proportional to the area of the loop and the rate of change of the magnetic field directed perpendicular to the loop. Thus, integrating the voltage that appears across the gap makes it possible to obtain information as to the variation in the magnetic field with time. However, with one loop it is possible to obtain the variation in the magnetic field perpendicular to that loop. To obtain the complete magnetic field in three-dimensional space,

9.2 Measurement of Electric Fields

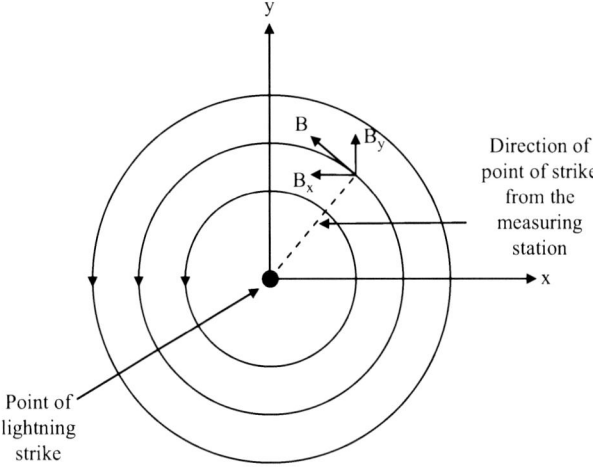

Fig. 9.5 On the surface of the ground the magnetic field lines from a lightning flash form circular loops around the strike point. At any point the direction of the magnetic field is tangential to the circle at that point. Thus, knowing the direction of the magnetic field makes it possible to determine the direction to the strike point from the point of observation (Figure created by author)

it is necessary to have three loops placed perpendicular to each other. This configuration yields the magnetic field variation in three perpendicular directions. Let us say that these components are $B_x(t)$, $B_y(t)$, and $B_z(t)$. Then the total magnetic field is

$$B(t) = \sqrt{B_x^2(t) + B_y^2(t) + B_z^2(t)}. \tag{9.7}$$

Now, if the lightning channel is vertical and if the ground is assumed to be perfectly conducting, following from Ampere's law, the magnetic field will form closed loops around the lightning channel (Fig. 9.5). In this case, the magnetic field at any point is parallel to the ground plane and it therefore has only x and y components. At any point the direction of the magnetic field makes an angle θ in the direction of x, where

$$\cos\theta = \frac{B_x(t)}{\sqrt{B_x^2(t) + B_y^2(t)}}. \tag{9.8}$$

Unfortunately, with this method it is not possible to measure constant or stationary magnetic fields because a voltage is induced in the loop only when the magnetic field changes. However, in the case of lightning flashes, this method works fine because, especially in return strokes, the magnetic field varies rapidly.

Today, this method is used routinely to measure the magnetic fields generated by lightning return strokes [2].

9.3 Electric Field Close to Lightning Channel

Many scientists have used the electric field as a vehicle to probe the activities of lightning flashes. Many of the processes described in Chap. 7 can be identified by the electromagnetic fields measured close to the lightning channel. For example, in close electric field records, it is possible to identify the preliminary breakdown, stepped or dart leaders, continuing currents, M-components, and K changes. It is also important to remember that many processes that take place during a lightning flash, especially inside the cloud, are not known. For this reason, measured electric fields may contain features that cannot be attributed to any of the processes identified in Chap. 7. Furthermore, in different publications the definition of positive (or negative) field may be different. In some publications, the positive direction of a field is defined as the outward normal to the Earth's surface. This is called the *physics convention*. According to this notation, a negative charge in a cloud produces a positive field at ground level and a negative return stroke produces a negative field change. In other publications, the atmospheric sign convention is used. According to that notation, a negative charge in a cloud produces a negative field at ground level, and a negative return stroke will produce a positive field change.

Recall from the analysis presented in Chap. 3 that the mathematical expression for the electromagnetic field generated by a lightning flash can be separated into three terms: static, induction, and radiation. The electric field produced by a lightning flash at any point is the sum of the contributions from the static, induction, and radiation. Very close to the channel (say, within approximately 5 km) the static term is the dominant one. Of course, the other two contributions are there, but they are overwhelmed by the static contribution. In the close electric field records, the electric field produced by the preliminary breakdown, stepped leader, and return stroke may appear either as in Fig. 9.6a or b, depending on the distance to the flash [3]. In this figure, the electric field is assumed to be positive if it is directed to ground, i.e., atmospheric sign convention. This convention is opposite to that used in Chap. 3 in performing theoretical calculations on the electric fields produced by lightning leaders and return strokes. Now, the static electric field at any point on ground as the leader approaches the ground consists of two terms, one produced by the removal of negative charge from the cloud and the other produced by the negative charge approaching the ground. Very close to the lightning channel the electric field is dominated by the one produced by the negative charge coming down along the stepped leader, and this produces an increasing negative electric field as in curve a of Fig. 9.6a. When the distance to the flash is larger, the dominant field component is produced by the removal of negative charge from the cloud charge center, and hence the preliminary breakdown and the stepped leader fields become

9.3 Electric Field Close to Lightning Channel

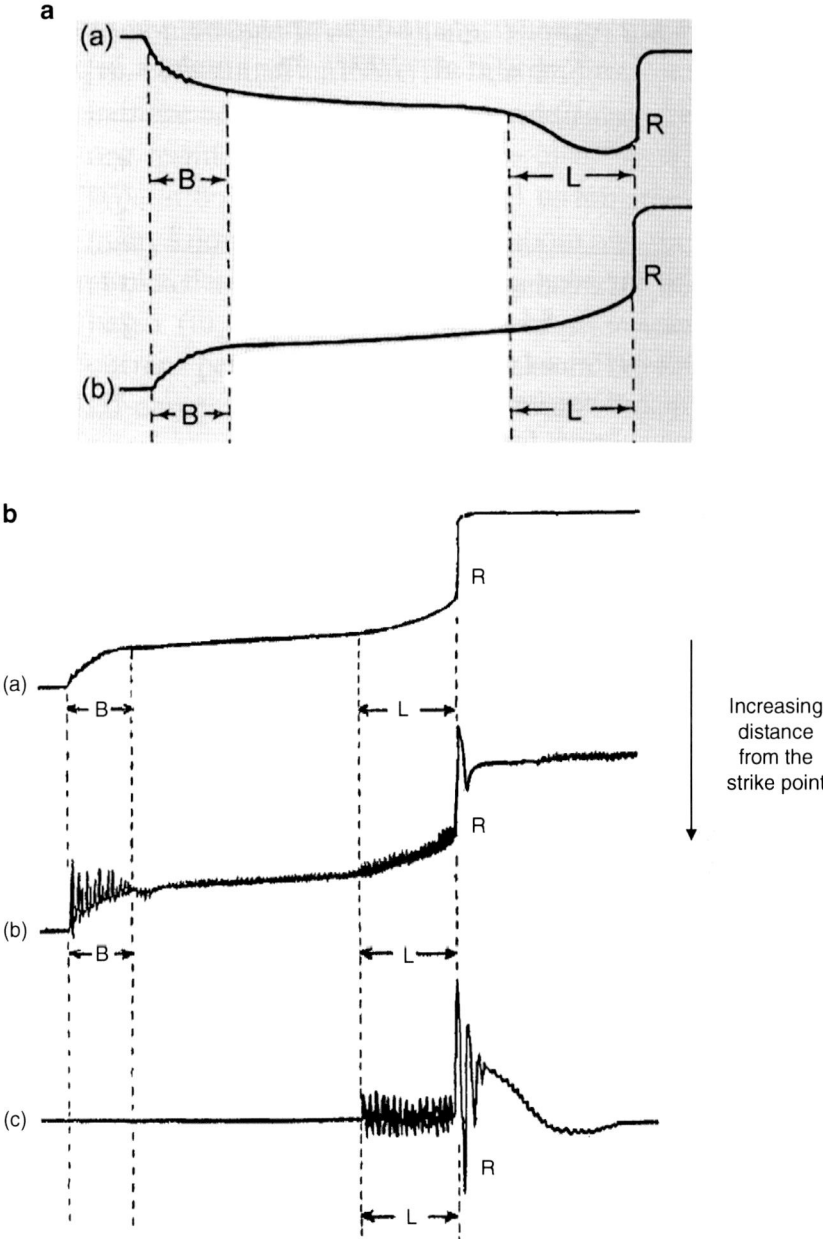

Fig. 9.6 (**a**) Appearance of electric field of a first return stroke, including preliminary breakdown, at (*a*) close and (*b*) intermediate distances. When the distance to the flash is small, the field from the preliminary breakdown (*B*) and the leader field is opposite to that of the return stroke (*R*). At intermediate distances the field of the preliminary breakdown and the field of the leader have the same polarity as those of the return stroke. At these distances the electric field is mainly electrostatic (Adapted from Clarence and Malan [3]; diagram based on atmospheric sign convention). (**b**) As the distance to the electric field increases, the radiation fields in the preliminary breakdown, leader, and return stroke become more dominant (Adapted from Clarence and Malan [3]; diagram based on atmospheric sign convention)

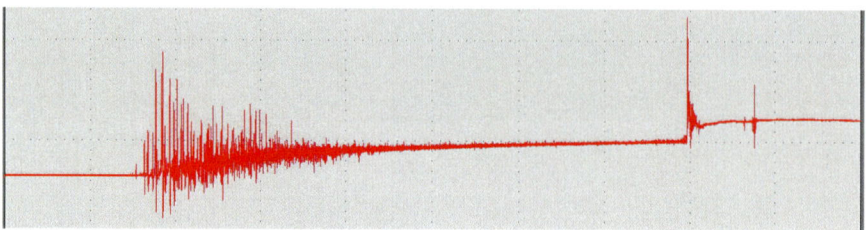

Fig. 9.7 First return stroke field at intermediate distance measured from a thunderstorm in a field experiment in Sweden [4]. Note the strong preliminary breakdown pulses. Observe also that the electrostatic field starts to increase at the initiation of the preliminary breakdown, indicating that the charge displacement in the lightning flash commences with the preliminary breakdown (diagram based on atmospheric sign convention)

positive, as in curve b of Fig. 9.6a. If the distance to the lightning flash is in the range of 5–10 km, then it is possible to see the pulse activity of the preliminary breakdown very clearly (Fig. 9.7). The waveform in Fig. 9.7 was captured in a study conducted in Sweden that is described in [4]. In Fig. 9.6 this pulse activity, which is mainly radiation, is hidden inside the large electrostatic field. This pulse activity is a characteristic feature of the preliminary breakdown that takes place in a cloud signaling the start of a ground flash (or, more correctly, the start of a stepped leader). In Fig. 9.6a, immediately after the preliminary breakdown, the negative electric field starts to increase, signaling the lowering of the negative charge toward the ground. Note that in Fig. 9.7, too, the electric field starts to increase together with the preliminary breakdown. The sign convention of the electric field in Fig. 9.7 is identical to that used in Fig. 9.6. As the stepped leader travels toward ground, the electric field increases continuously. Since the stepped leader travels in steps, the charge is also transported in a stepped manner. These pulses are usually difficult to see in close electric field records. Some pulses may be discernible very close in time to the return stroke (as in Fig. 9.7) because at this time the leader steps take place at heights closer to ground and therefore produce large electric field changes at ground level. When the stepped leader reaches the ground, the first return stroke is initiated. The first return stroke transports the negative charge brought down by the leader to the ground, and this generates a rapid increase in the electric field, as shown in Figs. 9.6 and 9.7. After the first return stroke, several dart leader return stroke sequences can take place in a flash, and this activity can generate an overall electric field change similar to that shown in Fig. 9.8 (the same sign convention as that of Figs. 9.6 and 9.7) [5]. Some subsequent strokes may give rise to continuing currents, and in Fig. 9.8 the third return stroke initiated a continuing current. The continuing current generates an electric field that increases continuously because of the continuous transport of charge to ground. Some records show the occurrence of M-components during the continuing current phase. An electric field record showing M-components during the continuing current stage is shown in Fig. 9.9. (Note that the sign convention here is opposite to that of Figs. 9.6, 9.7, and 9.8) [6]. These M-components generate a hook-shaped electric field on the electric field of the

9.3 Electric Field Close to Lightning Channel

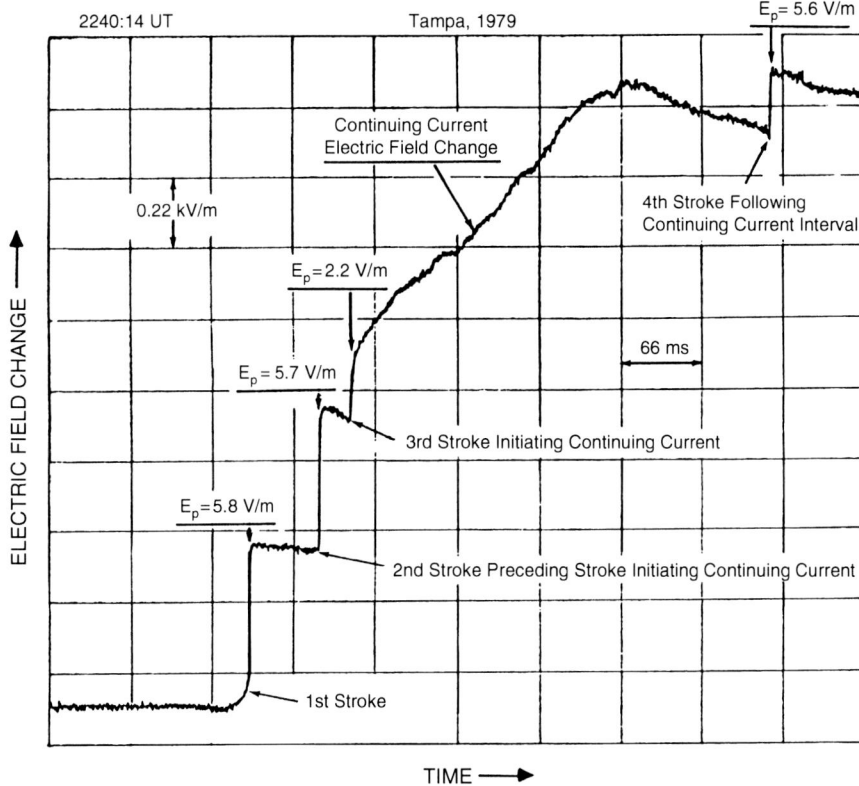

Fig. 9.8 Overall electric field change for a four-stroke lightning flash (at a distance of 6.5 km) with a long continuing current following the third stroke. The initial electric field (not resolved in the figure) normalized to 100 km are labeled E_p (Adapted from Rakov and Uman [5]; diagram based on atmospheric sign convention)

continuing current. They are marked M in the diagram. In a single continuing current period it is possible to observe several M-components. In close field records, K changes appear as a stepwise increase in the electric field. They may occur between return strokes or in the last stages of a lightning flash. Figure 9.9 shows the stepwise field changes produced by K changes.

Here, we have not illustrated the features of the magnetic field measured very close to the channel. The main features of the magnetic field very close to the channel are almost identical to the features of the current flowing in the lightning channel. See Chap. 8 for a description of the current measured at the channel base.

Figure 9.10 shows the electric field generated by a lightning flash at a distance of approximately 10–20 km [7]. This flash contained three return strokes. The features that could be identified easily are marked in the figure. Note that the flash starts with a preliminary breakdown. This preliminary breakdown region is usually approximately 1 ms long and contains a burst of pulses. The amplitude of the

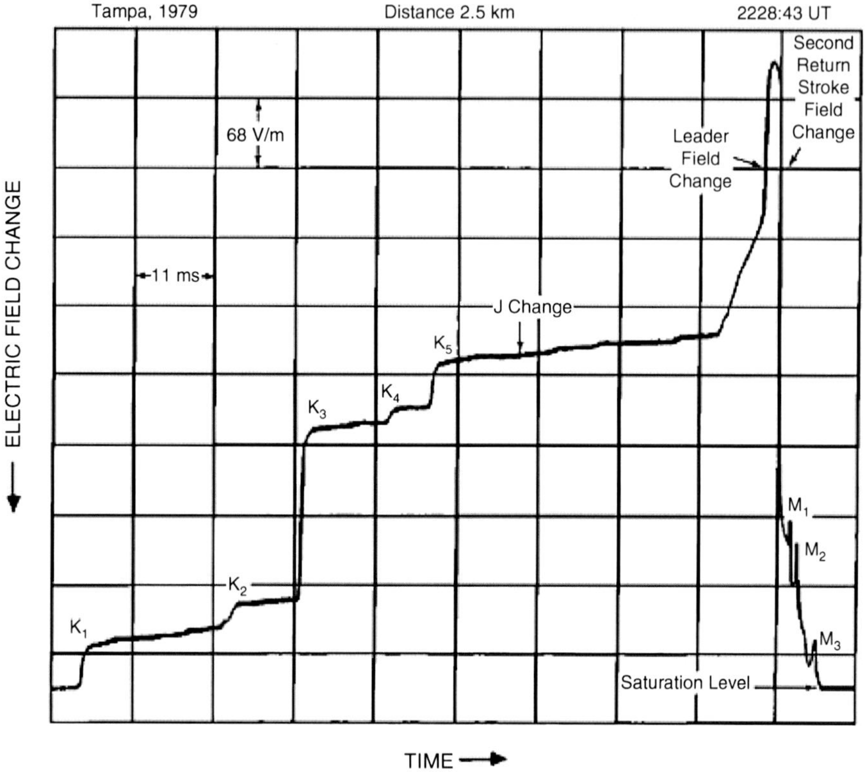

Fig. 9.9 Portion of an electric field record of a lightning flash showing electric field changes created by *K* changes. Note that the polarity of the positive electric field is opposite to that of Fig. 9.8 (Adapted from Thottappillil et al. [6]; diagram based on physics sign convention)

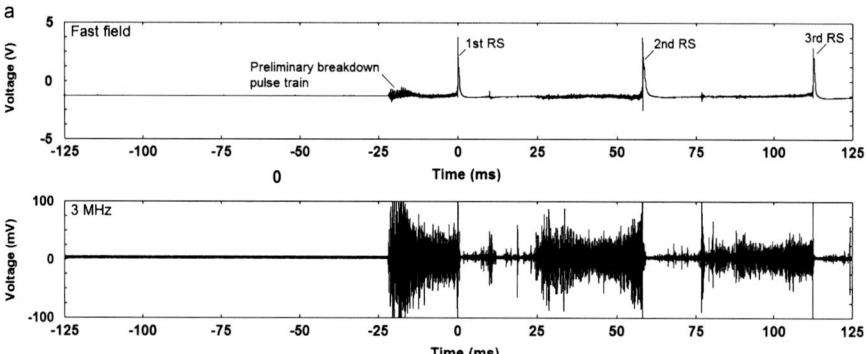

Fig. 9.10 Electric field of a three-return-stroke lightning flash at a distance of 10–20 km. *Upper diagram*: electric field; *lower diagram*: high-frequency [HF] radiation at 3 MHz. Note that, because of the shorter RC time constant of the measuring system, the static component decayed exponentially to zero. Observe the strong 3 MHz radiation associated with the leader stage of the return strokes and how it decreases rapidly to zero immediately after the return stroke (Adapted from Baharudin et al. [7]; diagram based on atmospheric sign convention)

9.3 Electric Field Close to Lightning Channel

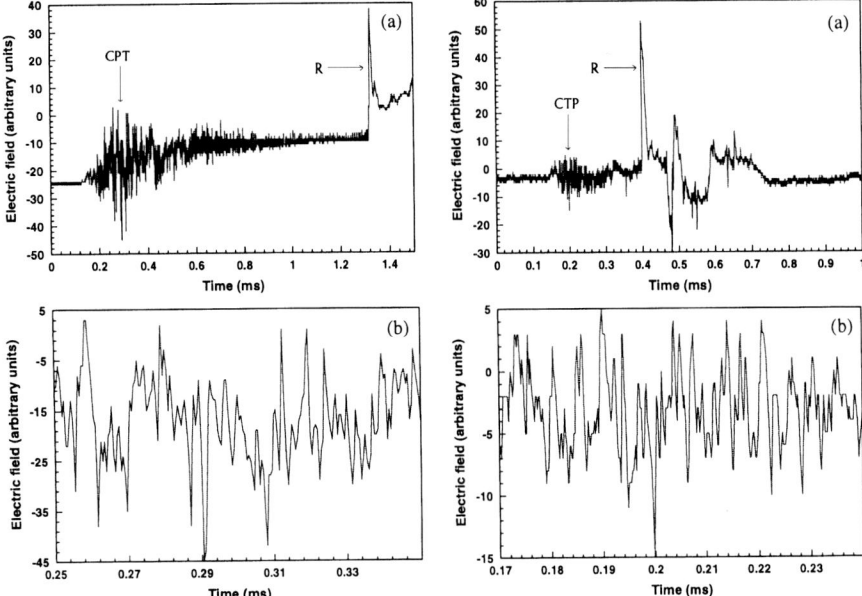

Fig. 9.11 Two examples (marked *a*) of chaotic pulse trains (marked *CPT*) preceding subsequent return strokes (*R*). The lower traces (marked *b*) show a portion of the pulse train in a faster time scale. A positive field corresponds to an upward deflection (Adapted from Gomes et al. [8]; diagram based on atmospheric sign convention)

largest pulses contained in the preliminary breakdown as compared to that of the main event of the discharge, the first return stroke, may vary from one geographical region to another. After the preliminary breakdown the stepped leader starts its journey toward the ground. Some records clearly show the pulses created by the stepped leader, especially within approximately 100 μs of the first return stroke. In general, the first return stroke is followed by several subsequent strokes. The bottom record shown in Fig. 9.10 is the 3 MHz radiation associated with the lightning flash. Note that 3 MHz radiation clearly indicates the cloud activity related to the origin of the leaders inside the cloud. Some field records show some electrical activity immediately before subsequent strokes and probably at the beginning of the dart leaders. The second subsequent stroke in Fig. 9.10 shows such activity. These pulse bursts are called *chaotic pulses* because they demonstrate no regular patterns. An example of a chaotic pulse burst together with the subsequent return stroke field is shown in Fig. 9.11 [8]. An expanded section of the chaotic pulse burst is also shown in the figure. The electrical activity in the cloud that follows return strokes can generate bursts of pulses with varying durations, and these are associated with K-change-type events in the cloud. Since continuing currents do not produce significant radiation fields (because they vary very slowly over time), they cannot be identified in radiation field records. The radiation fields

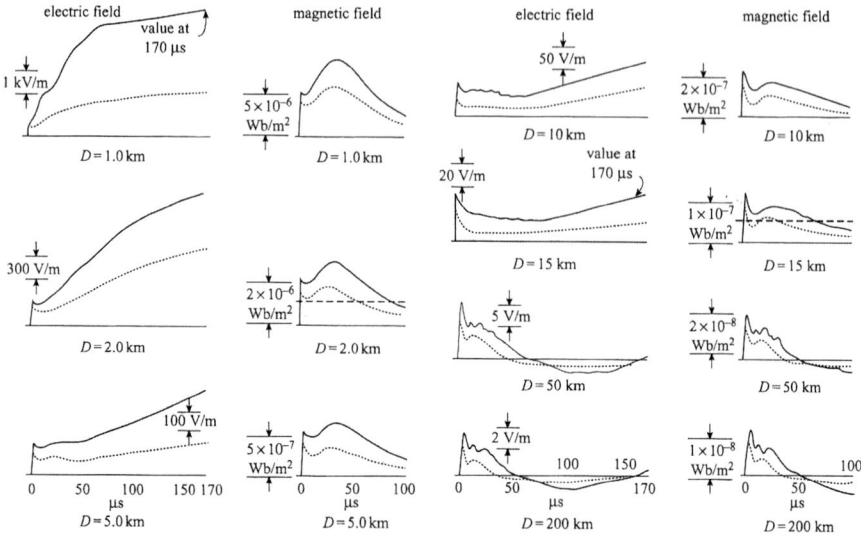

Fig. 9.12 Typical electric field intensity and magnetic flux density for first (*solid line*) and subsequent (*dotted line*) return strokes at different distances (Adapted from Lin et al. [9]; diagram based on atmospheric sign convention)

associated with K and M changes have not yet been investigated thoroughly in the literature. Some information on them is provided in Sect. 9.5.1.1.

Figure 9.12 shows how the return stroke electric and magnetic fields vary as a function of distance from the return stroke channel [9]. Observe how the fields vary with distance. At distances close to the channel the static electric field (the third term of Eq. 3.89 in Chap. 3) and the magnetic induction field (the second term of Eq. 3.90 in Chap. 3) are dominant, and at large distances the radiation fields take over.

9.4 Fine Structure of Radiation Fields from Stepped Leaders and Return Strokes

The electric field far from a channel, say, more than 50 km, is dominated by the radiation field. Many scientists have helped to identify the features of radiation fields from return strokes [9–12, among others]. One interesting fact about the radiation field generated by a transient event is that the integral of the radiation field is zero. The reason for this is as follows. A radiation field is proportional to the rate of change of the current associated with the transient process. Thus, if the radiation field is integrated with respect to time, one ends up with a quantity that contains the difference between the initial current and the final current of the transient process. Since this quantity is zero for a transient process (i.e., the system starts with zero current and ends up with zero current), the integral of the radiation field is zero. For this reason, pure radiation fields generated by transient processes such as return strokes are bipolar.

9.4 Fine Structure of Radiation Fields from Stepped Leaders and Return Strokes

The main features of radiation fields generated by negative return strokes are shown in Fig. 9.13a, b [10]. Several examples of radiation fields generated by positive return strokes are shown in Fig. 9.13c [11]. Let us consider the main features of these radiation fields.

As shown in Fig. 9.13a, b, broadband measurements of radiation fields of return strokes indicate that small field pulses with an amplitude in the range

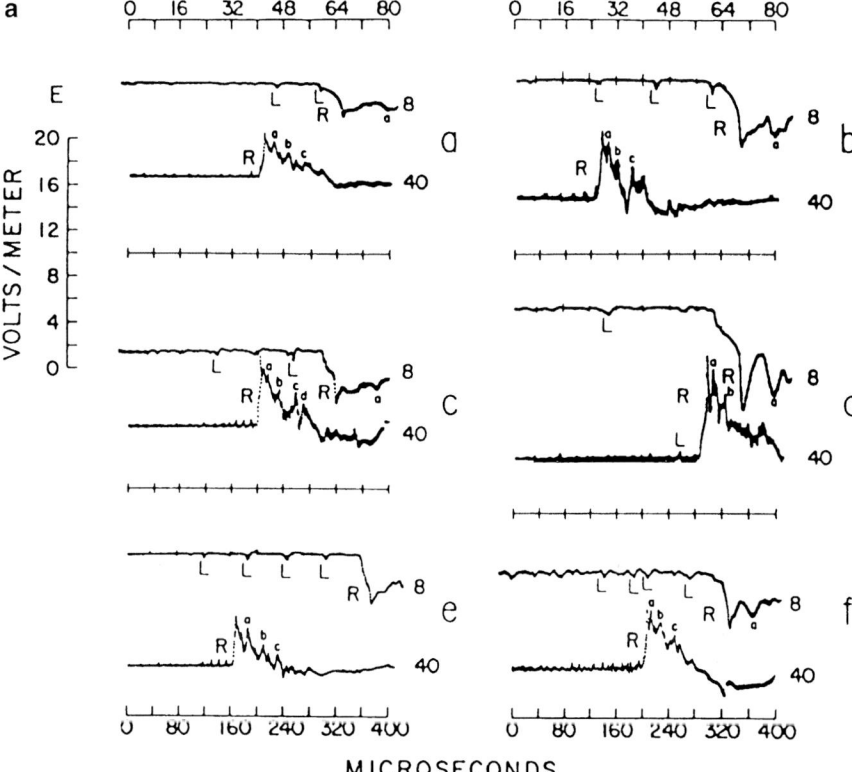

Fig. 9.13 (**a**) Typical radiation fields of first return strokes. Note the leader pulses that appear before the return stroke and the fine structure on the radiation fields marked *a*, *b*, and *c*. These variations are probably caused by the variation in current and geometry associated with branches of the return stroke channel (Adapted from Weidman and Krider [10]). The diagram is based on atmospheric sign convention, but note that the fast scale is inverted for clarity. (**b**) Sketches of shapes of electric radiation fields produced by (*a*) the first return stroke, (*b*) a subsequent return stroke preceded by a dart stepped leader, and (*c*) a subsequent stroke preceded by a dart leader. The small pulses characteristic of leader steps *L* are followed by a slow front *F* and an abrupt transition to peak *R*. The peak amplitudes are normalized to a distance of 100 km (Adapted from Weidman and Krider [10]; diagram based on atmospheric sign convention). (**c**) Electric radiation fields of positive return strokes observed in Sweden. The initial peak value (E_p) and the range (*D*) are given *below* each diagram. A positive field corresponds to an upward deflection. The time marked *S* denotes the arrival of the first ionospheric reflection (Adapted from Cooray [11]; diagram based on atmospheric sign convention)

150 9 Electromagnetic Fields of Lightning Flashes

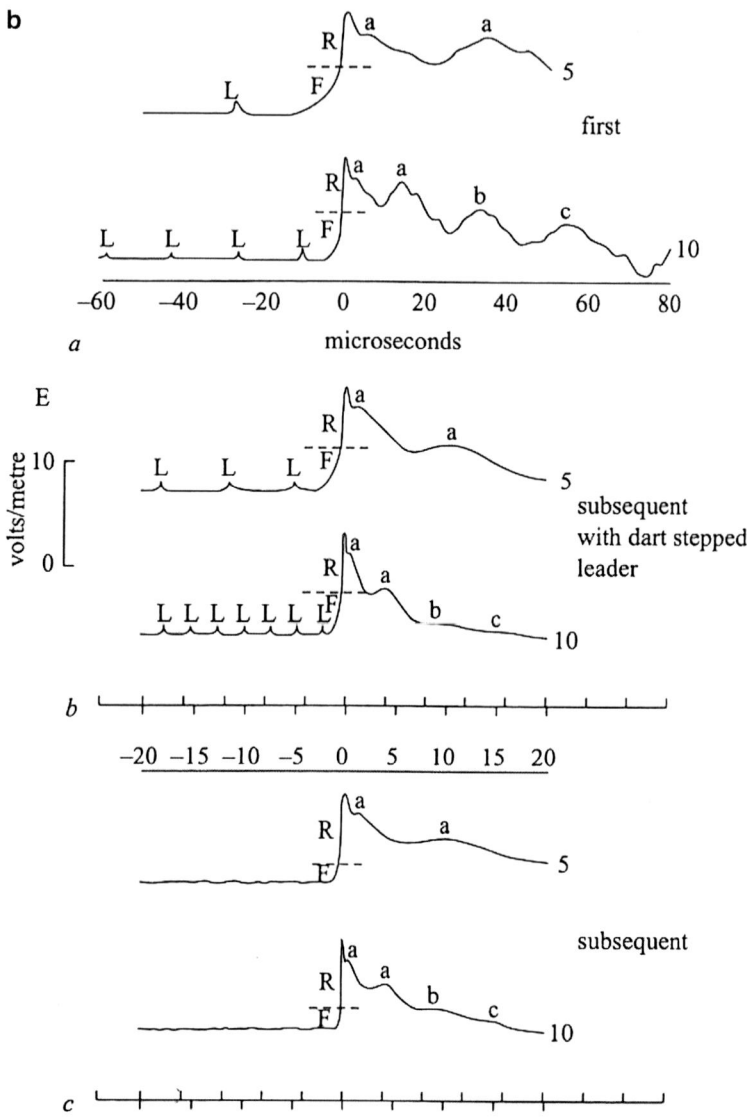

Fig. 9.13 (continued)

9.4 Fine Structure of Radiation Fields from Stepped Leaders and Return Strokes   151

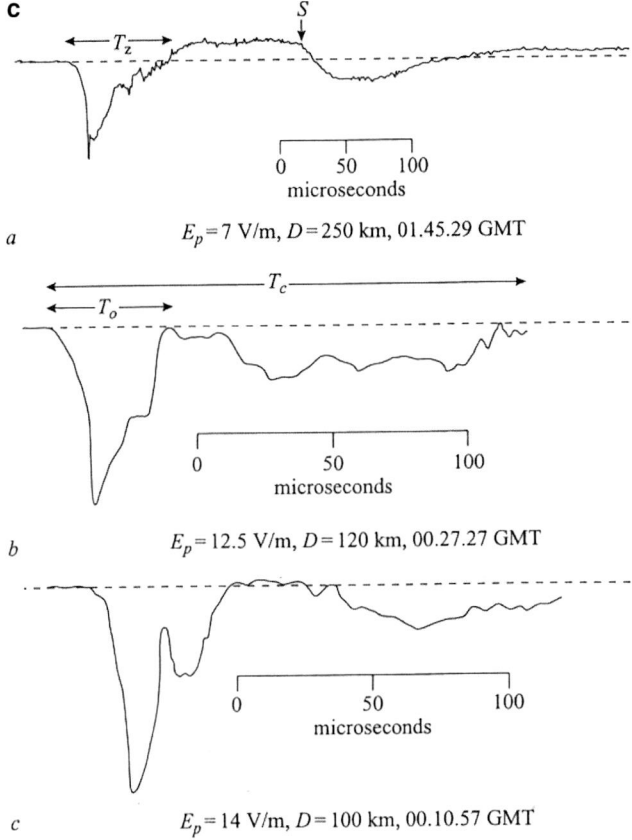

Fig. 9.13 (continued)

of approximately 0.5–1 V/m at 100 km occur in electric fields preceding return strokes. Several examples of such pulses in high resolution are shown in Fig. 9.14 [12]. Simultaneous optical and electric field measurements show that these radiation field pulses are produced by the processes taking place during the formation of leader steps. The time separation between these pulses immediately before the return strokes is approximately 10–15 µs. To observe the fine structure of radiation fields, it is necessary to measure the electric field from lightning flashes occurring over ground of good conductivity. The reason for this is that the finite conductivity of the ground will attenuate and absorb the high frequencies in the electromagnetic field (Chap. 12). This leads to the distortion and disappearance of the rapid features of electromagnetic fields with increasing distance. The best one can do in this connection is to measure the electromagnetic fields of lightning flashes striking the sea at a coastal station. Sea water is a much better conductor than soil because of its high salt content, and most fast features in the electromagnetic field are retained while they propagate over salt water. From the data gathered from such measurements one

Fig. 9.14 Time-resolved electric fields produced by stepping process of stepped leaders at distances of 20–30 km or less over salt water (Adapted from Kride et al. [12]; diagram based on atmospheric sign convention)

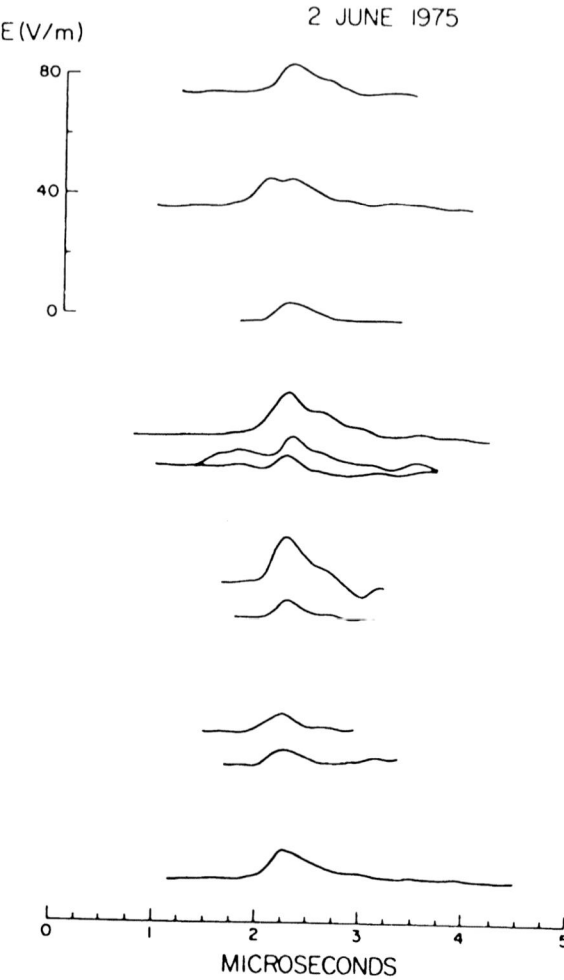

can observe that the rise time of the electric field pulses of a stepped leader is approximately 0.1 µs, and their duration is approximately 1 µs. The leader pulses immediately preceding the return stroke are *almost* unipolar (that means the overshoot has a small amplitude but longer duration), and their electric field time derivative normalized to 100 km is approximately 22 V/m/µs. The latter value is comparable to those of the electric field time derivatives of return strokes (see the sections below).

9.4.1 Slow Front and Fast Transition

As is apparent from the records shown in Fig. 9.13b, the initial rising part of a radiation field contains a slow front (a slowly rising portion of the electric field) followed by a fast transition [10]. The duration of the slow front is approximately 5 μs in negative first return strokes, approximately 10 μs in positive first return strokes, and approximately 0.5 μs in subsequent return strokes. The slow front is followed by a fast transition. The amplitude of the breakpoint (i.e., where the slow front ends and the fast transition begins) is approximately 50 % of the total peak in first return strokes (both positive and negative) and approximately 20 % in subsequent strokes. The 10–90 % rise time of the fast transition is similar in negative first and subsequent strokes, and the average value is approximately 0.1 μs. In radiation fields produced by positive return strokes, the corresponding rise time is approximately 0.2 μs [13].

9.4.2 Distribution of Peak Radiation Fields of Return Strokes

The peak of the radiation field of negative first return strokes normalized to 100 km (the values that would be observed at 100 km from the return stroke; observe that the radiation field decreases inversely with distance from the return stroke) lies in a range of 3–30 V/m, with a mean value lying somewhere between 5 and 10 V/m [5]. The mean of the initial peak of positive return strokes is larger than that of negative first return strokes by a factor of approximately two. The mean peak amplitudes of the radiation fields of subsequent return strokes are smaller than that of first return strokes by a factor of approximately two. The mean peak radiation field values of positive first return strokes is larger than the corresponding value of the negative first return strokes by a factor of approximately two [14].

9.4.3 Time Derivative of Radiation Field

The radiation field time derivative is a parameter that is difficult to measure accurately in land-based measurements owing to the strong attenuation of high frequencies in the signal by finitely conducting ground (for slowly varying fields, of course, the ground can be treated as a perfect conductor). As mentioned previously, to obtain accurate results or undistorted field signatures, the propagation path of the radiation field should be over sea water (high conducting surface) so that the propagation effects are negligible. The data available today on the characteristics of the time derivatives of the radiation field are obtained from measurements conducted with the latter approach. The average value of the peak radiation field derivative is approximately 30–40 V/m/μs when normalized to 100 km [15]. In the case of positive return strokes, the mean value of the peak radiation field derivative when normalized to 100 km is approximately 25 V/m/μs [13]. This is slightly less than the value corresponding to negative return strokes.

9.4.4 Subsidiary Peaks

After the initial peak, both negative first and subsequent strokes exhibit a shoulder within approximately 1 μs from the initial peak [10]. The decaying part of the first return stroke radiation field contains several subsidiary peaks whose separations are in the range of tens of microseconds. Subsequent strokes do not usually show these subsidiary peaks except in cases where the preceding stroke is the first stroke. Let us consider the possible reasons for these subsidiary peaks in the radiation field. Now, a change in the radiation field can be caused by a variation in the discharge current, a variation in the speed of propagation of the discharge, or a change in the geometry of the discharge channel. Subsidiary peaks may be caused by the variations in the current and the geometry as the return stroke front encounters branches in the leader channel. The subsequent return stroke fields are usually free of these subsidiary peaks, and, as mentioned in Chap. 7, subsequent return stroke channels usually do not contain branches.

9.4.5 Zero Crossing Time

The negative first return stroke waveform crosses the zero line in approximately 50–100 μs, whereas the corresponding values for the subsequent return strokes are 30–50 μs [16].

9.4.6 Ionospheric Reflections

It is important to mention that the radiation fields from lightning flashes travel not only along the ground but in a hemispherical region with the lightning channel at the center. The electromagnetic fields that propagate upward could be reflected from the ionosphere, and these ionospheric reflections could also contribute to the electric and magnetic fields measured at a given distance. Figure 9.15 shows how ionospheric reflections are produced. At any distance the measured field consists of the ground wave and ionospheric reflections. The separation in time between the ground wave and the ionospheric reflections decreases as the distance to the point of observation increases. Moreover, with increasing distance the ionospheric reflections become stronger because the reflection coefficient of the ionosphere increases as the angle of incidence increases. Thus, in electric and magnetic fields measured within approximately 50 km it is difficult to observe ionospheric reflections. However, field records obtained at distant observation points (say, at distances greater than approximately 100 km) clearly show these ionospheric reflections. The signature of the return stroke electromagnetic field

9.5 Electromagnetic Fields from Cloud Flashes

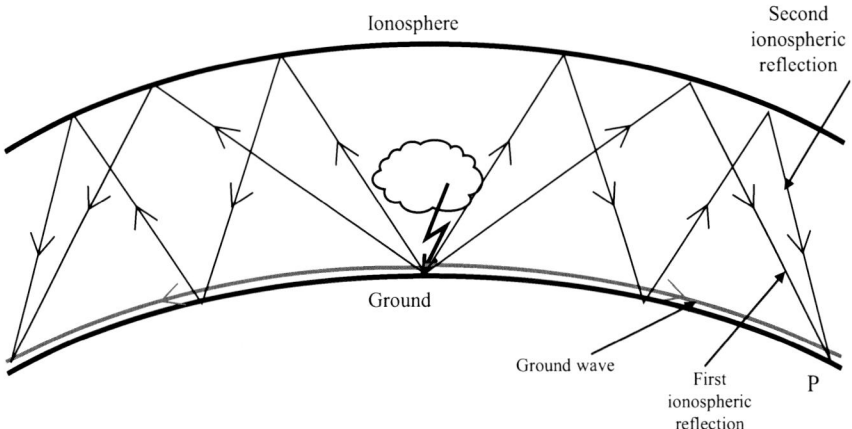

Fig. 9.15 Formation of ionospheric reflections. Assume that the electric field generated by the lightning flash is measured at point *P*. The signal at that point is due to the sum of electric fields arriving at that point along three different paths: first one along ground (ground wave), second one after reflecting once from ionosphere (first ionospheric reflection), and third one after reflecting twice from ionosphere (second ionospheric reflection) (Figure created by author)

without ionospheric reflections (i.e., the ground wave) is similar to that shown in Fig. 9.13a, b. An electric field from a return stroke distorted by ionospheric reflections is shown in Fig. 9.16 [17].

9.5 Electromagnetic Fields from Cloud Flashes

The electric fields of cloud flashes measured almost directly below the cloud can be very complicated to interpret because they are controlled by the unknown complex geometry of the discharge channel. Moreover, the exact nature of the physical process that gives rise to a particular feature of the electric field is hidden inside the cloud and cannot be observed directly. Thus, unlike the close field records of ground flashes, it is difficult to characterize the close fields of cloud flashes except to make broad conclusions concerning the direction (up or down) of the charge transfer. The close electric field is generally in agreement with the transport of negative charge to the upper regions of the cloud. However, since charge centers can be displaced in a horizontal direction because of wind shear and because significant horizontal movement of charges takes place during flashes, it is difficult to perform a detailed analysis and characterize the charge transfer process. Figure 9.17 shows an example of a close electric field generated by a cloud flash [18]. As illustrated in Chap. 3, the static electric field change produced by a cloud flash can have either positive or negative polarity depending on the distance to the point of observation. Close examination of these records shows the steplike

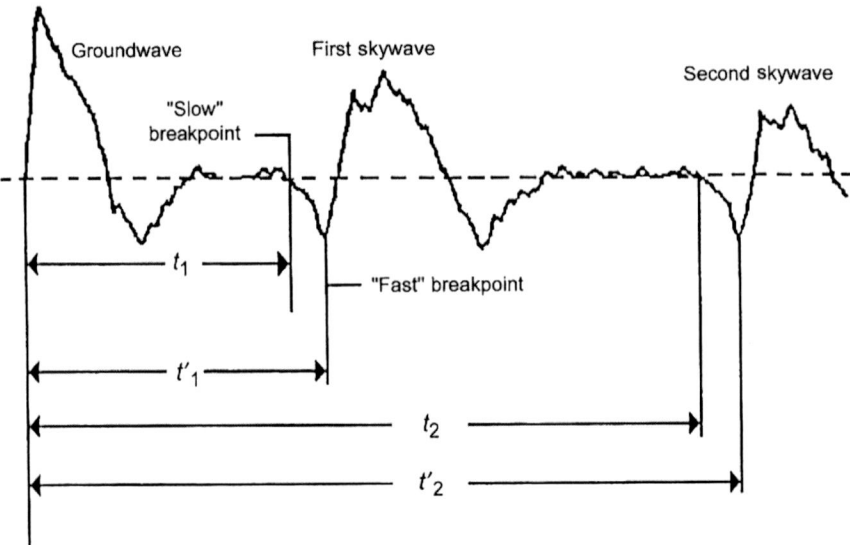

Fig. 9.16 Drawing of a typical radiation field waveform containing two ionospheric reflections. Time of arrival of ionospheric reflections marked by t_1 and t_2. For illustrative purposes the first and second ionospheric reflections are compressed in time by a factor of 2 relative to ground wave (Adapted from McDonald et al. [17]; see also Fig. 9.13c (waveform *a*)) (Figure based on atmospheric sign convention)

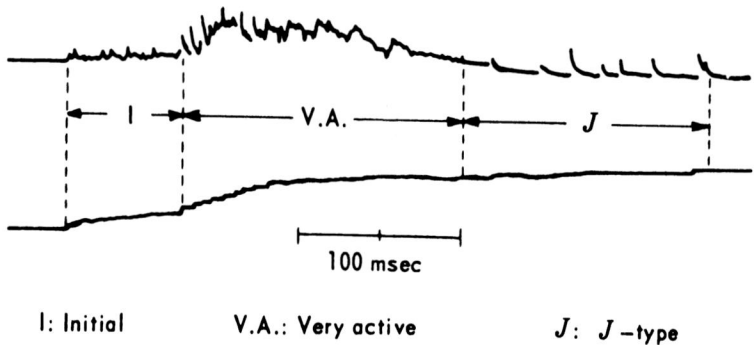

I: Initial   V.A.: Very active   J: J −type

Fig. 9.17 Typical electric field change of cloud flash. Upper and lower traces are recorded simultaneously by fast and slow antennas, respectively. In the fast antenna the RC constant is small, so that in the output of the system, most of the static field is removed; in the slow antenna the RC constant is large and the slowly varying static field is faithfully recorded (Adapted from Kitagawa and Brook [18]; diagram based on physics sign convention)

9.5 Electromagnetic Fields from Cloud Flashes

small electric field changes caused by K changes. These small field changes are barely discernible in the slow antenna record of Fig. 9.17 but are amplified in the fast antenna record.

9.5.1 Radiation Fields of Cloud Flashes

The radiation field from a cloud flash is more complicated but in principle is similar to that of a long-duration preliminary breakdown that occurs in ground flashes. However, in the case of preliminary breakdown in negative ground flashes, charges are transported towards ground, whereas the majority of cloud flashes transport negative charge upward. Thus, the polarity of the radiation field of cloud flashes is opposite to that of the preliminary breakdown pulses in negative ground flashes. A typical record of a radiation field generated by a cloud flash is shown in Fig. 9.18 [19]. The first 10 ms of a cloud flash is shown in Fig. 9.19 [20]. Note that the radiation field of a cloud flash consists of a burst of pulses separated by approximately 10–100 μs. The pulse activity is strong initially and decreases toward the end of the flash.

Individual radiation field pulses in a cloud flash can be either multipeaked or single-peaked. Examples of such pulses are shown in Fig. 9.20 [21]. Note that the multipeaked ones are strongly bipolar with several pulses superimposed on the rising part. The temporal features of these pulses are similar to pulses observed in the preliminary breakdown stage of ground flashes.

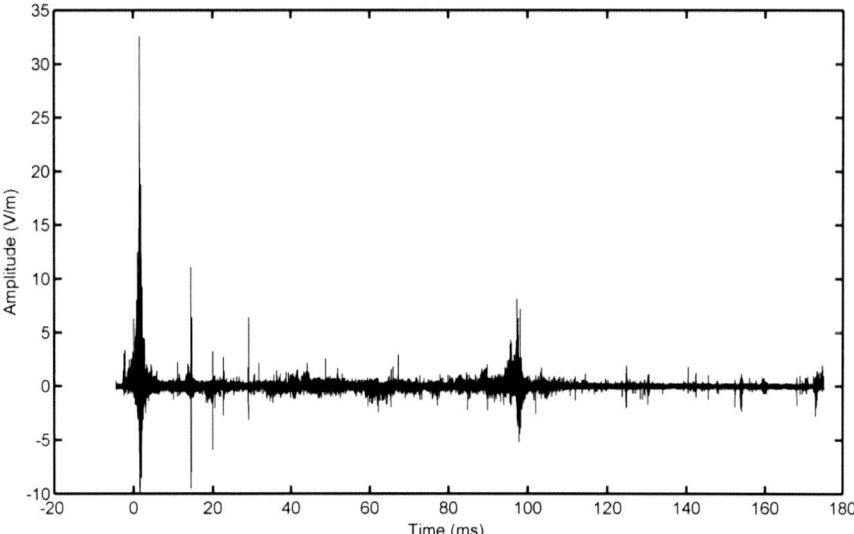

Fig. 9.18 Electric field of distant cloud flash. The electric field is pure radiation. Note that the electric field consists of a burst of electric radiation pulses (Adapted from Ahmad et al. [19]; diagram based on physics sign convention)

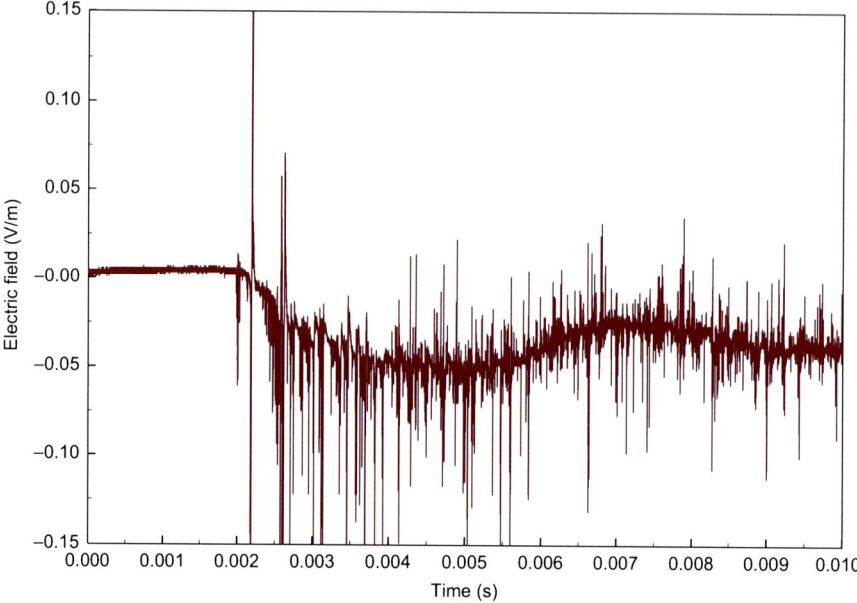

Fig. 9.19 First millisecond of cloud flash. Note intense pulse activity associated with cloud flash (Adapted from Sonnadara et al. [20]; diagram based on atmospheric sign convention)

Fig. 9.20 A cloud flash can produce two types of pulse. Some pulses are single-peaked, as in (**a**), while others are multipeaked, as in (**b**). They both occur in cloud flashes without any significant pattern to their occurrence. *Arrow*: starting point and ending point of pulses (Adapted from Villanueva et al. [21]; diagram based on physics sign convention)

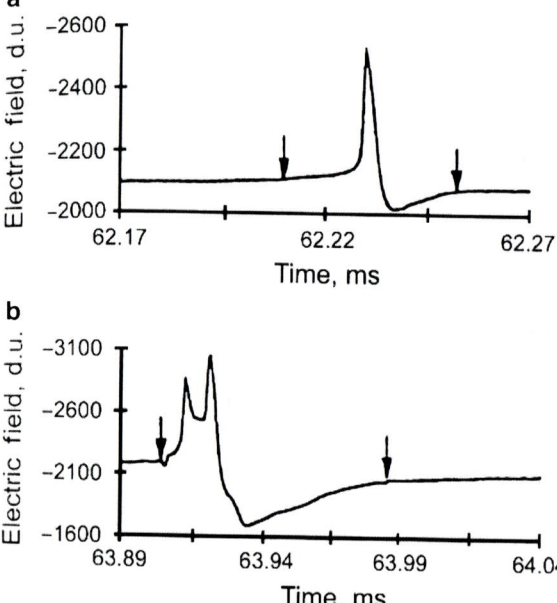

9.5 Electromagnetic Fields from Cloud Flashes

9.5.1.1 Radiation Field Produced by *K* changes and *M*-Components

Most of the information available concerning *K* changes and *M*-components are pertinent to close electric fields and, hence, the static electric field component of *K* changes and *M*-components. The information available at present on the radiation fields generated by *K* changes and *M*-components indicate that they are associated with one or several bipolar or almost bipolar radiation field pulses. Figure 9.21 shows one example each of *K* changes and *M*-components and the associated radiation fields [22]. The figure demonstrates that a *K* change that appears as a step in a close field may transform itself into one or several bipolar pulses when the distance is such that the radiation field is dominant. As depicted in Fig. 9.21, even at

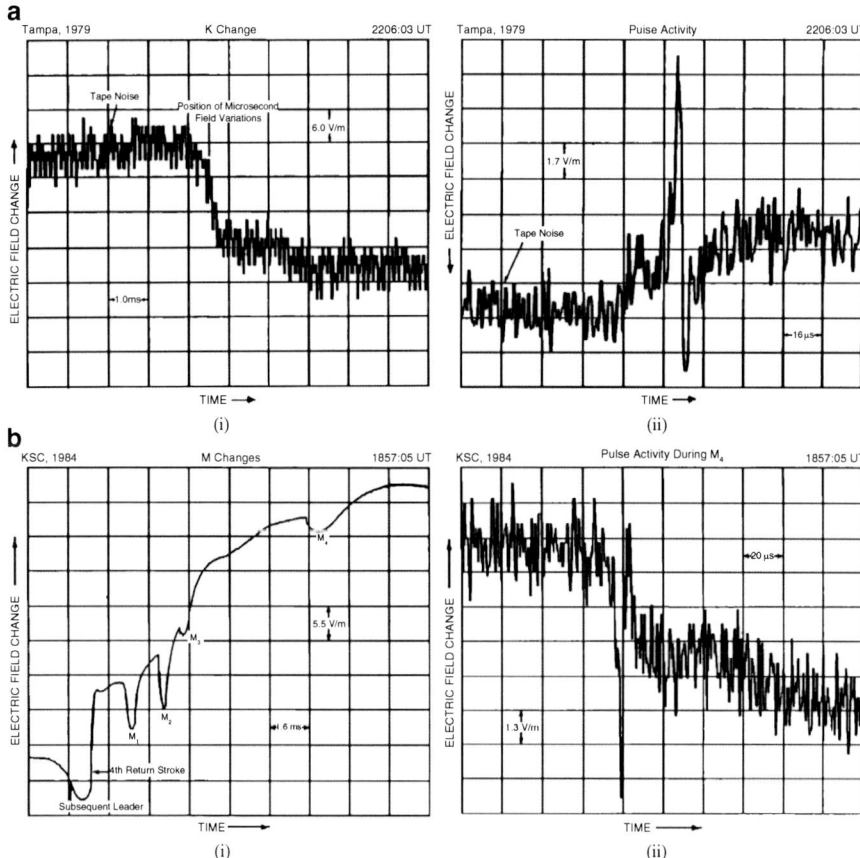

Fig. 9.21 (a) Example of *K* change that appears as step *i*) and corresponding radiation field (*ii*). This means that at great distances the field change due to a *K* change appears as a pulse of radiation as in (*ii*). (b) Example of M-components that appear as hooks in close field *i*) and radiation field associated with one M-component (*ii*). This shows that when measured far from the source, M-components appear as a pulse as in (*ii*) or as a burst of pulses (Adapted from Rakov et al. [22]; diagram based on atmospheric sign convention)

close distances it might be possible to see the pulse structure if measurements are taken with high resolution. As the distance increases, the steplike electric field, which is mainly electrostatic, disappears, leaving behind the bipolar pulses.

9.5.1.2 Isolated Electrical Breakdowns

Even though the preliminary breakdown process is associated with ground flashes, it is an electrical event that takes place inside the cloud (in principle, it is a cloud discharge). Some thunderstorms give rise to electrical breakdowns similar to preliminary breakdowns but without any connection to ground flashes [23]. Qualitatively, there is no difference between the radiation fields of a preliminary breakdown process and isolated breakdown events. The duration of the pulse burst is approximately 1 ms, and the duration of the individual pulses associated with the isolated electrical breakdown process is approximately tens of microseconds. These breakdown events might be generated by a process very similar to the process that initiates positive and negative ground flashes, but for some reason they fail to proceed all the way to a ground flash. Figure 9.22 shows two examples of isolated electrical breakdowns, one with positive polarity (identical to the preliminary breakdown in negative ground flashes) and the other with negative polarity. The temporal features of the individual pulses in the isolated breakdown events are very similar to the radiation field pulses generated by cloud flashes.

9.6 Compact Cloud Flashes

In general, cloud flashes give rise to a burst of pulses, as shown in Figs. 9.18 and 9.19. However, scientists have also observed electrical activity in clouds that leads to the generation of a single pulse. Such pulses are known in the literature as *narrow bipolar pulses*. The electrical event that generates this type of pulse is called a *compact cloud flash*. At present, the exact physical process associated with compact cloud flashes is not known. However, it has been suggested that these compact discharges are a result of relativistic electron avalanches created by the acceleration of energetic electrons, generated by cosmic rays, in thundercloud fields. The close electric fields of compact cloud flashes show that the main electrical activity is not preceded by any leader discharges, adding support to the foregoing hypothesis. A typical radiation field pulse generated by a compact cloud flash is shown in Fig. 9.23 [24]. Note that its duration is approximately 20 μs. These are the strongest

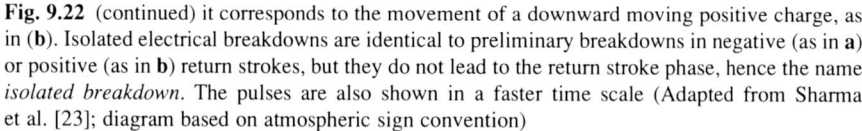

Fig. 9.22 (continued) it corresponds to the movement of a downward moving positive charge, as in (**b**). Isolated electrical breakdowns are identical to preliminary breakdowns in negative (as in **a**) or positive (as in **b**) return strokes, but they do not lead to the return stroke phase, hence the name *isolated breakdown*. The pulses are also shown in a faster time scale (Adapted from Sharma et al. [23]; diagram based on atmospheric sign convention)

9.6 Compact Cloud Flashes

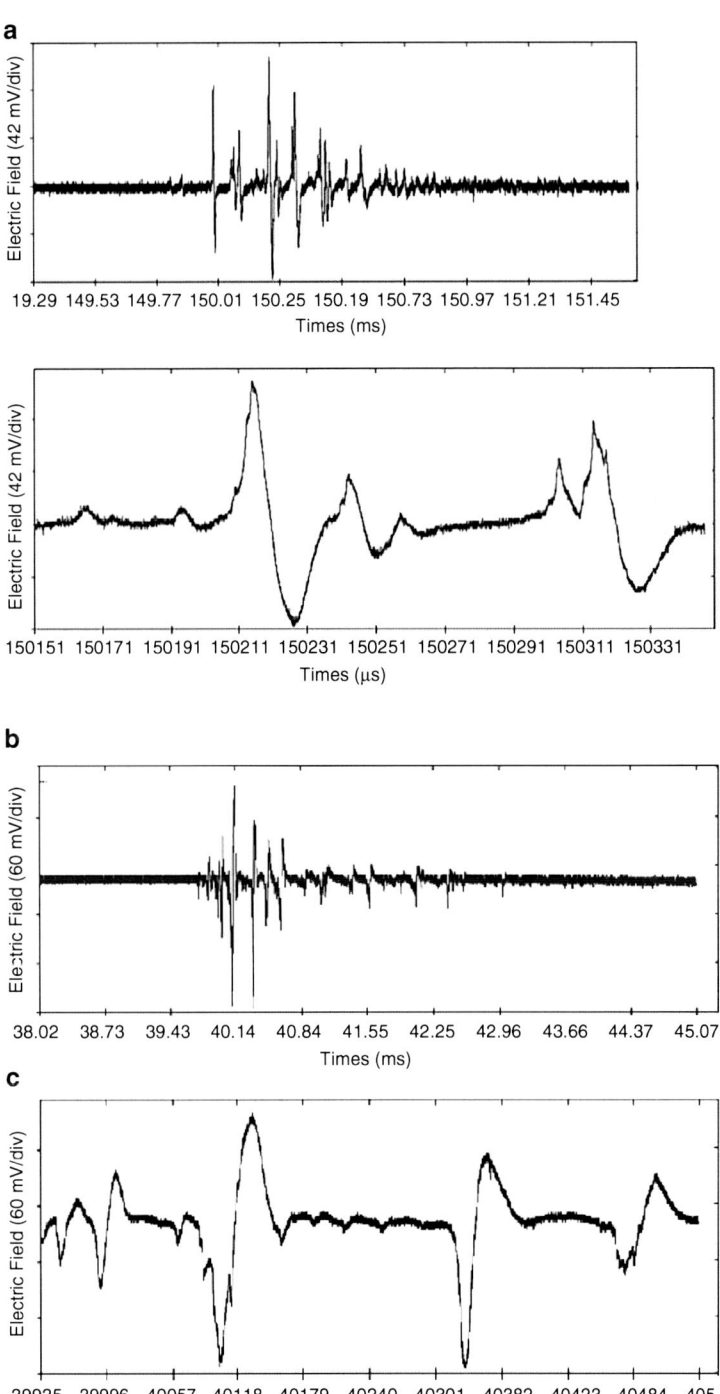

Fig. 9.22 Pulses associated with isolated electric breakdowns. Sometimes an isolated breakdown corresponds to the movement of downward moving negative charge ward, as in (**a**); in other cases,

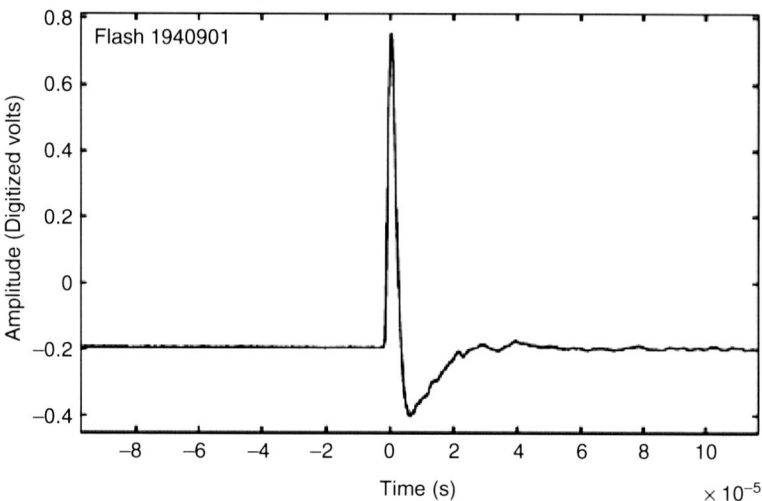

Fig. 9.23 Example of a narrow bipolar pulse observed from thunderstorm in Malaysia. Total duration of pulse is approximately 20 μs; width of initial half cycle (i.e., the time to zero crossing from the beginning of the pulse) is approximately 5 μs (Adapted from Ahmad et al. [24]; diagram based on atmospheric sign convention)

generators of radio noise during lightning flashes, and their amplitudes when normalized to 100 km are comparable to those of return strokes.

9.7 Electromagnetic Spectrum

During thunderstorms the cracking noise coming out of a radio receiver, say, tuned to 1 MHz, is caused by the 1 MHz frequency content of the electromagnetic field of lightning flashes. Thus, the frequency content of lightning electromagnetic fields is of interest whenever there is an interest in understanding the response of a tuned circuit or a system where induced currents (or voltages) can oscillate at a given frequency, i.e., resonance frequencies. The frequency of oscillation of the currents and voltages of a system are governed by the physical dimensions of the system. The smaller the dimensions, the higher the resonance frequency.

The radiation fields generated by various events in ground and cloud flashes can be categorized as impulses or transients. Depending on the event under consideration, they may vary in amplitude and duration. Two methods can be used to obtain the frequency content of the electromagnetic radiation of lightning flashes. The first method is to record the electromagnetic field in high resolution and then use Fourier transformation to obtain the frequency content. In this procedure, one selects the event of interest in the lightning flash, for example, stepped leader pulses, return stroke radiation fields, or chaotic pulses, and studies the frequency content of the individual events. The other method is to use radio receivers tuned to

9.7 Electromagnetic Spectrum

different frequencies so that the spectral amplitude of the lightning flash at these frequencies can be measured. In this method, it is not possible to separate individual events in the flash. The measured frequencies correspond to the entire flash. Of course, the same results can be obtained by Fourier transforming the radiation field of the entire flash measured in high resolution. However, the higher the frequency of interest, the smaller should be the time interval between the digital sampling used in recording the signal. Thus, the method requires a significant amount of computing power and memory to estimate the spectrum of a complete cloud flash, including high frequencies.

Figure 9.24 shows the frequency spectrum of several individual events in lightning flashes [25]. In these studies the spectrum is obtained by Fourier transformation of the recorded electric fields. Figure 9.25 shows the frequency spectrum of lightning flashes obtained using narrow band recorders together with the frequency spectrum obtained by Fourier transforming the complete radiation fields generated by cloud flashes [19]. Note that, as expected, the two spectra are similar.

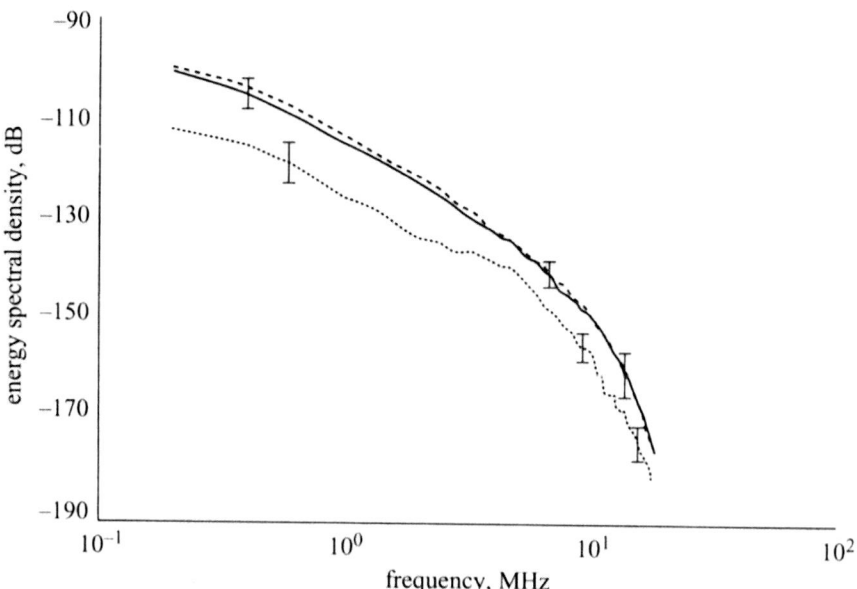

Fig. 9.24 Average spectra of 74 first return strokes (*solid line*) and 55 subsequent strokes (*dashed line*). *Dotted line*: average spectrum of 18 pulses generated by cloud flashes (Adapted from Willett et al. [25])

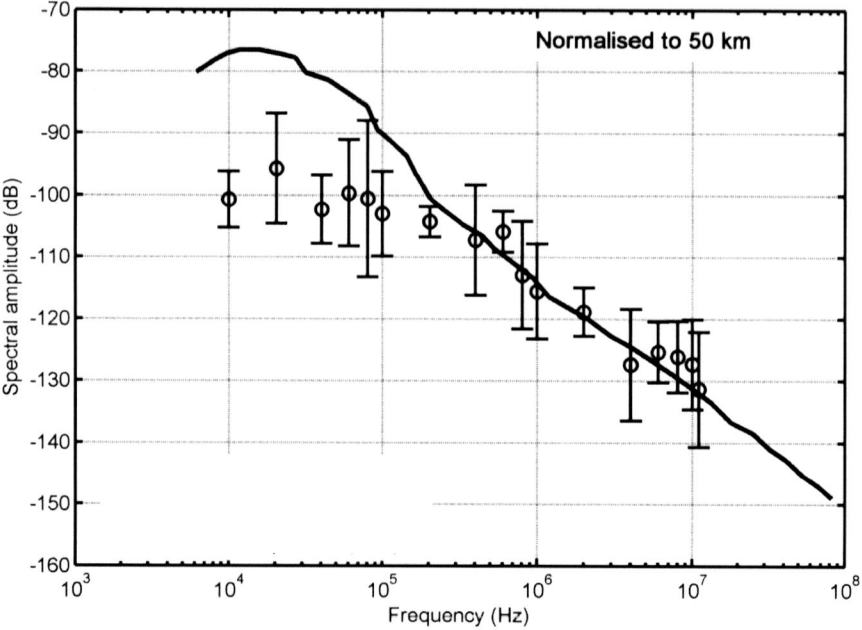

Fig. 9.25 Average spectral amplitude of long-duration cloud flashes (*open circle*) and standard deviation (*vertical bars*) as compared to narrow band spectrum of Horner and Bradley [26], as normalized by Le Vine [27] and as presented by Sonnadara et al. [20], which is shown by the *solid line* (Adapted from Ahmad et al. [19])

References

1. Fernando M, Cooray V (2012) Measurement of electromagnetic fields. In: Cooray V (ed) Lightning electromagnetics. IET Publishers, London
2. Krider EP, Noggle RC, Uman MA (1976) A gated wide band direction finder for lightning return strokes. J Appl Meteorol 15:301–306
3. Clarence ND, Malan DJ (1957) Preliminary discharge processes in lightning flashes to ground. Q J Roy Meteorol Soc 83:161–172
4. Esa MRM, Ahmad MR, Cooray V (2014) Wavelet analysis of the first electric field pulse of lightning flashes in Sweden. Atmos Res 138:253–267
5. Rakov V, Uman M (1990) Long continuing currents in negative lightning ground flashes. J Geophys Res 95:5455–5470
6. Thottappillil R, Rakov VA, Uman MA (1990) K and M changes in close lightning ground flashes in Florida. J Geophys Res 95:18631–18640
7. Baharudin ZA, Fernando M, Ahmad NA, Mäkelä JS, Rahman M, Cooray V (2012) Electric field changes generated by the preliminary breakdown for the negative cloud-to-ground lightning flashes in Malaysia and Sweden. J Atmos Sol-Terr Phys 84–85:15–24
8. Gomes C, Cooray V, Fernando M, Jayaratne C (1998) Chaotic pulse trains associated with negative subsequent stokes. Proceedings of the international conference on lightning protection, Birmingham, England, 1998

References

9. Lin YT, Uman MA, Tiller JA, Brantley RD, Beasley WH, Krider EP, Weidman CD (1979) Characterization of lightning return stroke electric and magnetic fields from simultaneous two-station measurements. J Geophys Res 84:6307–6314
10. Weidman CD, Krider EP (1980/1982) The fine structure of lightning return stroke wave forms. J Geophys Res Lett 7:955–958, Correct J Geophys Res 87:7351
11. Cooray V (1984) Further characteristics of positive radiation fields from lightning in Sweden. J Geophys Res 84:11807–11815
12. Krider EP, Weidman CD, Noggle RC (1977) The electric fields produced by lightning stepped leaders. J Geophys Res 82:951–960
13. Cooray V, Fernando M, Gomes C, Sorenssen T (2004) The fine structure of positive lightning return-stroke radiation fields. Trans IEEE (EMC) 46:87–95
14. Cooray V, Lundquist S (1982) On the characteristics of some radiation fields from lightning and their possible origin in positive ground flashes. J Geophys Res 87:11203–11214
15. Krider EP, Letenturier C, Willett JC (1996) Submicrosecond field radiated during the onset of first return strokes in cloud-to-ground lightning, J. Geophys Res 101(D1):1589–1597
16. Cooray V, Lundquist S (1985) Characteristics of the radiation fields from lightning in Sri Lanka in the tropics. J Geophys Res 90:6099–6109
17. McDonald TB III, Uman MA, Tiller JA, Beasley WH (1979) Lightning location and lower-ionospheric height determination from two-station magnetic field measurements. J Geophys Res 84:1727–1734
18. Kitagawa N, Brook M (1960) A comparison of intracloud and cloud-to-ground lightning discharges. J Geophys Res 65:1189–1201
19. Ahmad NA, Baharudin Z, Fernando M, Cooray V (2011) Radiation field spectra of long duration cloud flashes, in broadband and HF radiation from cloud flashes and narrow bipolar pulses, PhD thesis of N. A. Ahmad, University of Uppsala, Uppsala, 2011
20. Sonnadara U, Cooray V, Fernando M (2006) The lightning radiation field spectra of cloud flashes in the interval from 20 kHz to 20 MHz. Trans IEEE (EMC) 48:234–239
21. Villanueva Y, Rakov VA, Uman MA, Brook M (1994) Microsecond-scale electric field pulses in cloud lightning discharges. J Geophys Res 99:14353–14360
22. Rakov VA, Thottappillil R, Uman MA (1992) Electric field pulses in K and M changes of lightning ground flashes. J Geophys Res 97:9935–9950
23. Sharma SR, Cooray V, Fernando M (2008) Isolated breakdown activity in Swedish lightning. J Atmos Solar-Terr Phys 70(8):1213–1221
24. Ahmad NA, Fernando M, Baharudin ZA, Cooray V, Ahmad H, Abdul Malek Z (2010) Characteristics of narrow bipolar pulses observed in Malaysia. J Atmos Sol-Terr Phys 72:534–540
25. Willett JC, Bailey JC, Letenturier C, Krider EP (1990) Lightning electromagnetic radiation field spectra in the interval from 0.2 to 20 MHz. J Geophys Res 95:20367–20387
26. Horner F, Bradley PA (1964) The spectra of atmospherics from near lightning discharges. J Atmos Sol-Terr Phys 26:1155–1166
27. Le Vine DM (1987) Review measurements of the RF spectrum of radiation from lightning. Meteorol Atmos Phys 37:195–204

Chapter 10
The Return Stroke – How to Model It

10.1 Introduction

To calculate the electromagnetic fields generated by return strokes, it is necessary to know the spatial and temporal variation of the return-stroke current along the channel. Unfortunately, only the return-stroke velocity and the current generated at the channel base can be measured directly, and the way in which the return-stroke current varies along the channel must be extracted indirectly. For example, the magnitude of the current and its wave shape at different heights can in principle be extracted by studying the optical radiation. Unfortunately, the exact relationship between the return-stroke-generated optical radiation and the return-stroke current parameters are not known, and therefore only qualitative inferences can be made on the return-stroke current. The information thus obtained indicates that the return-stroke peak current decreases with height while the rise time of the current increases.

The difficulty of measuring current parameters along a lightning channel directly made it necessary for the lightning research community to make reasonable assumptions concerning how the return-stroke current varies along the channel. These assumptions are generally known as *return-stroke models*. Such models specify how the return-stroke current varies as a function of height along the channel. In the literature, there are many categories of return-stroke models. The main categories are as follows:

(a) Physics models incorporating basic physics to describe each stage of the return stroke [1]. Models of this kind utilize the basic conservation laws, namely, conservation of charge, conservation of momentum, and conservation of energy, in combination with the equations of fluid dynamics and thermodynamics, to evaluate the temporal and radial variation of the channel temperature, channel pressure, channel radius, acoustic parameters, optical radiation, and electrical properties of a channel such as electron density and conductivity. The input parameters of the model are the temporal variation of the current in the channel, the temperature, the pressure, and radius of the channel at the initiation of discharge. Actually, these are not really return-stroke models in that they cannot specify how the return-stroke current varies along the channel as a function of space and time. They can only be used to study how the channel properties vary at a given point on the discharge channel if the current input to that channel section is known. However, these physical models can be combined with other return-stroke model types (described subsequently) to give a reasonable description of the physical parameters of the return-stroke channel.

(b) Electromagnetic models where the return stroke is modeled using Maxwell's equations [2, 3]. In electromagnetic models the lightning channel is represented by a perfectly conducting wire of finite radius located (vertically) over a ground plane (Fig. 10.1). The current or voltage at the ground end of the wire is given as a boundary condition, and Maxwell's equations are solved for the particular geometry and boundary conditions using either the method of moments or the finite difference technique to obtain the distribution of the current along the conductor. Since the speed of propagation of electromagnetic signals along a conductor located in air is equal to the speed of light, in models of this kind the electrical constants of the medium in which the conductor is located or the electrical characteristics of the conductor itself are modified in different ways to obtain a wave speed comparable to that of a return stroke. The techniques employed for this purpose are (1) a resistive wire in air above ground, (2) replacing air above the conducting boundary (i.e., ground) by a dielectric medium, (3) coating the wire with a dielectric material, (4) loading the wire with additional distributed series inductances, and (5) considering two parallel wires having additional distributed capacitances between them. It is doubtful whether Maxwell's equations alone can describe the development of a return stroke. The optical radiation and thunder signal generated by a return stroke

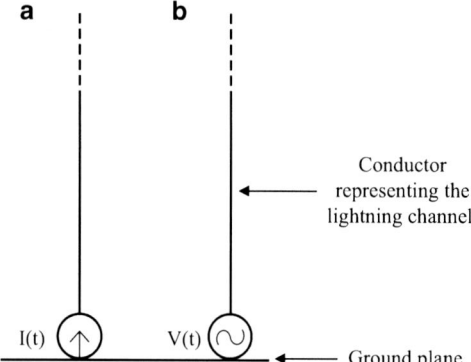

Fig. 10.1 In electromagnetic models the lightning channel is treated as a vertical conductor located over a perfectly conducting ground plane. The initiation of the return stroke is represented by switching on a current source (as in (*a*)) or a voltage source (as in (*b*)) that varies as a function of time at the base of the channel. The resulting current signature along the channel as a function of time is obtained by solving Maxwell's equations (Figure created by author based on information available in the literature)

10.1 Introduction

demonstrate very clearly that the physical parameters of the channel such as pressure and temperature vary rapidly with time during a return stroke. These effects cannot be described by Maxwell's equations alone. Even to include the ionization processes in the channel requires using the physics of electrical discharges, which cannot be explained by pure Maxwell's equations, either.

(c) Transmission line models where the return-stroke channel is treated as a transmission line and use the corresponding transmission line equations to solve for the current at different points along the channel [4]. These models assume that the leader channel can be considered a transmission line with specific inductance (L), capacitance (C), and resistance (R) per unit length. The models can also account for the variation of L, C, and R along a channel. Two types of transmission line models exist in the literature. The first type describes the return stroke as an injection of a current waveform from the ground end of a vertical transmission line located over a ground plane (Fig. 10.2). The solution of transmission line equations provides a description of the temporal variation of the current as a function of time and height along the transmission line. The second type of model considers the leader channel as a charged vertical transmission line located over a ground plane (Fig. 10.3). The initiation of the return stroke is represented by connecting this charged transmission line to ground either directly or through an impedance. This generates a wave of ground potential propagating along the transmission line, and the current associated with this potential wave is the return-stroke

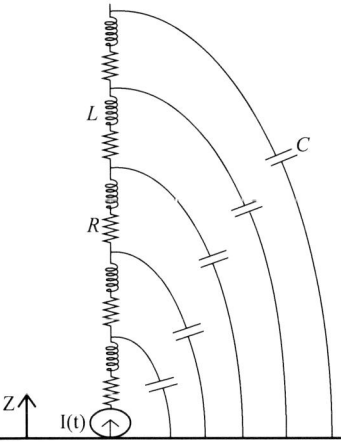

Fig. 10.2 In one type of transmission line representation, the lightning channel is assumed to be a vertical transmission line with per-unit parameters L, C, and R. In this representation, C is the capacitance of the channel section with respect to ground. The return stroke is initiated by injecting a current pulse, i.e., $I(t)$, at the base of the transmission line. The current distribution along the channel as a function of time is obtained by solving transmission line equations with specified boundary conditions. In more sophisticated models, the resistance per unit length of any given channel section can be assumed to vary with time or as a function of the current flowing through the channel section (Figure created by author)

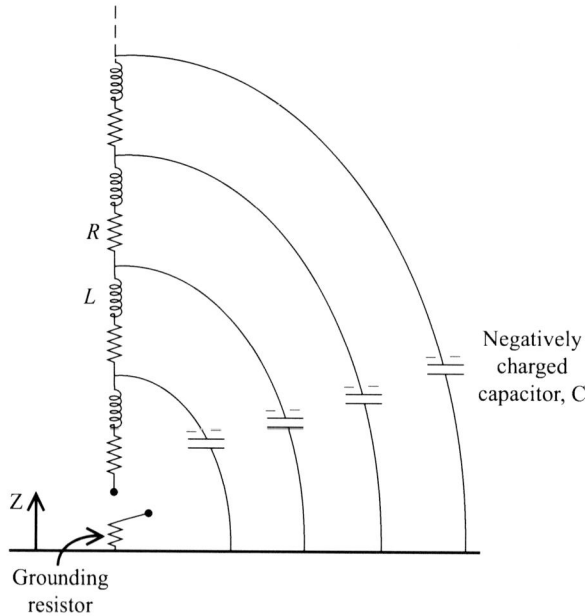

Fig. 10.3 Second approach to utilizing the transmission line theory to represent return strokes. In this case, the leader channel is represented by a charged transmission line. The return stroke is initiated by connecting the transmission line to ground across a resistor. The resistor represents the grounding impedance. The connection of the charged transmission line to ground gives rise to the propagation of a potential wave (and an associated current waveform) along the line. The signature of the potential and the current as a function of time at any point on the line is obtained by solving transmission line equations. This procedure is more physically appealing than the procedure illustrated in Fig. 10.2 (Figure created by author)

current. The temporal and spatial variation of the current in this case can be obtained by numerically solving the transmission line equations. In more specialized transmission line models, the channel resistance is allowed to vary as a function of return-stroke current. The connection between the channel resistance and the return-stroke current is obtained from the physics models described earlier.

(d) Engineering models that combine basic ideas of all of the previously described three model types and merge them with available experimental data in creating a model [5]. The most successful and simplest return-stroke models belong to this category. These models are used frequently in calculating electromagnetic fields in engineering applications. Engineering return-stroke models can be divided into three types: current-generation models, current-propagation models, and current-dissipation models. In this chapter, the basic concepts used in these engineering models are described.

Some of the concepts used in engineering models are physically sound while others are not. Unfortunately, however, in contrast to the study of physics, where a model is rejected if it is inconsistent with known physical laws, even if the model can produce results in agreement with certain observations, the current state of the art in lightning research is to accept the simplest model capable of generating signatures similar to measured lightning electric and magnetic fields as valid and forget about the physical inconsistencies inherent in the model. This tendency of the lightning research community, probably motivated by lightning protection engineers who require models that can be implemented easily using computer software, has actually led to a decline in the progress of lightning return-stroke models. With this note of warning, let us consider the basic ideas that have been put forward to model return strokes. First, let us summarize the basic processes that take place during a return stroke that are of interest in lightning return-stroke models.

10.2 Return-Stroke Process

The stepped leader transports negative charge from cloud to ground. It contains a central core where the temperature is on the order of tens of thousands of degrees. The central core is so thin that the charge on the leader cannot be accommodated within the central core because it would lead to very high electric fields at its surface that would be large enough to generate electrical discharges such as streamers propagating from the central core. As the charge on the central core increases, these streamer discharges carry this charge away from the central core and deposit it around the central core. This region where most of the charge of the stepped leader resides is called the *corona sheath* (Chap. 7). The radius of the corona sheath can be calculated approximately by realizing that the electric field in air at standard atmospheric pressure and temperature cannot exceed 3×10^6 V/m without causing an electrical breakdown leading to electrical discharges. Thus, the outer radius of a stepped leader close to ground can be estimated by assuming that at the outer surface of the corona sheath the electric field is approximately 3×10^6 V/m. Of course, the breakdown electric field appropriate to the atmospheric pressure at the height of interest must be used in evaluating the radius of the leader channel at elevated heights. This consideration gives the radius of a typical stepped leader close to ground as approximately 10 m. The radius of the central core of a stepped leader is approximately 1 mm, and since it is a good conductor, it is approximately at the potential of the cloud.

During the attachment of the stepped leader to a grounded object, the stepped leader is usually met by a connecting leader issuing from the grounded object. The potential of the tip of the connecting leader is almost at ground potential (zero potential), and, as mentioned earlier, the tip of the stepped leader is at cloud potential. The cloud potential corresponding to a typical stepped leader could be approximately 50 MV (Chap. 14). When the connecting leader and the stepped leader meet each other, the potential of the tip of the stepped leader changes to ground potential. The process is slightly more complicated when the details are taken into account, but this treatment is sufficient to describe the action of a return stroke. This sudden change in potential leads to a rapid transfer of charge from the stepped leader tip to ground along the connecting leader. This is because the charge on the stepped leader is bounded there by the cloud potential, and when the potential is relaxed, the charge is released to ground. This transfer of charge from the tip in the vicinity of the point of connection between the stepped leader and the connecting leader creates an electric field that forces the charges on the leader channel slightly ahead of the contact point to transfer to ground. This action of transferring charge to ground proceeds upward, and the transfer of charge to ground from points at and below the upward moving potential discontinuity creates a current that flows along the lightning channel. The upward moving potential discontinuity that releases the charge of the stepped leader is the return-stroke front, and the current propagating to ground from points below the return-stroke front is the return-stroke current. What happens during the return stroke to the charge deposited in the leader channel can be easily understood by imagining the following simple experiment. Fill a cylinder that is closed at one end with sand. Now, holding the open end of the cylinder with one hand, turn the cylinder upside down. When the hand is released, the sand starts to fall and as time progresses, sand from higher and higher regions will be drawn toward ground. The sand propagates downward while the action goes up.

Experimental data show that return-stroke channels are no more than a few centimeters in diameter. The question, then, is how a charge located over a region that is several meters in radius is transferred to ground. This removal of charge from the corona sheath takes place as follows. When the core of the leader channel changes from cloud potential to ground potential, the presence of the corona charge induces a charge of opposite polarity on the return-stroke channel (i.e., positive in the case of negative return strokes), and this induced charge creates a strong radial electric field around the central core. This electric field generates positive streamers (recall that the discharges that carried away the leader charge to the corona sheath are negative streamers) that propagate out and neutralize the negative charge in the corona sheath. A consequence of this is that return strokes remove positive charges from ground and deposit them in the corona sheath of the stepped leader. In effect, this is equivalent to transferring negative charge from the corona sheath to ground. This process is pictorially depicted in Fig. 10.4.

10.3 Current-Generation-Type Return-Stroke Model

The description given in the previous section provides a physical picture that is being used to construct current-generation models. This physical picture is as follows [5]. A return stroke is a wave of ground potential (or zero potential) that travels upward along the stepped leader channel. As this potential wave reaches any point on the stepped leader channel, it generates positive streamers that travel to the negative corona sheath and neutralizes the negative charge located there. This neutralization of the corona sheath gives rise to a current pulse of negative charge

10.3 Current-Generation-Type Return-Stroke Model

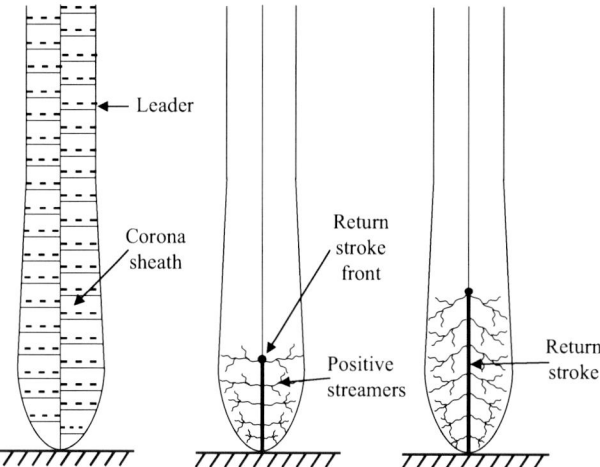

Fig. 10.4 The stepped leader channel is negatively charged and the bulk of the charge resides on the corona sheath. As the return stroke surges up the channel, it extends positive streamers into the corona sheath and neutralizes the charge on the corona sheath. The charge necessary to neutralize the charge on the corona sheath is provided by the ground (Figure created by author)

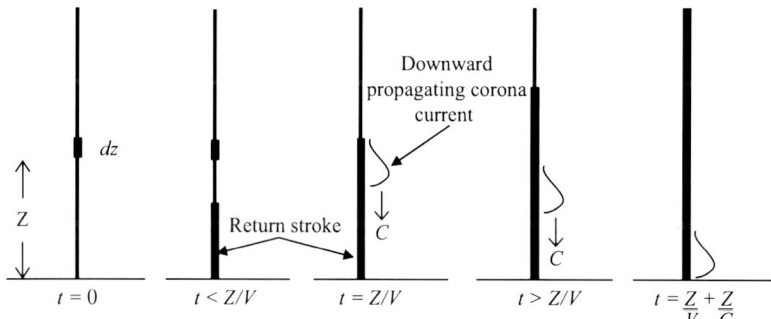

Fig. 10.5 Principle of current-generation-type return-stroke models. Consider a channel element dz on the leader channel. The return stroke takes time z/v to reach this channel element. In the preceding statement, v is the speed of the return stroke and is assumed to be uniform along the channel. When the return-stroke front reaches this element, it neutralizes the charge in the corona sheath associated with this element. This neutralization process gives rise to a current pulse (consisting of a negative charge). This current pulse propagates to ground along the return stroke channel at the speed of light c. It takes time z/c to reach ground. In other words, this current pulse reaches ground after a time of $z/v + z/c$ from the beginning of the return stroke. Since the current pulse consists of a negative charge (i.e., it is a negative current), it effectively transfers the negative charge from the corona sheath to ground (Figure created by author)

that travels to ground along the core of the stepped leader. Figure 10.5 illustrates how a small channel section on the leader channel contributes to this process.

Let us describe the physical process described earlier mathematically. Let us denote by v the speed of propagation of the return-stroke front, and let us denote by

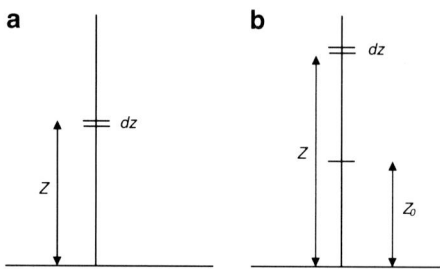

Fig. 10.6 (a) Geometry relevant to calculation of return-stroke current at ground level in current-generation-type return-stroke models. An expression is first written for the current at ground level due to the current element dz. The total current at ground level is obtained by summing the contribution (using integration) from all channel elements from ground level to top of channel. (b) Geometry relevant to calculation of return-stroke current at given height Z_0 in current-generation-type return-stroke models. An expression is first written for the current at height Z_0 due to the current element dz. The total current at height Z_0 is obtained by summing the contribution (using integration) from all channel elements from height Z_0 to top of channel (Figure created by author)

$\rho(z)$ the positive charge deposited by the subsequent return stroke per unit length of the corona sheath located at height z. This is identical in magnitude to the negative charge removed by the return stroke from the same unit channel section. Consider a channel element of length dz located at height z along the leader channel (Fig. 10.6a). If the speed of the return stroke is constant, then the return-stroke front will reach this point at time z/v. Let us denote the current resulting from the neutralization of a unit length of the corona sheath at height z by $i_c(z, t)$. We call this the *corona current*. We have assumed that this current is a function of z because the charge deposited by the return stroke on the leader channel may vary with height (Chap. 14). We have also assumed that this current is a function of time because neutralization of the corona sheath by positive streamers may take some time. Recall that the release of this charge, and hence the corona current associated with that channel element, is initiated at time $t = z/v$. We will also assume that this corona current travels to ground at a speed equal to the speed of light. We denote it by c. In fact, this downward propagation speed of the corona current may not be exactly equal to the speed of light, but since the return-stroke channel below its front is hot and conducting, it is not a very bad assumption. Thus, the corona current, once released at height z, takes a time of z/c to reach ground. Therefore, mathematically we can describe the current reaching ground as a result of the neutralization of the channel element of length dz located at height z by $i_c(z, t - z/v - z/c)dz$. In writing down this expression we have assumed that the return stroke is initiated at $t = 0$. This expression indicates that the current from the channel element at height z appears at ground level after a time delay of $z/v + z/c$ from the beginning of the return stroke (the first term describes the time taken by the return-stroke front to reach height z, while the second term describes the time taken by the corona current to reach ground) . Now we are ready to write down an expression for the current reaching ground during the return stroke. It is obtained by summing up

10.3 Current-Generation-Type Return-Stroke Model

the contribution of corona currents generated by all channel elements in the return-stroke channel. Recall that $i_c(z, t)$ is the corona current released by a channel of unit length located at height z. Thus, the corona current released by a channel length of dz is given by $i_c(z, t)dz$. Therefore, the current at ground level is given by

$$i(0, t) = \int_0^{z_m} i_c\left(z, t - \frac{z}{v} - \frac{z}{c}\right) dz. \tag{10.1}$$

Now we have an expression for the current at ground level. But what is the upper limit z_m of integration in the preceding equation? Well, it depends on how far into the time we would like to estimate the current. Say that we would like to estimate the current at ground level from time zero to some time, say, t. Since it will take some time for the return-stroke front to propagate to a given point on the channel and it will take more time for the current released from that point to appear at ground level, we must estimate the highest point on the channel that can contribute to the return-stroke current at ground level at time t. It is not difficult to see that the height of this point, z_m is given by solving the equation

$$\frac{z_m}{v} + \frac{z_m}{c} = t, \tag{10.2}$$

that is,

$$z_m = \frac{t}{\left\{\frac{1}{v} + \frac{1}{c}\right\}}. \tag{10.3}$$

This is why the upper limit of the integral is set at z_m. Now, how do we obtain the return-stroke current at any other point of the channel? We apply the same principle. Let us assume that we would like to calculate how the return-stroke current changes at a point located at height z_0. First we can see that only the channel elements located above this height (i.e., z_0) contribute to the current at this point. Let us again consider a channel element of length dz located at height z (Fig. 10.6b). Point z is located above z_0. That is, $z > z_0$. As previously, the corona current released by the current element dz is $i_c(z, t)dz$. This current is released at time z/v. This current will take a time of $(z - z_0)/c$ to reach point z_0. Thus, the contribution to the current at point z_0 by the channel element dz located at height z is $i_c(z, t - z/v - (z - z_0)/c)dz$. Now, remember that the current at point z_0 starts only after the return-stroke front has passed this point. This happens when $t = z_0/v$. Thus, as previously, the total current at point z_0 is given by

$$i(z_0, t) = \int_{z_0}^{z_{0m}} i_c\left(t - \frac{z_0}{v} - \frac{(z - z_0)}{v} - \frac{(z - z_0)}{c}\right) dz \quad \text{valid for} \quad t \geq z_0/v. \tag{10.4}$$

Using the same argument as we used earlier, we see that the equation to be satisfied by the upper limit of the integral z_{0m} is

$$(z_{0m} - z_0)\left\{\frac{1}{v} + \frac{1}{c}\right\} = t - \frac{z_0}{v}. \tag{10.5}$$

This yields

$$z_{0m} = \frac{t - \frac{z_0}{v}}{\left\{\frac{1}{v} + \frac{1}{c}\right\}} + z_0. \tag{10.6}$$

When z_0 goes to zero, this equation reduces to 3. Now, except for a few addition details, we have an expression for how the return-stroke current changes at any point on the return-stroke channel as a function of time. But now, if we attempt to calculate the integrals to evaluate the current, we come up against a difficulty because we still do not know what $i_c(z,t)$ is. It is not that easy to obtain this information using basic physics. We can safely say that the positive streamers will take some time to completely neutralize the charge at any given point. Let us say that this time is τ. This also means that the duration of the corona current is of the same order of magnitude as τ. This we can appreciate by writing (note that the corona current is zero when $t < z/v$)

$$i_c(z,t) = i_0(z)e^{-(t-z/v)/\tau} \quad t > z/v. \tag{10.7}$$

The preceding equation says that the current will go to zero in a time on the order of τ and the amplitude of the current (defined by $i_0(z)$) is a function of height. Now, recall that the corona current at a given height was caused by the charge removed by the return stroke from a unit length at that height. From the preceding equation we see directly that the amount of charge released from a unit channel section at height z (or the amount of positive charge deposited by the return stroke on a unit channel length at height z) is

$$\rho(z) = \tau i_0(z). \tag{10.8}$$

We have obtained this by integrating Eq. (10.7) from time z/v to infinity. The set of equations given above completely describe the variation of the return-stroke current with height. Recall that in deriving all the equations we have assumed that the return-stroke speed is constant. How should we treat the problem if the return-stroke speed is not a constant but varies along the channel, which indeed is the experimental observation? To introduce this, we must understand how to obtain the average return-stroke speed between two points on the channel. For example, consider a point at height z. What is the average return-stroke speed along the channel from ground to height z? Let us denote the return-stroke speed that varies with height by $v(z)$. To obtain the average speed, we must divide the channel from zero to z into very small elements and assume

10.3 Current-Generation-Type Return-Stroke Model

that the speed is constant along each of these small elements. But of course, it may vary from one element to another. To obtain the average speed, we first calculate the total time necessary for the return-stroke front to reach level z by adding the elementary time intervals taken by the return-stroke front to traverse each element. When the total length of the channel is divided by this time, we obtain the average return-stroke speed. Mathematically this can be written as

$$v_{av}(z) = \frac{z}{\int_0^z \frac{d\xi}{v(\xi)}}. \tag{10.9}$$

In the preceding equation, $v_{av}(z)$ is the average speed of the return stroke over the channel length from 0 to z. Thus, the time taken by the return-stroke front to move from ground to height z is given by $z/v_{av}(z)$. In the same way, the time taken by the return-stroke front to travel from point z_0 to z is $z/v_{av}(z) - z_0/v_{av}(z_0)$. Once this is understood, it is a simple matter to change the preceding equations to the case of variable return-stroke speed. The current at ground level is then given by

$$i(0,t) = \int_0^{z_m} i_c\left(t - \frac{z}{v_{av}(z)} - \frac{z}{c}\right) dz, \tag{10.10}$$

$$z_m = \frac{t}{\left\{\frac{1}{v_{av}(z_m)} + \frac{1}{c}\right\}}. \tag{10.11}$$

However, now we have an additional problem. We cannot directly obtain the value of z_m from the preceding equation by solving it analytically. But this equation can be easily solved numerically using a computer to obtain z_m for any given time t. Similarly, the total current at a point z_0 on the return-stroke channel is given by

$$i(z_0,t) = \int_{z_0}^{z_{0m}} i_c\left(t - \frac{z_0}{v_{av}(z_0)} - \left\{\frac{z}{v_{av}(z)} - \frac{z_0}{v_{av}(z_0)}\right\} - \frac{(z-z_0)}{c}\right) dz \text{ valid for } t \geq z_0/v_{av}(z_0). \tag{10.12}$$

Using the same argument as we used earlier, we see that the equation to be satisfied by the upper limit of the integral is

$$\left\{\frac{z_{0m}}{v_{av}(z_{0m})} - \frac{z_0}{v_{av}(z_0)}\right\} + \frac{(z_{0m} - z_0)}{c} = t - \frac{z_0}{v_{av}(z_0)}. \tag{10.13}$$

This equation must also be solved numerically using a computer to obtain the value of the upper limit of the integral.

10.3.1 Selection of Input Parameters in Current-Generation-Type Return-Stroke Models

We can see from the set of equations derived in the previous section that the input parameters of current-generation-type return-stroke models are $v(z)$, $\rho(z)$, and τ. Note also that the channel base current is a function of these three parameters. If the channel base current is given or measured, then it can be used as an input parameter instead of one of the preceding three. In other words, the input parameters of current-generation-type return-stroke models are any three of the following set of parameters: $i(0,t)$, $v(z)$, $\rho(z)$, and τ. Several researchers have introduced engineering return-stroke models belonging to the current-generation type. They were introduced by Heidler [6], Diendorfer and Uman [7], Cooray [8], Cooray et al. [9], and Cooray and Rakov [10]. In the models introduced by Heidler [6] and Diendorfer and Uman [7], the input parameters are $i(0,t)$, $v(z)$, and τ. In the model introduced by Cooray and Rakov, they were $\rho(z)$, $v(z)$, and τ. As an example, let us consider how Cooray and Rakov [10] selected the model parameters in their model.

The experimental data show that the return-stroke velocity decreases with height, and close to ground its speed is approximately 1.5×10^8 m/s. The decrease in the velocity with height can be represented by an exponential function with a decay height of approximately 3,000 m. That is,

$$v(z) = v_0 e^{-z/\lambda}, \tag{10.14}$$

with $v_0 = 1.5 \times 10^8$ m/s and $\lambda = 3,000$ m. This generates an average velocity over the first kilometer of approximately 10^8 m/s, which is close to the values measured. The value of τ may be roughly related to the time taken by positive streamers to travel the radius of the corona sheath. Since the speed of streamers is approximately $2 \times 10^5 - 10^6$ m/s, the neutralization of the corona sheath is achieved in a matter of several microseconds. Thus, the value of τ is on the order of a few microseconds. The value selected by Cooray and Rakov was 1 μs. In Chap. 14, a procedure for evaluating the distribution of the positive charge per unit length deposited by the return-stroke channel is described. In the same chapter, an expression for this parameter is also given. Copying directly from that chapter we find that

$$\rho(z) = a_0 I_p + \frac{I_p(a + bz)}{1 + c_0 z + dz^2}, \tag{10.15}$$

where z is the coordinate directed along the channel with its origin at ground level, $\rho(z)$ is the charge deposited by the subsequent return stroke per unit length of the leader channel (in C/m), and I_p is the return-stroke peak current. The values of the parameters of this equation are $a_o = 5.09 \times 10^{-6}$, $a = 1.325 \times 10^{-5}$, $b = 7.06 \times 10^{-6}$, $c_o = 2.089$, and $d = 1.492 \times 10^{-2}$. It is important to point out

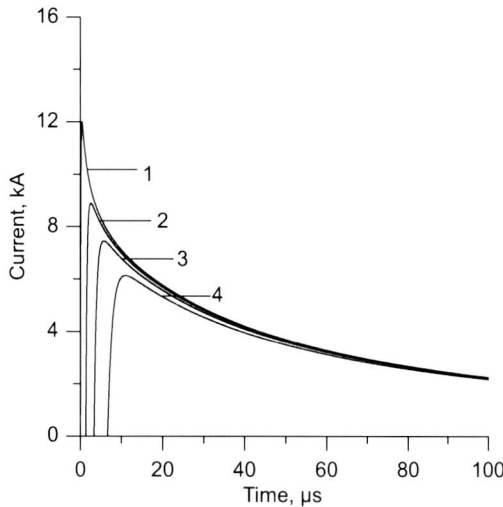

Fig. 10.7 Current waveform at different heights as predicted by Cooray and Rakov [10] return-stroke model. Note that the current amplitude decreases with height while the current rise time increases. Also, the current starts at any given height only after the return stroke has reached that height. This is why current waveforms start at different times in the diagram. (*1*) 0 m. (*2*) 200 m. (*3*) 500 m. (*4*) 1,000 m

that, because of the spread in the experimental data used to derive this relationship, for a given return-stroke current the values of these constants may vary by approximately 30 % from one return stroke to another. The expression given in Eq. (10.15) was used by Cooray and Rakov to represent the distribution of positive charge deposited by a return stroke. Note that Cooray and Rakov's model was originally introduced to represent subsequent return strokes, and therefore the constants of Eq. (10.15) that are appropriate for subsequent return strokes are used. The predicted channel base return-stroke current and the return-stroke current at different points along the channel as predicted by this model for a typical subsequent return stroke are shown in Fig. 10.7. Observe that the model predicts a current waveform that decreases its amplitude with height while the rise time of the current increases.

10.4 Current-Propagation-Type Return-Stroke Models

In the current-propagation models, the return-stroke channel is assumed to be a medium of propagation for the current injected into the channel at the channel base. Different models of this type differ by the way in which they treat the attenuation and dispersion of the current as it propagates along the channel.

The simplest, but not very physically reasonable, model assumes that the current propagates along the return-stroke channel at a constant speed without attenuation or dispersion. Such a model was introduced by Uman and McLain [11], and it is known as a *transmission line model*. This name should not be

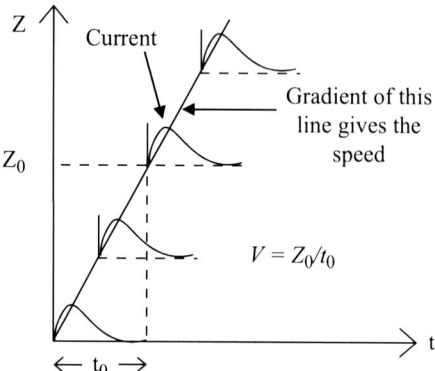

Fig. 10.8 Current at different levels as predicted by transmission line model. In the model, it is assumed that the return stroke can be represented by a current pulse propagating upward along the leader channel without attenuation or distortion at uniform speed. Thus, the same current appears at different heights but with a delay depending on the speed of the return stroke. For example, the current at height Z_0 is initiated at time t_0, where $t_0 = Z_0/v$. Thus, the gradient of the slanted straight line in the figure gives the speed of propagation. Note that the assumption made in this model is identical to what happens when a current pulse is injected into a lossless and uniform transmission line (Figure created by author based on information available in literature)

confused with the transmission line models described previously. It is called a transmission line model because the physical scenario assumed in the model is identical to the propagation of a current pulse along a uniform and lossless transmission line. Let us assume that the current injected at the channel base is denoted by $i_b(t)$. If the current does not change its amplitude or shape as it propagates along the channel, then at any given level the same current that was injected at the channel base will appear, but with a delay that is given by the time taken by the return stroke to travel (Fig. 10.8). Thus, the return-stroke current at any height z is given by

$$i(z,t) = i_b(t - z/v) \quad \text{for} \quad t > z/v. \qquad (10.16)$$

Note that z/v is the time taken by the return-stroke front (or the current front) to reach level z. Observe that, according to this model, the return stroke does not deposit any charge along the stepped leader channel. The current injected at ground level appears as it is in the cloud after the propagation delay. In effect, it transports positive charge from ground to cloud without affecting any charge being deposited on the channel by the leader. This is one reason why the model is not physically reasonable.

10.4 Current-Propagation-Type Return-Stroke Models

The transmission line model has become very popular in the lightning community because it predicts a simple relationship between the peak current and the peak electric radiation field measured at a distance D as follows:

$$E_p = \frac{\mu_0 I_p v}{2\pi D}. \tag{10.17}$$

This equation provides a means to obtain the peak return-stroke current from the measured radiation fields if the velocity of the return stroke is known.

In other models, the current amplitude is assumed to decrease with height. That is, as the current travels upward, part of the associated charge is deposited along the channel, leading to a decrease in current amplitude with height. This makes these models physically more reasonable than the transmission line model discussed previously. Let us assume that the way in which the current amplitude decreases with height is given by $A(z)$. Furthermore, assume that the return-stroke speed is uniform along the channel. Then the return-stroke current at any height is given by

$$i(z,t) = A(z) i_b(t - z/v) \quad \text{for} \quad t > z/v. \tag{10.18}$$

If the return-stroke speed is not uniform, then the parameter z/v in the preceding equation should be replaced by $z/v_{av}(z)$, where $v_{av}(z)$ is the average return-stroke speed along the channel length from 0 to z. Once the return-stroke current at any level is specified, then it is a simple matter to calculate the return-stroke electric fields from it. In a model introduced by Nucci et al. [12], the return-stroke current amplitude is assumed to decrease exponentially with height. That is,

$$A(z) = e^{-z/\lambda}, \tag{10.19}$$

where λ is the decay height constant of the model (set to 2 km in the model). In a model introduced by Rakov and Dulzon [13], the return-stroke current amplitude was assumed to decrease linearly with height, i.e.,

$$A(z) = \left(1 - \frac{z}{H}\right), \tag{10.20}$$

where H is the height of the leader channel.

These models can be complicated further by assuming that not only the current amplitude but also the current shape and the return-stroke speed change with height. A model of this general nature was introduced by Cooray and Orville [14]. The return-stroke current at level z in such a general current-propagation model is given by

$$I(z,t) = A(z) F\{z, t - z/v_{av}(z)\} \quad t > z/v_{av}(z), \tag{10.21}$$

where $A(z)$ is a function that represents the attenuation of the peak current and $F(z, t)$ describes the wave shape of the current at height z. We can define the function $F(z, t)$ as follows:

$$F(z,t) = \int_0^t I_b(\tau) R(z, t-\tau) d\tau, \qquad (10.22)$$

where $I_b(t)$ is the channel base current and $R(z, t)$ is a function that describes how the shape of the current waveform is modified with height. However, this operation itself leads to the attenuation of the current, and if we would like to represent the attenuation only by the factor $A(z)$, then we must normalize this function to the peak current at the channel base. Let t_p be the time at which the peak of the function defined in Eq. (10.22) is reached. Then the normalization can be performed as follows:

$$F(z,t) = I_p \frac{\int_0^t I_b(\tau) R(z, t-\tau) d\tau}{\int_0^{t_p} I_b(\tau) R(z, t-\tau) d\tau}, \qquad (10.23)$$

where I_p is the peak current at the channel base. The parameters utilized by Cooray and Orville in their model are as follows:

$$v = v_1 \left\{ a e^{-z/\lambda_1} + b e^{-z/\lambda_2} \right\}, \qquad (10.24)$$

$$A(z) = \left\{ c e^{-z/\lambda_3} + d e^{-z/\lambda_4} \right\}, \qquad (10.25)$$

$$R(z,t) = \frac{e^{-t/\tau(z)}}{\tau(z)}. \qquad (10.26)$$

Cooray and Orville assumed three types of variation for $\tau(z)$ in their model:

$$\tau(z) = \eta z, \qquad (10.27)$$

$$\tau(z) = 1 - e^{-z/\lambda_5}, \qquad (10.28)$$

$$\tau(z) = 1 - e^{-(z/\lambda_6)^2}. \qquad (10.29)$$

The numerical values of the parameters used in the model are $v_1 = 2.2 \times 10^8$, $a = 0.5$, $b = 0.5$, $\lambda_1 = 100.0$ m, $\lambda_2 = 5{,}000$ m, $c = 0.3$, $d = 0.7$, $\lambda_3 = 100$ m,

10.5 Analytical Expression for Channel Base Current

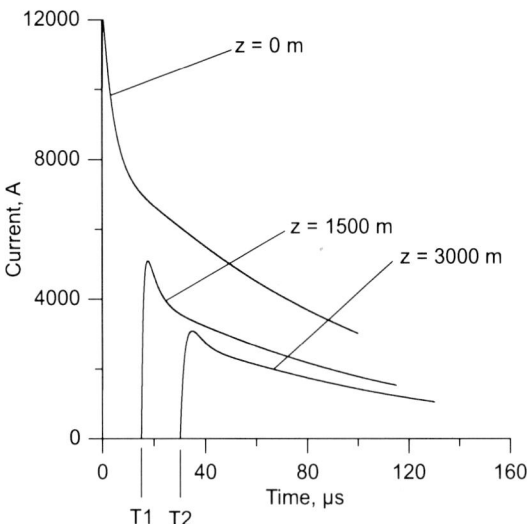

Fig. 10.9 Return-stroke current at several locations along channel according to Cooray and Orville model [14]. In the calculation, the return-stroke speed v is kept constant at 10^8 m/s and $\eta = 7 \times 10^{-10}$ s m^{-1}; $T1$ and $T2$: times at which current at heights 1,500 and 3,000 m is initiated (i.e., $T1 = 1,500/v$, $T2 = 3,000/v$). Note that the rise time of the current waveform increases with height

Table 10.1 Parameters of first and subsequent return-stroke currents

	I_1 kA	τ_{11} μs	τ_{12} μs	n_1	I_2 kA	τ_{21} μs	τ_{22} μs	n_2
First return stroke	28	1.8	95	2				
Subsequent return stroke	10.7	0.25	2.5	2	6.5	2	230	2

$\lambda_4 = 3,000$ m, $\lambda_5 = 500$ m, $\lambda_6 = 250$–500 m, and $\eta = 7 \times 10^{-10}$ s m^{-1}. Figure 10.9 shows the return-stroke current at several heights along the channel as predicted by this model.

10.5 Analytical Expression for Channel Base Current

In current-propagation models and in some of the current-generation models, the return-stroke current at the channel base is used as an input parameter. In many models, the return-stroke current is analytically represented by the equation [15].

$$I_b(0,t) = \frac{I_1}{\eta_1}\left(\frac{t}{\tau_{11}}\right)^{n_1}\frac{e^{-t/\tau_{12}}}{1+\left(\frac{t}{\tau_{11}}\right)^{n_1}} + \frac{I_2}{\eta_2}\left(\frac{t}{\tau_{21}}\right)^{n_2}\frac{e^{-t/\tau_{22}}}{1+\left(\frac{t}{\tau_{21}}\right)^{n_2}}, \qquad (10.30)$$

$$\eta_m = \exp\left\{-(\tau_{m1}/\tau_{m2})(n_m\tau_{m2}/\tau_{m1})^{1/n_m}\right\}. \qquad (10.31)$$

The parameters used to represent first and subsequent stroke currents are given in Table 10.1.

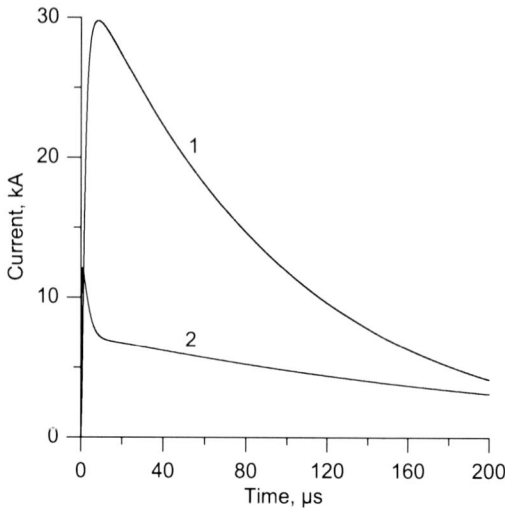

Fig. 10.10 Waveform that can be used to represent (1) first return-stroke current and (2) subsequent return-stroke current. Analytical expression for the current waveform is given in Eq. (10.30). In many return-stroke models, these current waveforms are used to simulate the return-stroke current at the channel base (Figure created by author)

These two current waveforms (i.e., first and subsequent) are depicted in Fig. 10.10.

References

1. Cooray V (2003) Mathematical modelling of return strokes. In: Cooray V (ed) Lightning flash. IET Publishers, London
2. Baba Y, Rakov V (2012) Electromagnetic models of lightning return strokes. In: Cooray V (ed) Lightning electromagnetics. IET Publishers, London
3. Moini R, Sadeghi SH (2012) Antenna models of lightning return strokes: an integral approach based on the method of moments In: Cooray V (ed) Lightning electromagnetics. IET publishers, London
4. De Conti, Silveira AFH, Visacro S (2012) Transmission lines of lightning return stroke. In: Cooray V (ed) Lightning electromagnetics. IET Publishers, London
5. Cooray V (2012) Return stroke models for engineering applications. In: Cooray V (ed) Lightning electromagnetics. IET Publishers, London
6. Heidler F (1985) Travelling current source model for LEMP calculation In: Proceedings of the 6th international symposium on EMC, vol 29F2, Zurich, Switzerland, pp 157–162
7. Diendorfer G, Uman MA (1990) An improved return stroke model with specified channel base current,". J Geophys Res 95:13621–13644
8. Cooray V (1998) Predicting the spatial and temporal variation of the electromagnetic fields, currents, and speeds of subsequent return strokes. IEEE Trans Electromagn Compat 40:427–435
9. Cooray V, Montano R, Rakov V (2004) A model to represent first return strokes with connecting leaders. J Electrost 40:97–109
10. Cooray V, Rakov V (2011) Engineering lightning return stroke models incorporating current reflection from ground and finitely conducting ground effects. Proc IEEE (EMC) 53:773–781

11. Uman MA, McLain DK (1969) Magnetic field of lightning return stroke. J Geophys Res 74:6899–6910
12. Nucci CA, Diendorfer G, Uman MA, Rachidi F, Ianoz M, Mazzetti C (1988) On lightning return stroke models for LEMP calculations. In: 19th international conference on lightning protection, Graz, Austria
13. Rakov VA, Dulzon AA (1991) A modified transmission line model for lightning return stroke field calculation. In: Proceedings of the 9th international symposium on electromagnetic compatibility, vol 44H1, Zurich, Switzerland, pp 229–235
14. Cooray V, Orville RE (1990) The effects of variation of current amplitude, current risetime and return stroke velocity along the return stroke channel on the electromagnetic fields generated by return strokes. J Geophys Res 95(D11):18617–18630
15. Delfino F, Procopio R, Rossi M, Rachidi F, Nucci CA (2007) An algorithm for the exact evaluation of the underground lightning electromagnetic fields. IEEE Trans (EMC) 49(2):401–411

Chapter 11
Experimental Data and Theories on Return Stroke Speed

11.1 Introduction

As discussed in Chap. 7, a return stroke is a potential discontinuity that propagates from ground to cloud along the leader channel. The potential ahead of the return stroke front is at cloud potential and the potential behind the return stroke front is close to the ground potential. A similar process also happens if a perfectly conducting vertical wire is charged to a given potential and the end of the wire is connected to ground. In this case, too, a potential discontinuity propagates along the conductor. The speed of propagation of the potential discontinuity in the case of the perfectly conducting wire is equal to the speed of light in air. On the other hand, the speed of propagation of the return stroke front is considerably less than the speed of light, and to date there is no consensus among researchers as to the cause of this speed reduction in the return stroke front. However, return stroke speed is an important parameter in return stroke models, and almost all the return stroke models available at present use return stroke speed as an input parameter.

Several researchers have attempted to derive the return stroke speed using the return stroke current as an input. In this chapter we will present the assumptions of these models and the corresponding results obtained. However, before we proceed further let us summarize the information currently available on return stroke speed as gathered from experimental observations.

11.2 Experimental Observations

Several experimental techniques can be used to measure return stroke speed. The classical method is to use a streak camera in which the image of the return stroke is recorded on a rapidly moving film. Since the image of the return stroke at different times is recorded on different parts of the film, knowing the speed of the film

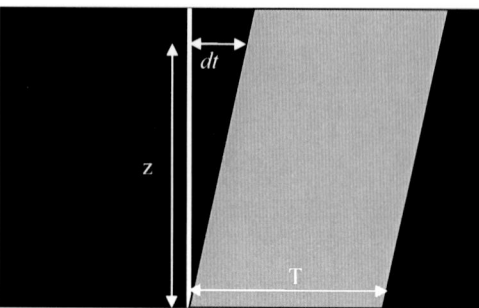

Fig. 11.1 Estimating the speed of the return stroke using a streak camera. Assume that the return stroke moves upward at a speed v and the return stroke channel remains luminous for a time T. Because of the movement of the film in the streak camera, the image of the return stroke channel is spread out in time (*gray portion of image*). Superimposing this image with the image of a stationary camera (*bright straight line*) makes it possible to estimate the speed of the return stroke. According to the foregoing picture, the speed v of the return stroke is given by $v = z/dt$. Note that at $t = 0$, both the still picture and the streak picture coincide (figure created by author)

makes it possible to obtain the information on the return stroke speed by comparing the streak image with an image taken by a stationary camera (Fig. 11.1). The second method for measuring the return stroke speed is to isolate small sections of the lightning channel using horizontal slits and, using photo multipliers, record the optical signal of the return stroke through these slits. Since the slits observe different parts of the channel, it is possible to create a time development of the channel from the recorded light signals. The speed of the return stroke can also be obtained using interferometric techniques that can track the radiation sources during the development of the return stroke. Finally, the fast video cameras available today can record up to 1 million frames per second or more and the use of these cameras allows one to resolve the development of the return stroke channel and, hence, measure its speed of development.

Recent measurements of return stroke speeds from natural lightning return strokes were taken, among others, by Idone and Orville [1] and Mach and Rust [2]. According to these studies, the average speed of a return stroke over the first kilometer or so of the return stroke channel is approximately 1–1.5×10^8 m/s. Close to the ground end (channel segment less than 500 m) the average return stroke speed was approximately 1.5–2.0×10^8 m/s, with a slight tendency for the subsequent return stroke speed to be higher than that of the first return stroke. All these studies show a clear tendency for the return stroke speed to decrease with height. Since the major portion of the return stroke channel is hidden inside the cloud, the average return stroke speed over longer lengths of the return stroke channel can be obtained only through interferometric studies. According to the interferometric studies conducted by Shao et al. [3], for a single first return stroke, the average speed from ground to the point of initiation of the stepped leader is approximately 5×10^7 m/s. For subsequent return strokes they found that the average speeds along the channel lengths of 10–15 km range from 0.5×10^8–10^8 m/s.

No measurements are currently available on the variation in return stroke speed over the first hundred meters or so of natural lightning flashes. However, a few such measurements are available from triggered subsequent return strokes. These data seem to indicate that the return stroke speed increases initially, reaches a peak, and then decreases [4].

Researchers have also investigated experimentally whether there is a relationship between the return stroke speed and the corresponding return stroke peak current. Except for a study conducted by Idone et al. [5] in New Mexico, experimental observations have not found a correlation between return stroke speed and the corresponding peak current. As we will see in the next section, many of the theories that attempt to derive the return stroke speed from fundamental principles predict that the return stroke speed is correlated with the return stroke peak current and the speed should increase with increasing return stroke current. In this respect, the theory seems to be in conflict with the experimental data. Let us now consider the basic assumptions of these theories.

11.3 Theories of Return Stroke Speed

11.3.1 Theories of Lundholm and Wagner

Both Lundholm [6] and Wagner [7] visualized the return stroke as a step current pulse of amplitude I_p propagating along the leader channel at a constant speed v (Fig. 11.2). This model, which of course is a gross simplification of the return stroke, was launched in an attempt to discover the relationship between the return stroke current and its speed. The treatment of Lundholm was as follows: an upward moving current pulse generates an electric field at the surface of the return stroke channel. Lundholm showed, using Maxwell's equations, that the

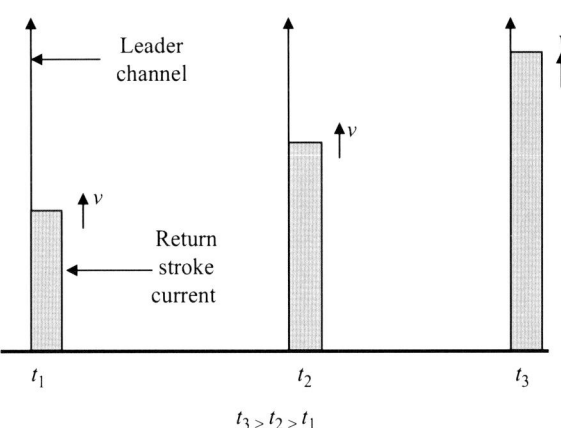

Fig. 11.2 Current in return stroke channel according to Lundholm's [6] model. According to this model, the return stroke can be simulated by a step current pulse propagating along the return stroke channel at constant speed v

component of this field parallel to the surface of the channel and at a point just below the return stroke front is given approximately by

$$E_z = \frac{I_p}{4\pi\varepsilon_o c}\left[\frac{c^2 - v^2}{cv^2\left(t - \frac{z}{v}\right)}\right] = I_p R \qquad t > z/v, \qquad (11.1)$$

where I_p is the peak current, R is the resistance per unit length of the return stroke channel, t is the time, and c is the speed of light. To proceed further, he assumed the validity of Toepler's law. Let us now see what Toepler's law is. After studying the resistance of spark channels Toepler [8] came to the following conclusion. If $i(t)$ is the current flowing along the spark channel of length d, its resistance is given by

$$r(t) = \frac{kd}{\int_0^t i(t)dt}. \qquad (11.2)$$

Thus, the resistance per unit length of the discharge channel (or spark channel) is given by

$$R = \frac{k}{Q}, \qquad (11.3)$$

where R is the resistance per unit length of the spark channel, Q is the charge that has passed through the discharge, and k is Toepler's spark constant. This equation is known as Toepler's law. Lundholm assumed that this law is valid for lightning return strokes. Since $Q = I_p \times (t - z/v)$ for the situation under consideration, the combination of Eqs. 11.1, 11.2, and 11.3 provides a relationship between I_p and v, namely,

$$v = \frac{c}{\sqrt{1 + \frac{4\pi\varepsilon_o c^2 k}{I_p}}}. \qquad (11.4)$$

The basics of Wagner's treatment are as follows. Consider the return stroke as a transmission line with inner and outer conductor radii a and b. A charge per unit length ρ is deposited on this transmission line during the leader stage. The energy of the system is electrostatic. As the return stroke surges up through the transmission line releasing the bound charge, the energy of the system changes from electrostatic to magnetic energy. Consequently, the energy E_u released per unit length of the return stroke channel is

$$E_u = \frac{\rho^2}{2C} - \frac{LI_p^2}{2}, \qquad (11.5)$$

11.3 Theories of Return Stroke Speed

where L and C are the inductance and capacitance per unit length, respectively, I_p is the peak current, and ρ is the linear charge density of the return stroke channel. Observe that if the speed of propagation v of the return stroke is equal to the speed of light c, then $E_u = 0$. This is indeed the case since $c = \sqrt{(LC)^{-1}}$ and $I_p = \rho v$. When $v \neq c$, the preceding equation gives the energy that must be absorbed by the front of the return stroke. From experiments conducted in the laboratory, it was observed that an energy of 0.2 J/m/A is required to raise a spark to a conducting state. Then the energy E_u per unit length that must be absorbed by the front of the wave to bring the channel to a conductivity that will support a current of I_p amperes is equal to $0.2\,I_p$ J. Substituting this into Eq. (11.5) and after some mathematical manipulations we arrive at the equation

$$v = \frac{c}{\sqrt{1 + \dfrac{0.4}{LI_p}}}. \tag{11.6}$$

The value of the inductance per unit length was calculated by assuming a (the radius of the inner core) $= 0.03$ m and b (the radius of the current return conductor) $= 180$ m. Unfortunately, Wagner did not provide any justification for the value assumed for the latter parameter. The return stroke speed as a function of peak current as predicted by the preceding equation is depicted in Fig. 11.3.

The derivations of Lundholm and of Wagner could be criticized on several points. Both assumed the current to be a step function, which is an oversimplification. Furthermore, because of the assumption of an instantaneous rise in the current at the front, any relationship between the rise time of the return stroke current and

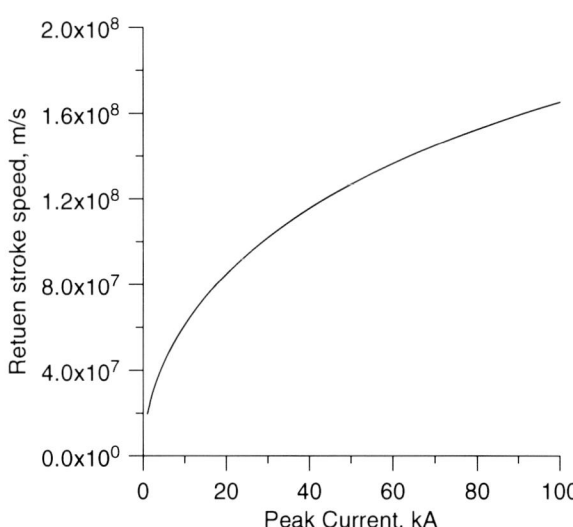

Fig. 11.3 Return stroke speed as function of return stroke peak current according to Wagner's [7] model

the speed is lost. The validity of Lundholm's derivation is based on the assumed validity of Toepler's Law. This law characterizes the temporal variation in the channel resistance in the early stages of a spark discharge and may not be applicable to return strokes since they propagate along highly conducting channels already thermalized by leaders. Wagner's simulations depend on the radius of the current return conductor (i.e., the radius of the outer cylinder), which does not exist in practice. These simplifications cast doubt on the quantitative validity of both Lundholm's and Wagner's results. Nonetheless, the expressions derived may still provide a qualitative description of the relationship between the return stroke current and the speed.

11.3.2 Rai's Theory

Rai [9] envisaged the return stroke front as an upward moving equipotential surface that separates a highly ionized gas region below and a neutral gas region with low temperature (virgin air) above. Based on the work of Albright and Tidman [10], he assumed that the speed of propagation v of the potential discontinuity at the wavefront of the return stroke was given by

$$v = \frac{eE\left\{1 + \frac{T_o}{\eta}\right\}}{m\chi}, \qquad (11.7)$$

where m and e are respectively the electron mass and charge, χ is the elastic scattering frequency of electrons, T_o is the electron temperature of the channel, E is the electric field at the wavefront, and η is the ionization potential of the constituent gas. If σ is the conductivity of the return stroke channel in the vicinity of its tip, then Eq. (11.7) can be written as

$$v = \frac{e\left\{1 + \frac{T_o}{\eta}\right\}J}{m\chi\sigma}. \qquad (11.8)$$

If the current at the front and the radius of the cross section of the return stroke are known, then the speed can be calculated from the preceding expression. For example, if the current is a step function of amplitude I_o, then the return stroke speed is given by

$$v = \frac{e\left\{1 + \frac{T_o}{\eta}\right\}I_p}{m\chi\sigma A}. \qquad (11.9)$$

11.3 Theories of Return Stroke Speed

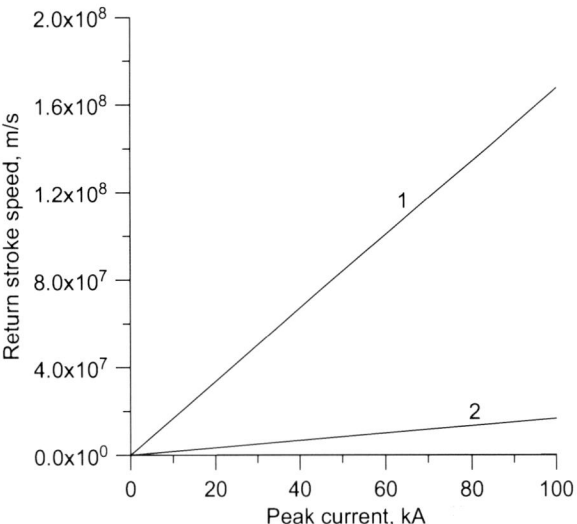

Fig. 11.4 Return stroke speed as function of return stroke peak current according to Rai's [9] model. In the calculation, the return stroke current waveform is simulated by a step. *Curve 1* corresponds to 10 S/m and *curve 2* to 100 S/m

Note that in the preceding equation, A is the cross-sectional area of the return stroke channel. The speed of the return stroke as a function of peak current as predicted by this equation is shown in Fig. 11.4. In the calculation, the following numerical parameters given by Rai were used: $\chi = 0.4 \times 10^{13}\,\mathrm{s}^{-1}$, $\eta = 17\,\mathrm{eV}$, $T_0 = 3 \times 10^4\,\mathrm{K}$, $A = 3.14 \times 10^{-6}\,\mathrm{m}^2$ (i.e., 1 mm radius). Results are obtained for $\sigma = 10$ S/m and $\sigma = 100$ S/m. Note that to obtain reasonable results, the conductivity of the channel at the return stroke tip must be approximately 10 S/m. However, since the leader channel is at a temperature of approximately 10,000–20,000 K, the conductivity at the tip of the return stroke channel could be approximately 10^4 S/m. Unfortunately, Eq. (11.9) predicts rather low velocities if this value is used in that equation.

In his original calculation, Rai [9] assumed that the amplitude of the current step at the return stroke front was equal to the magnitude of the current flowing at that time at the return stroke channel base, i.e., at ground level. Based on this assumption, he derived an expression for the variation in the return stroke speed along the return stroke channel. Unfortunately, there is no justification for assuming that the current at ground level and the current at the return stroke front have the same magnitude, unless the incorrect assumption is made that the speed of propagation of the information along the return stroke channel is infinite.

Before proceeding further some comments are appropriate on the basic assumptions of the model. First, Eq. 11.6 is valid as long as the medium ahead of the front is virgin air. In reality, the return stroke propagates through the core of the leader channel, which is highly conducting because of its elevated temperature. Second, the theory is only applicable if there is a current discontinuity at the return stroke front (otherwise the current density in Eq. 11.9 goes to zero, making the speed equal to zero as well); however, such a current discontinuity is not physically possible.

11.3.3 Cooray – Model I

The procedure used by Cooray [11, 12] to calculate the return stroke speed is based on the following facts. Return strokes propagate along the channels of either the stepped or the dart leaders. Cooray [11, 12] assumed that, with the exception of the very tip, these leader channels can be treated as arc channels in air with an axial potential gradient of the order of 2–10 kV/m. Thus, to induce a significant change in the current in the leader channel, the minimum field required is of the same order of magnitude as this axial electric field. In the analysis, it is of course assumed that the initial radius of the return stroke is smaller than that of the leader core. During the return stroke phase the current in the leader channel starts to increase, and the first significant change in the current takes place at the tip of the return stroke. Thus, the return stroke should maintain a field at the tip that is of the order of the axial field along the leader channel. When this condition is satisfied, electrons accelerate toward the tip of the return stroke, causing an increase in the current (Fig. 11.5). On the basis of this reasoning, Cooray [11, 12] assumed that the field at the tip of the return stroke was equal to the potential gradient of the leader channel.

Let us consider how this assumption can be used to evaluate the return stroke speed as a function of height. First, it is necessary to calculate the electric field at the tip of the return stroke. The electric field at the return stroke front can be calculated

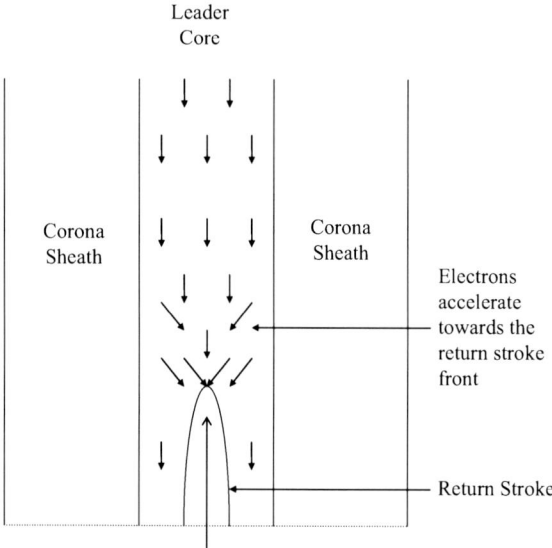

Fig. 11.5 Pictorial depiction of concept used by Cooray [11, 12] to connect return stroke speed to return stroke current. In the model, it is assumed that the electric field at the tip of the return stroke is of the order of the external field that exists along the core of the leader channel. When this condition is satisfied, electrons accelerate toward the return stroke front, enhancing ionization and leading to an increase in the current at the tip of the return stroke (Adapted from Cooray [14])

11.3 Theories of Return Stroke Speed

using the classical dipole method used frequently by lightning researchers or by using the electromagnetic fields of accelerating charges as introduced recently by Cooray and Cooray [13]. Both methods give identical results for the value of the field, even though the final expressions for the field are completely different.

According to the results obtained through this procedure, the main conclusion that one can make is that the return stroke speed is determined primarily by three parameters: the rise time of the return stroke current, its peak value, and the potential gradient of the leader channel. The return stroke speed decreases with increasing rise time and with decreasing current peak. Moreover, the return stroke speed increases as the potential gradient of the channel decreases. Since the potential gradient of the channel decreases with increasing conductivity, the model predicts higher speed in highly conducting channels. One disadvantage of this model is that it is empirical in nature and does not incorporate the physics of the breakdown process at the tip of the channel. Irrespective of these shortcomings, the theory can be used to draw some qualitative conclusions concerning the return stroke speed. Let us assume that the return stroke current is a ramp function and that it propagates along the return stroke channel at constant speed v. Consider a situation where the location of the return stroke front is several tens of meters above the ground, so that the contribution to the electric field at the return stroke front by the image current in the perfectly conducting ground is negligible. In this case, it can be shown that the electric field at the return stroke front E_f is given approximately given by

$$E_f = \frac{1}{4\pi\varepsilon_0}\left(\frac{1}{v^2} - \frac{1}{c^2}\right) G \ln\left[\frac{Z_0}{R}\right], \tag{11.10}$$

where G is the gradient of the ramp (i.e., A/s), R is the radius of the return stroke channel, and Z_0 is the height of the return stroke front from ground level. In the case of real lightning currents, the equivalent gradient can be obtained from the ratio I_p/t_r, where I_p is the peak current and t_r is the rise time of the current. Thus, the preceding equation can be written as

$$v = \frac{1}{\sqrt{\frac{4\pi\varepsilon_0 E_f}{\left\{\frac{I_p}{t_r}\right\}\ln\left[\frac{Z_0}{R}\right]} + \frac{1}{c^2}}}. \tag{11.11}$$

This equation shows that the speed of propagation of the current pulse is determined by the ratio I_p/t_r. Only if the rise time of the current were kept constant while changing the return stroke peak current would one find the return stroke speed to increase with the return stroke current. Since the average derivative of the current depends only very weakly on the return stroke peak current (see the next section), any dependence of return stroke speed on return stroke peak current must be weak. Thus, any experimental scatter in the data can make the relationship between the peak current and the return stroke speed random.

11.3.4 Cooray – Model II

Observe that three of the theories presented in the previous sections assume that the current at the return stroke front rises to its peak value instantaneously. In other words, these theories do not allow for a return stroke current that rises from zero to its peak value at the front of the return stroke. For this reason, any connection between the return stroke speed and the current rise time is lost in the theory. This could be why these theories cannot account for the apparent absence of a relationship between the return stroke current and the return stroke speed. On the other hand, Cooray Model I takes into account the finite rise time of the current, but, as was mentioned in the previous section, the theory is empirical in nature and does not incorporate the physics of electrical discharges in its formulation.

To remove this disadvantage of the foregoing theories, Cooray [14] developed a theory in which the physical development of the discharge channel is taken into account in evaluating the return stroke speed. The basic foundation of the theory is as follows. If the spatial and temporal variations in the return stroke current and the channel radius are given as inputs, then the temporal variation in the electric field can be calculated at any point along the axis of the return stroke. For example, the electric field calculated at a point located at 100 m in a subsequent return stroke for three current waveforms is shown in Fig. 11.6a. The corresponding current waveforms are shown in Fig. 11.6b. Note that as the return stroke front approaches the point of observation, the electric field rises rapidly and reaches a peak a fraction of a microsecond after the passage of the return stroke front. After that, the electric field continues to decrease with time. This electric field depends both on the rise time of the current and the return stroke speed. Note that the peak electric field decreases as the speed of the return stroke and the current rise time increase. Since the current flowing at that point is given as an input, the energy released per unit length during the return stroke can be calculated at any given point on the channel. The energy calculated thus depends on the return stroke speed and the rise time of the return stroke current. For example, Fig. 11.7 shows the energy released over the first 80 μs of the return stroke for three current waveforms. In the calculation, the return stroke speed is kept constant at 1.5×10^8 m/s. This technique can be used, for example, to calculate the total energy released in any given section of the channel below the return stroke front.

Now, it is reasonable to think that the return stroke speed is controlled by the electric field at the front or the tip of the return stroke. This field in turn is controlled by the charge distribution below the return stroke front. However, calculations show that a significant contribution to this field comes only from the first 100 m or so of the channel section below the return stroke front. In other words, the return stroke speed is controlled by the physical processes taking place in the first 100 m or so of the channel below the front.

So far, all the results and conclusions presented in this section are based purely on Maxwell's equations. Let us summarize the results by the following equation: let U represent the energy released in the first 100 m of the channel below a return stroke front. Since this is the section of the channel that controls the return stroke speed, the physical processes taking place here with the aid of released

11.3 Theories of Return Stroke Speed

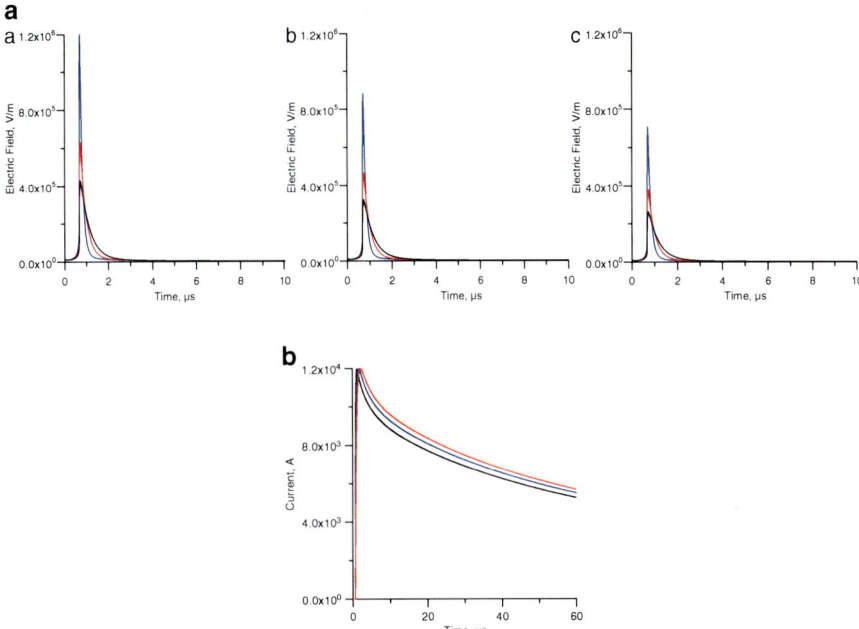

Fig. 11.6 (a) Electric field at point located 100 m above ground as function of time. (a) Electric fields corresponding to $v = 1.0 \times 10^8$ m/s and current rise times of 0.5 µs (curve with highest amplitude), 1.0 µs (curve with intermediate amplitude), and 1.5 µs (curve with lowest amplitude). (b) Electric fields corresponding to $v = 1.5 \times 10^8$ m/s and current rise times of 0.5, 1.0, and 1.5 µs. (c) Electric fields corresponding to $v = 2.0 \times 10^8$ m/s and current rise times of 0.5, 1.0, and 1.5 µs. In the calculations, the channel base peak current was kept constant at 12 kA (Adapted from Cooray [14]). (b) (i) The three current waveforms at the channel base current used in the calculation. The rise times of the channel base current were 0.5 µs (curve with lowest amplitude tail), 1.0 µs (curve with intermediate amplitude tail), and 1.5 µs (curve with highest amplitude tail). The speed of the return stroke was 1.5×10^8 m/s (Adapted from Cooray [14])

energy U should provide the correct conditions for the return stroke front to propagate at a specified speed. Since this energy is a function of the peak current, current rise time, and return stroke speed, let us write it as

$$U = f(I_p, t_r, v), \qquad (11.12)$$

where I_p is the return stroke peak current, t_r is the current rise time, and v is the return stroke speed. The next step is to combine the preceding equation with the fluid dynamics and thermodynamics of the electrical discharges.

To simplify the analysis, we assume that the first return stroke current front can be represented by a linear ramp (Fig. 11.8). Of course, this is a crude approximation, but it is a better approximation than a step function. With this approximation the current immediately below the return stroke front can be represented by a ramp function. The gradient of the ramp varies with the return stroke peak current and

Fig. 11.7 Energy dissipated over first 100 μs at point located 100 m above ground as function of time. (i) $I_p = 12$ kA; $v = 1.5 \times 10^8$ m/s; current rise time $= 0.5$ μs (curve with lowest amplitude). (ii) $I_p = 12$ kA; $v = 1.5 \times 10^8$ m/s; current rise time $= 1.0$ μs (curve with intermediate amplitude). (iii) $I_p = 12$ kA; $v = 1.5 \times 10^8$ m/s; current rise time $= 1.5$ μs (curve with highest amplitude) (Adapted from Cooray [14])

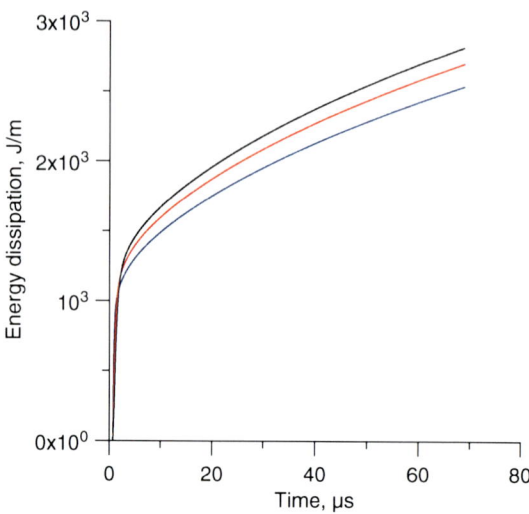

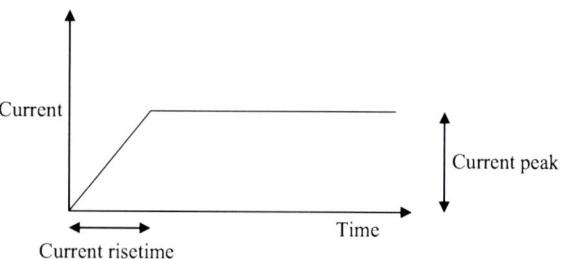

Fig. 11.8 Idealized first return stroke current waveform assumed in analysis of Cooray [14]. Note that the gradient of the rising part of the current is given by I_p/t_r, where I_p is the peak current and t_r is the rise time

rise time. Now, the energy that will be dissipated in the creation of a 1-m discharge channel that supports a ramp current can be calculated using the theory developed by Braginski [15]. According to that theory, the radius of a spark channel transporting a ramp current is given by

$$r(t) = 0.93 \times 10^{-3} \rho_0^{-1/6} i^{1/3} t^{1/2}, \qquad (11.13)$$

where $r(t)$ is the radius of the channel at time t in meters, t is the time in microseconds, i is the current in the channel at that instant in kA, and ρ_0 is the density of air at atmospheric pressure (1.29×10^{-3} g/cm³). Since Braginski [15] assumed that the conductivity of the high temperature channel remained more or less constant around 10^4 S/m, the channel resistance and, hence, the energy dissipation in the channel can be calculated as a function of time. In the present study, the energy released over the 100 m of channel below the front (through which a ramp current is propagating) is obtained. Let us denote this calculated energy by U_B. Since it is also a function of the peak current, current rise time, and return stroke speed, let us write it as

11.3 Theories of Return Stroke Speed

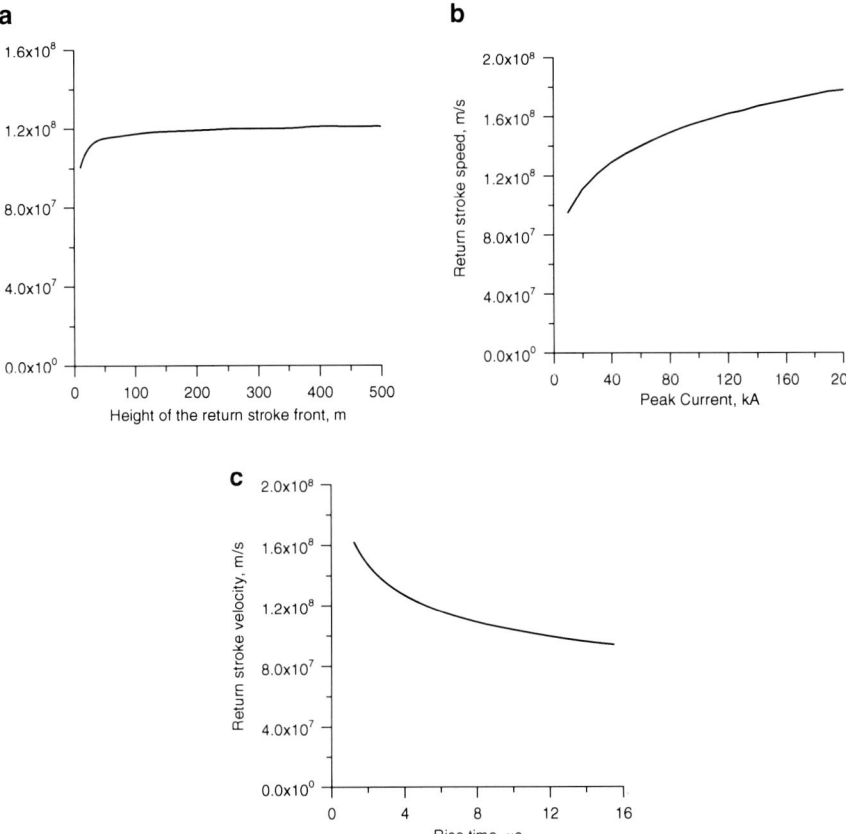

Fig. 11.9 Variation in return stroke speed as function of (**a**) height of return stroke front, (**b**) return stroke peak current, and (**b**) return stroke current rise time. In (**a**), rise time = 0.5 μs and $I_p = 30$ kA. In (**b**), rise time = 0.5 μs. In (**c**), $I_p = 30$ kA (Adapted from Cooray [14])

$$U_B = f_b(I_p, t_r, v). \quad (11.14)$$

Note that the foregoing parameter depends on the return stroke speed only indirectly because the integration time to calculate the energy dissipation over the first 100 m of the channel below the return stroke front depends on the speed of propagation of the return stroke front. Cooray [14] equated $f(I_p, t_r, v)$ to $f_b(I_p, t_r, v)$, which gave rise to an expression for the return stroke speed as a function of its peak current. The results obtained thus show that the front speed is related to the height of the front, return stroke peak current, and return stroke rise time.

Let us consider how the return stroke speed changes with these parameters. Figure 11.9 shows how the return stroke speed varies as a function of the height of the return stroke front, peak current, and current rise time. First, observe that the return stroke speed increases as the return stroke front surges upward through the channel. The speed at a given height increases with increasing return stroke current and with decreasing return stroke rise time. Of course, recall that the return stroke current rise time may also vary along the channel, and this will also add another variable to the return stroke speed. These results still show that the return stroke speed is a strong function of the return stroke current. How do we explain the apparent lack of relationship between these two parameters? According to Cooray [14], the previous models did not take into account the variation in return stroke current rise time with return stroke current peak. By measuring the rise time of currents measured by Berger on San Salvatore [16], Cooray [14] showed that the return stroke peak current and return stroke current rise time are correlated to each other and that the return stroke current rise time increases with increasing peak current. The relationship obtained by Cooray [14] is shown in Fig. 11.10. When this variation was included in the model, the return stroke speed remained more or less the same with increasing current. The results are shown in Fig. 11.11. Thus, the theory managed to explain the reason for the apparent lack of correlation between the return stroke current and return stroke speed.

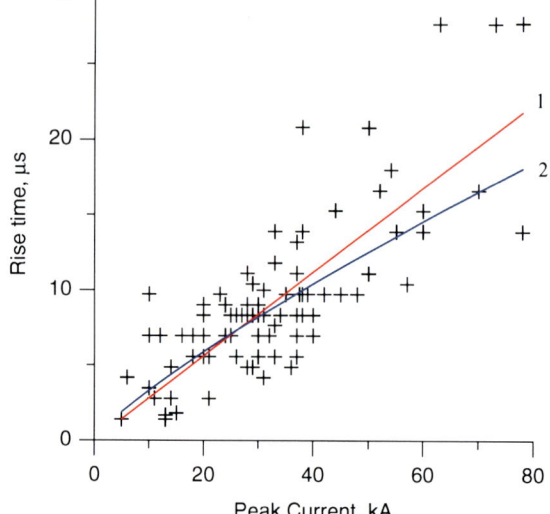

Fig. 11.10 Peak current and corresponding rise time of first return stroke waveforms measured by Berger [16]. The best fit lines, both linear (*curve 1*) and power (*curve 2*) regression, are also depicted in the figure (Adapted from Cooray [14])

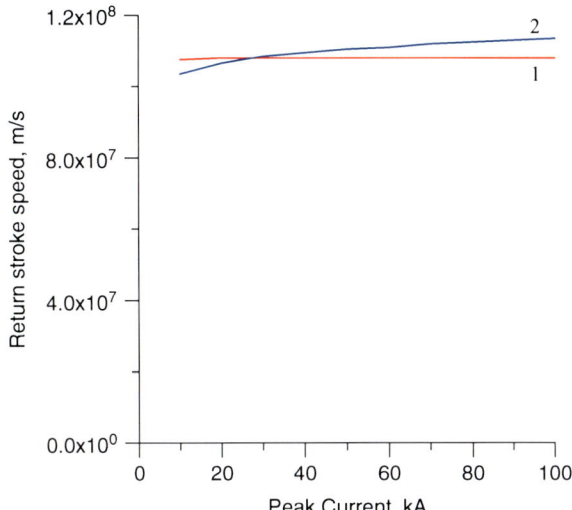

Fig. 11.11 Variation in return stroke velocity as function of peak current if return stroke current rise time is assumed to increase according to curve 1 of Fig. 10.10 (curve 1) or according to curve 2 of Fig. 10.10 (Curve 2) (Adapted from Cooray [14])

References

1. Idone VP, Orville R (1982) Lightning return stroke velocities in the thunderstorm research program. J Geophys Res 87:4903–4916
2. Mach DM, Rust WD (1993) Two-dimensional velocity, optical risetime, and peak current estimates for natural positive lightning return strokes. J Geophys Res 98:2635–2638
3. Shao XM, Krehbiel PR, Thomas RJ, Rison W (1995) Radio interferometric observations of cloud-to-ground lightning phenomena in Florida. J Geophys Res 100:2749–2783
4. Olsen RC, Jordan DM, Rakov VA, Uman MA, Grimes N (2004) Observed one dimensional return stroke propagation speeds in the bottom 170 m of a rocket-triggered lightning channel. Geophys Res Lett 31:L16104–L16107
5. Idone PI, Orville RE, Hubert P, Barret L, Eybert-Berard A (1984) Correlated observations on three triggered lightning flashes. J Geophys Res 89:1385–1394
6. Lundholm R (1957) Induced overvoltage-surges on transmission lines and their bearing on the lightning performance at medium voltage networks. PhD dissertation, Royal Institute of Technology (KTH), Sweden
7. Wagner CF (1963) Relationship between return stroke current and velocity of the return stroke. AIEE Trans 80:609–617
8. Toepler M (1906) Zur kenntnis der gesetz der gleitfunkenbildung. Ann Phys D, 4 folge 21:193–222
9. Rai J (1978) Current and velocity of the return lightning stroke. J Atmos Terr Phys 40:1275–1285
10. Albright NW, Tidman DA (1972) Ionizing potential waves and high voltage break down streamers. Phys Fluid 15:86
11. Cooray V (1993) A model for the subsequent return strokes. J Electrost 30:343–354
12. Cooray V (2014) Return stroke speed models. In: Cooray V (ed) The lightning flash, 2nd edn. IET Publishers, London

13. Cooray V, Cooray G (2010) The electromagnetic fields of an accelerating charge: applications in lightning return-stroke models. IEEE Trans (EMC) 52:944–955
14. Cooray V (2014) Power and energy dissipation in negative lightning return strokes. Atmos Res 149:359–371
15. Braginski SI (1958) Theory of development of a spark channel. Sov Phys JEPT 34(7):1068–1074
16. Berger K (1972) Methods and results of lightning records at Monte San Salvatore from 1963–1971. Bull Schweiz Elektrotech ver 63:21403–21422

Chapter 12
Propagation Effects Caused by Finitely Conducting Ground on Lightning Return Stroke Electromagnetic Fields

12.1 Introduction

Consider a radio antenna located over ground and transmitting at a given frequency. If the ground is perfectly conducting, the amplitude of the radio wave generated by the antenna decreases inversely with distance as one moves away from the antenna, i.e., the signal decreases as $1/r$, where r is the distance from the antenna to the point of observation. However, if the conductivity of the ground is finite, then the amplitude of the radio wave decreases much more rapidly than the inverse distance. The higher the frequency of the wave, the higher the rate of decrease of the radio signal with distance. This attenuation of the radio signal or the electromagnetic wave by finitely conducting ground is called *propagation effects*. Attenuation of the electromagnetic wave results from the absorption of energy from the electromagnetic field by the finitely conducting ground. The higher the frequency of the electromagnetic field, the higher the amount of energy absorbed by the ground. Let us analyze this a bit further. Consider an electromagnetic field generated by a radio antenna tuned to a given frequency. If the ground is perfectly conducting, at any given point on the ground the electric field is perpendicular to the ground surface and the magnetic field is in the azimuthal direction (Fig. 12.1a). The direction of energy flow or the Poynting vector (i.e., $\mathbf{E} \times \mathbf{H}$) of the electromagnetic field at that point is directed parallel to the ground. Now consider the electromagnetic field of a radio antenna located over finitely conducting ground. In this case, the magnetic field has the same direction as before, but the electric field is inclined to the surface of the ground (Fig. 12.1b). That is, there is a component of the electric field parallel to the ground. This component is called the *horizontal electric field*. The Poynting vector in this case is directed towards the ground (a component of which is directed into the ground), indicating that energy is absorbed from the electromagnetic field by the ground. With increasing frequency or with decreasing ground conductivity the angle between the vertical and the direction of the

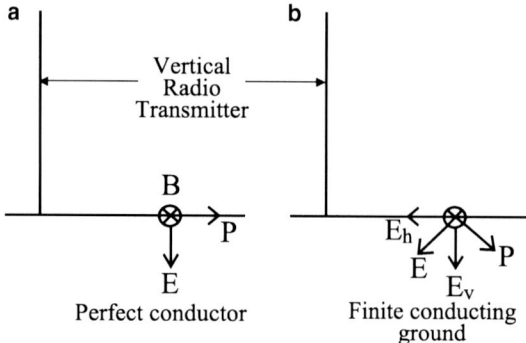

Fig. 12.1 (a) The electric field vector **E** over perfectly conducting ground is perpendicular to the ground surface, and the Poynting vector **P** [which is equal to $\frac{1}{\mu_0}(\mathbf{E} \times \mathbf{B})$] is parallel to the surface. The magnetic field **B** is azimuthal and goes into the plane. (b) Over finitely conducting ground the electric field vector is inclined to the ground surface (the magnetic field has the same direction), and the Poynting vector is directed towards the ground (a component of which is directed into the ground). Thus, in the case of finitely conducting ground, energy is absorbed by the ground from the electromagnetic field (Figure created by author)

electric field increases, i.e., the horizontal electric field increases. Thus, the energy absorbed by the ground increases with increasing frequency and with decreasing conductivity.

In the case of lightning-generated electromagnetic fields, which contain all the frequencies by differing amounts, the higher frequencies are selectively attenuated by ground, and as a result the fast or rapid features of the electromagnetic field disappear as the distance of propagation increases. The effect of this selective attenuation of higher frequencies is to increase the rise time and decrease the amplitude of the electromagnetic field (Sect. 12.3). Interestingly, in the case of lightning generated electric fields the Poynting vector rotates with time having a direction towards the ground at short times and becoming more and more parallel to ground with increasing time.

Knowledge concerning the characteristics of electromagnetic fields generated by lightning flashes is of importance in evaluating the interaction of these electromagnetic fields with electrical networks and in the remote sensing of lightning current parameters from the measured fields. However, as described previously, electromagnetic fields generated by lightning change their character as they propagate over the ground surface because of selective attenuation of high-frequency signals by finitely conducting ground. Thus, the peak and rise time of lightning-generated electric fields and electric field time derivatives measured at a given distance from the lightning channel may deviate more or less from the values that would be present over perfectly conducting ground depending on the distance of propagation and the conductivity of the ground.

In this chapter a simple procedure to evaluate propagation effects on electromagnetic fields generated by lightning return strokes is described.

12.2 Theory

In Chap. 3 we derived an expression for an electromagnetic field generated by a Hertzian dipole located above a perfectly conducting ground. This electric field is given by

$$de_z(j\omega,\rho) = \frac{I(j\omega)dz}{2\pi\varepsilon_o}\left(\frac{2-3\sin^2\theta}{j\omega R^3} + \frac{2-3\sin^2\theta}{cR^2} + j\omega\frac{\sin^2\theta}{c^2 R}\right)e^{-j\omega R/c}. \quad (12.1)$$

See Fig. 12.2 for the relevant geometry. Now, when the ground is finitely conducting with a conductivity denoted by σ, a correction term must be introduced into the preceding equation. This correction term is a function of the height of the dipole, its frequency, the distance to the point of observation, and the ground conductivity. With this correction term the corresponding electric field over finitely conducting ground is given by

$$de_z(j\omega,\rho) = \frac{I(j\omega)dz}{2\pi\varepsilon_o}\left(\frac{2-3\sin^2\theta}{j\omega R^3} + \frac{2-3\sin^2\theta}{cR^2} + j\omega\frac{\sin^2\theta}{c^2 R}\right)e^{-j\omega R/c} + S(z,\rho,\omega,\sigma). \quad (12.2)$$

An expression for this correction term was derived by Sommerfeld in 1906 [1] and is given in terms of an infinite integral that contains oscillating terms in the integrand. This oscillating nature of the integrand demands significant computational power in the numerical calculation of these integrals. Several scientists, notably Norton [2] and Bannister [3], attempted to derive an analytical function that can approximate Sommerfeld's integral. The results of these studies showed that for a very good approximation the function S can be written as

$$S = \frac{I(j\omega)dz}{2\pi\varepsilon_o}\frac{j\omega\sin^2\theta}{c^2 R}\frac{1}{2}\{(R_v+1) + (1-R_v)a(z,\rho,j\omega,\sigma) - 2\}e^{-j\omega R/c}. \quad (12.3)$$

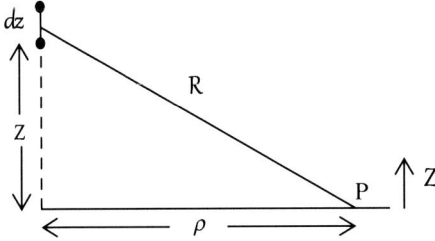

Fig. 12.2 Geometry relevant to parameters in field expressions of dipole. A dipole of length dz is located at height z above the ground plane. The distance to the point of observation P from the dipole is R, and the horizontal distance to the point of observation from the dipole is ρ (Figure created by author)

Substituting this into Eq. 12.2 gives the vertical electric field at ground level at point P due to a dipole located at height z as

$$de_z(j\omega,\rho) = \frac{I(j\omega)dz}{2\pi\varepsilon_o}\left(\frac{2-3\sin^2\theta}{j\omega R^3} + \frac{2-3\sin^2\theta}{cR^2} + j\omega\frac{\sin^2\theta}{c^2R}f(z,\rho,j\omega,\sigma)\right)e^{-j\omega R/c}. \tag{12.4}$$

In the preceding equations,

$$f(z,\rho,j\omega,\sigma) = \frac{1}{2}\{(R_v+1) + (1-R_v)a(z,\rho,j\omega,\sigma)\}, \tag{12.5}$$

$$R_v = \frac{\cos\theta - \Delta_1}{\cos\theta + \Delta_1} \tag{12.6}$$

$$\Delta_1 = \frac{k_o}{k_1}\left(1 - \frac{k_o^2}{k_1^2}\sin^2\theta\right)^{1/2}, \tag{12.7a}$$

$$k_o = \omega^2\mu_o\varepsilon_o \tag{12.7b}$$

$$k_1 = -j\omega\mu_o(\sigma + j\omega\varepsilon_o\varepsilon_r) \tag{12.7c}$$

$$\cos\theta = z/R \tag{12.7d}$$

$$\eta = -\frac{j\omega R}{2c\sin^2\theta}[\cos\theta + \Delta_1]^2, \tag{12.8}$$

$$a(z,\rho,j\omega,\sigma) = 1 - j(\pi\eta)^{1/2}e^{-\eta}\,erfc\left(j\eta^{1/2}\right). \tag{12.9}$$

In these equations, *erfc* stands for the complementary error function. The function $a(z,\rho,j\omega,\sigma)$ is the attenuation function corresponding to a dipole at height z over homogeneous and finitely conducting ground of surface impedance Δ_1. In the above equation ε_r is the relative dielectric constant of the soil, ε_o is the relative permittivity of free space and μ_o is the magnetic permeability of free space. These expressions can be calculated numerically with little difficulty. Note also that when the dipole is at ground level, i.e., $z=0$, $f(z,\rho,j\omega,\sigma)$ reduces to a.

Now, the lightning channel can be divided into a large number of elementary channel sections, each of which can be treated as an elementary dipole. Therefore, the total electric field generated by the return stroke can be obtained by summing the contribution from each dipole, taking into account the time delays. Hence, the total electric field due to a lightning return stroke over finitely conducting ground at a horizontal distance ρ can be written as

$$e_z(j\omega,\rho) = \int_0^H \frac{I(j\omega)}{2\pi\varepsilon_o}\left(\frac{2-3\sin^2\theta}{j\omega R^3} + \frac{2-3\sin^2\theta}{cR^2} + j\omega\frac{\sin^2\theta}{c^2R}f(z,\rho,j\omega,\sigma)\right)e^{-j\omega R/c}dz. \tag{12.10}$$

If the spatial and temporal variations in the return stroke current is known, then $I(j\omega)$ can be obtained through Fourier transformation.

12.2 Theory

Numerical evaluation of the integral in Eq. 12.10 gives the vertical electric field in the frequency domain (i.e., the electric field corresponding to a given frequency ω), and the time domain electric field can be obtained through inverse Fourier transformation. However, this calculation still involves a significant amount of numerical computation. To simplify the calculations further, Cooray and Lundquist [4] simplified this equation using the following arguments. As far as the propagation effects are concerned, the section of the electromagnetic field that is of interest is the part occurring within the first few microseconds. If the speed of propagation of the return stroke front is approximately 10^8 m/s, then the length of the channel that contributes to the radiation field during this time would be no larger than a few hundred meters. Thus, in the preceding equation, the attenuation function $f(z, \rho, j\omega, \sigma)$ can be replaced by $f(0, \rho, j\omega, \sigma)$, the attenuation function corresponding to a dipole located at ground level. The result is

$$e_z(j\omega, \rho) = \int_0^H \frac{I(j\omega)}{2\pi\varepsilon_o} \left(\frac{2 - 3\sin^2\theta}{j\omega R^3} + \frac{2 - 3\sin^2\theta}{cR^2} + j\omega \frac{\sin^2\theta}{c^2 R} a(\rho, j\omega, \sigma) \right) e^{-j\omega R/c} dz. \tag{12.11}$$

Note that $f(0, \rho, j\omega, \sigma)$ is equal to $a(\rho, j\omega, \sigma)$. This equation can be inverse Fourier transformed directly into the time domain, giving the result

$$E_z(t, \rho) = E_{z,s}(t, \rho) + E_{z,i}(t, \rho) + \int_0^t E_{z,r}(t - \tau, \rho) A(\rho, \tau, \sigma) d\tau, \tag{12.12}$$

where $A(\rho, \tau, \sigma)$ is the inverse Fourier transformation of $a(\rho, j\omega, \sigma)$. In this equation, $E_{z,s}(t, \rho)$, $E_{z,i}(t, \rho)$, and $E_{z,r}(t, \rho)$ are the static, induction, and radiation field components, respectively, of the electric fields generated by the return stroke over perfectly conducting ground. These field components are given by

$$E_{z,s}(t, \rho) = \int_0^H \frac{dz}{2\pi\varepsilon_o} \left\{ \frac{2 - 3\sin^2\theta}{R^3} \int_0^t i(z, \tau - R/c) d\tau \right\}, \tag{12.13}$$

$$E_{z,i}(t, \rho) = \int_0^H \frac{dz}{2\pi\varepsilon_o} \frac{2 - 3\sin^2\theta}{cR^2} i(z, t - R/c), \tag{12.14}$$

$$E_{z,r}(t, \rho) = \int_0^H \frac{dz}{2\pi\varepsilon_o} \frac{\sin^2\theta}{c^2 R} \frac{\partial i(z, t - R/c)}{\partial t}. \tag{12.15}$$

According to the preceding equations, only the radiation field term is disturbed by propagation effects while the static and induction terms remain intact. In cases where the distance to the point of observation is so large that it is only the radiation field that is of interest, Eq. 12.12 reduces to

$$E_z(t,\rho) = \int_0^t E_{z,r}(t-\tau,\rho) A(\rho,\tau,\sigma) d\tau. \tag{12.16}$$

As mentioned previously, in Eq. 12.16, $A(\rho,t,\sigma)$ is the inverse Fourier transformation of $a(\rho,j\omega,\sigma)$. To use Eq. 12.16, the function $A(\rho,t,\sigma)$ must be obtained for each conductivity and distance of interest by inverse Fourier transformation of $a(\rho,j\omega,\sigma)$. On the other hand, Wait [5] derived an analytical approximation for $A(\rho,t,\sigma)$ that can be used in Eq. 12.16 to further reduce the computational time. The analytical approximation to $A(\rho,t,\sigma)$ derived by Wait is given by

$$A(\rho,t,\sigma) = \frac{d}{dt}\left(1 - \exp\left(-\frac{t^2}{4\zeta^2}\right) + 2\beta(\varepsilon_r + 1)\frac{Q(t/2\zeta)}{t}\right), \tag{12.17}$$

with

$$Q(x) = x^2(1-x^2)\exp(-x^2), \tag{12.18}$$
$$\beta = 1/\mu_o \sigma c^2, \tag{12.19}$$
$$\zeta^2 = \rho/2\mu_o \sigma c^3. \tag{12.20}$$

The third term inside the brackets of Eq. 12.17 takes account approximately of the displacement current in the ground. In many cases of practical interest, this term can be neglected, and the attenuation function in those cases is given by

$$A(\rho,t,\sigma) = \frac{d}{dt}\left(1 - \exp\left(-\frac{t^2}{4\zeta^2}\right)\right). \tag{12.21}$$

Note that if the preceding expression is used in Eq. 12.16 to calculate the propagation effects, then the predicted propagation effects depend only on the parameter ρ/σ.

To illustrate the propagation effects, assume that the radiation field over perfectly conducting ground can be represented by a step function. Then the radiation field at different distances and conductivities are given by

$$E_z(t,\rho) = 1 - \exp\left(-\frac{t^2}{4\zeta^2}\right). \tag{12.22}$$

The results corresponding to 50-km, 100-km, and 200-km propagation over finitely conducting ground of 0.001 S/m are depicted in Fig. 12.3a. Observe how the rise time of the step increases with increasing distance. Note that in this figure

12.2 Theory

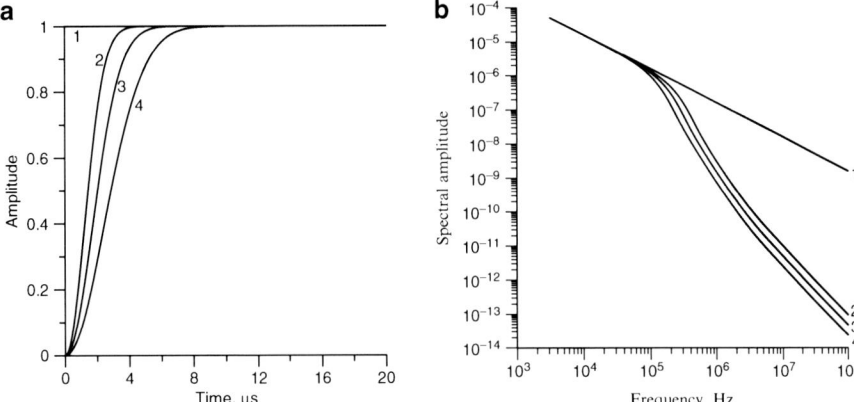

Fig. 12.3 (**a**) Effects of propagation on electric radiation field step as it propagates over various distances along finitely conducting ground. Observe that the field amplitude is normalized by removal of the inverse distance dependence of the radiation field. (1) Perfect conductor. (2) 50-km propagation over ground of 0.001 S/m conductivity. (3) 100-km propagation over ground of 0.001 S/m conductivity. (4) 200-km propagation over ground of 0.001 S/m conductivity. Note how the rise time of the step (which is zero over perfectly conducting ground) increases with increasing distance of propagation (Figure created by author). (**b**) Spectrum of electric fields shown in Fig. 12.3b. Note how the higher frequencies are attenuated with respect to lower frequencies as the signal propagates along finitely conducting ground (Figure created by the author)

the inverse distance reduction in the radiation field with distance is removed to illustrate the effects of propagation. In other words, if there were no propagation effects, then all the waveforms would resemble step functions with unit amplitude. The frequency spectrum corresponding to each waveform is shown in Fig. 12.3b. Note how the propagation effects remove the high frequencies from the signal and as a result the rise time of the signal increases.

Equation 12.12 can be used to calculate the propagation effects if the distance to the point of observation is larger than approximately 1 km. If the distance is less than that, then Eq. 12.10 must be used to obtain accurate results. However, even at these distances Eq. 12.12 can still generate reasonable results if the conductivity of the ground is higher than approximately 0.01 S/m. If the distance to the point of observation is less than approximately 200 m, then it is necessary to solve Sommerfeld's integrals to obtain accurate results. Even at these distances Eq. 12.10 can still provide a reasonable estimation of propagation effects. However, it is important to point out that at points of observation in the vicinity of the channel, the propagation effects modify only the electric field derivative. The propagation effects on the electric field signature itself are not very significant. Thus, if the propagation effects are needed only for the electric field, then the simplified equations presented earlier can be used to estimate them, even at distances close to the lightning channel.

12.3 Illustration of Propagation Effects

As explained earlier, propagation effects are caused by the selective attenuation of the higher frequencies in the electromagnetic field by finitely conducting ground. Thus, the fast features of lightning return stroke electromagnetic fields, such as the rise time, are more sensitive to propagation effects than the slow variations, such as the duration of the electromagnetic field. For the same reason, the electric field derivative is more sensitive to propagation effects than the electric field itself. For example, Fig. 12.4 shows how the electric field in the vicinity of the channel is modified by propagation effects. Observe that propagation effects do not modify the electric field (or the modifications are insignificant) in the vicinity of the channel.

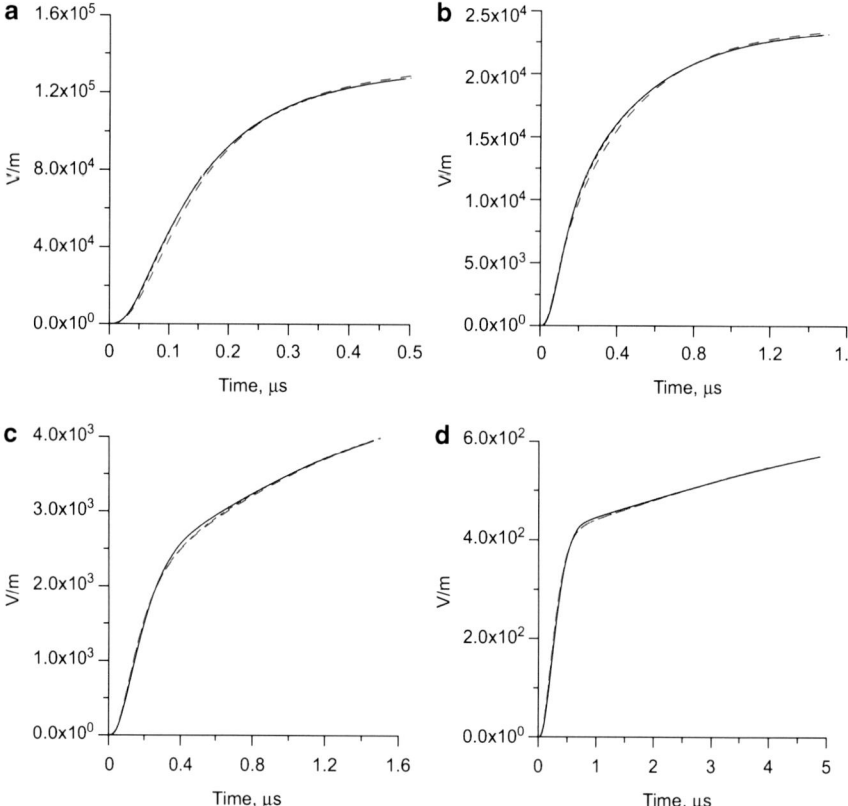

Fig. 12.4 Vertical electric field at ground level at (**a**) 10 m, (**b**) 50 m, (**c**) 200 m, and (**d**) 1000 m from lightning channel. *Solid line*: field obtained from Sommerfeld's equation; *short-dash line* (*blue*): Bannister approximation; *long-dash line* (*red*): Norton approximation. The conductivity of the ground is 0.001 S/m, and the relative dielectric constant is 5. Note that the equations of Bannister and Norton represent good approximations to Sommerfeld's exact integrals (Adapted from Cooray [6])

12.3 Illustration of Propagation Effects

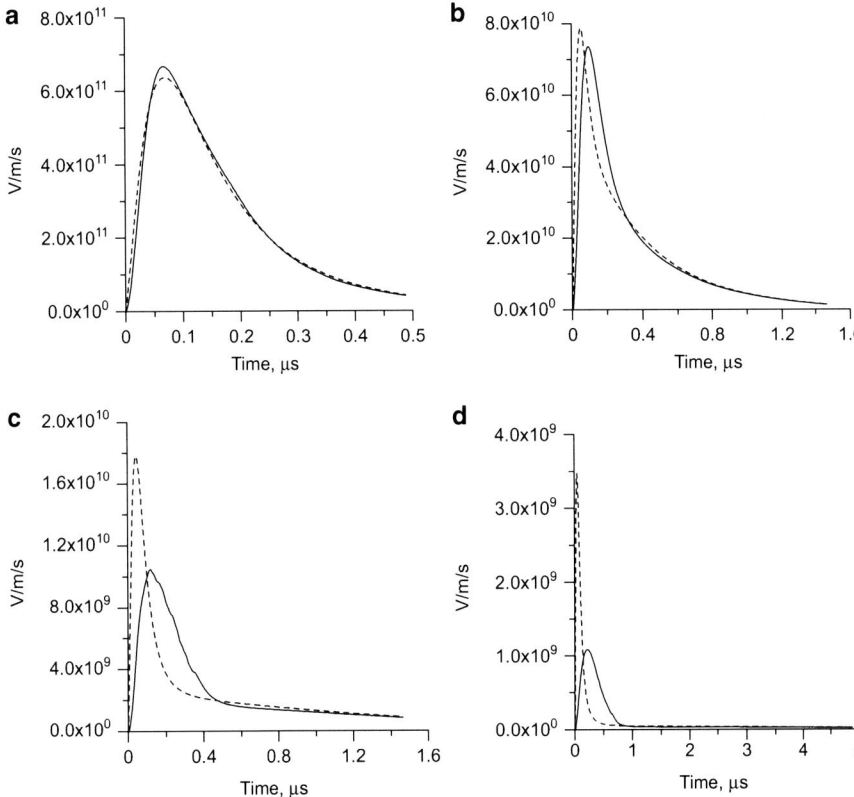

Fig. 12.5 Time derivative of vertical electric field at ground level at (**a**) 10 m, (**b**) 50 m, (**c**) 200 m, and (**d**) 1000 m from lightning channel. *Solid line*: time derivative of electric field over finitely conducting ground calculated using Sommerfeld's integrals; *dotted line*: time derivative of electric field over perfectly conducting ground. The conductivity of the ground is 0.001 S/m, and the relative dielectric constant is 5 (Adapted from Cooray [6])

On the other hand, Fig. 12.5 shows how the derivative of the electric field within approximately 1 km from the channel is modified by propagation effects. Note that propagation effects can significantly modify the peak amplitude of the electric field derivative. It is of interest to note that even very close to the return stroke channel the main contribution to the electric field derivative comes from the radiation field. In the radiation fields the electric field derivative and the magnetic field derivative have the same time signature, and they are related through the equation $dE/dt = c\, dB/dt$, where c is the speed of light. Thus, the amount of attenuation of the peak of the magnetic field derivative is the same as the amount of attenuation of the electric field derivative. Figure 12.6 shows the how the peak amplitude of the magnetic field derivative decreases with distance over finitely conducting ground.

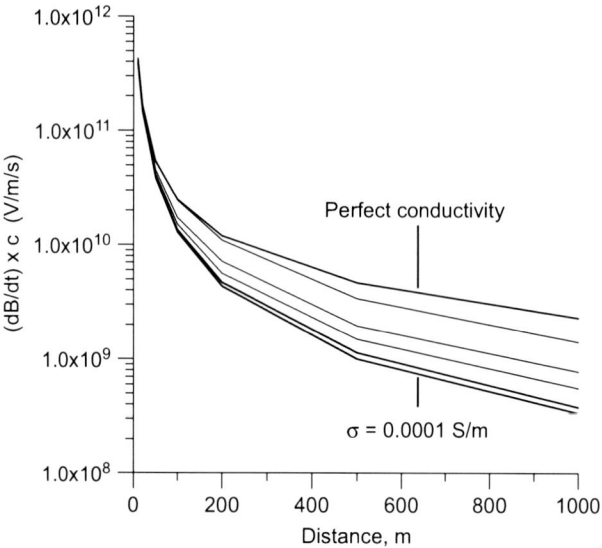

Fig. 12.6 Peak amplitude of magnetic field time derivative multiplied by speed of light in free space as function of distance for several conductivities. The results are shown for perfectly conducting ground: $\sigma = 0.01$ S/m, $\sigma = 0.001$ S/m, $\sigma = .0005$ S/m, $\sigma = 0.0002$ S/m, $\sigma = 0.0001$ S/m. Note that the peak amplitude at a given distance decreases as the conductivity decreases from infinity (perfect conductivity) to 0.0001 S/m. Since the derivative of the electric field is related to the derivative of the magnetic field through the equation $dE/dt = c\,dB/dt$ for distances larger than approximately a few tens of meters, where c is the speed of light, the results can be used to estimate the attenuation of dE/dt over finitely conducting ground (Adapted from [7])

For the reason described earlier, the data in Fig. 12.6 can also be used to obtain the attenuation of the electric field derivative at different distances and conductivities.

Figure 12.7 shows the long-distance propagation effects on the first return stroke radiation field, obtained using the model introduced by Nucci et al. [8] (Chap. 10), as it propagates over distances of 100 and 200 km over ground of 0.001 S/m conductivity. For reference, the radiation field that would be present over perfectly conducting ground is also shown in the figure. Observe how the rise time of the radiation field increases and its amplitude decreases with increasing distance. Observe also that at these distances the zero crossing time of the radiation field is not affected much by the propagation effects. Note again that in this figure the inverse distance reduction in the radiation field with distance is removed to illustrate the effects of propagation. In other words, if there were no propagation effects, all the waveforms would resemble a waveform over perfectly conducting ground.

The data shown so far are based on a theoretical calculation of propagation effects. In 1997, Cooray et al. [9] conducted an experiment in Denmark to study propagation effects experimentally. In the experiment, the electric fields from

12.3 Illustration of Propagation Effects

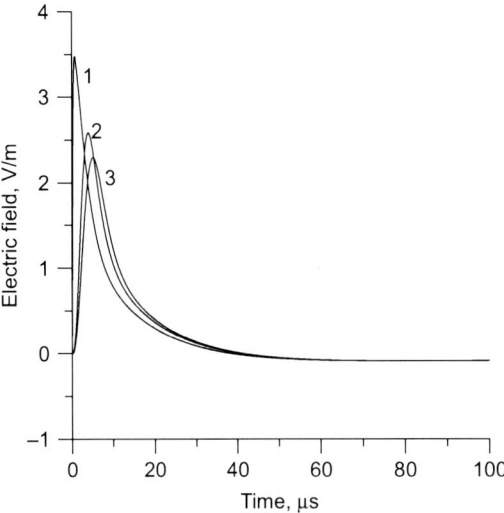

Fig. 12.7 Effect of long-distance propagation of radiation field of lightning return stroke as simulated by model of Nucci et al. [8]. (1) Perfect conductor. (2) 100-km propagation over ground of 0.001 S/m conductivity. (3) 200-km propagation over ground of 0.001 S/m conductivity. Note that the rise time of the radiation field increases and the peak amplitude of the radiation field decreases as the signal propagates over finitely conducting ground. Note that in the diagram the inverse distance dependence of the peak amplitude of the radiation field is removed to illustrate the effects of propagation. In other words, if the ground were perfectly conducting, then all the waveforms would have identical wave shapes and amplitudes. Note also that the zero crossing time of the radiation field is not modified significantly by propagation effects (Figure created by author)

lightning return strokes striking the sea were measured simultaneously at two stations, one located on the coast and the other situated 250 km inland. The propagation path of the electromagnetic fields from the strike point to the coastal station was over salt water, with the exception of the last 10 50 m. Since salt water is a good conductor, the propagation effects experienced by the radiation fields measured at the coastal station were negligible. Thus, the radiation field measured at the coastal station can be assumed to represent the electric radiation field that would be present over perfectly conducting ground. Figure 12.8 shows the electric radiation fields measured at the two stations and the radiation field calculated using Eq. 12.12, assuming the radiation field measured at the coastal station represented the one over perfectly conducting ground. Since the distance to the lightning flashes was more than 50 km from the coastal station, the measured portion of the field was pure radiation. Thus, only the third term in Eq. 12.12 is needed in the analysis. Note how the peak amplitude decreases and the rise time increases with propagation effects. The results also demonstrate that Eqs. 12.12 and 12.16, based on the simplified approximations, can be used with reasonable accuracy to calculate propagation effects.

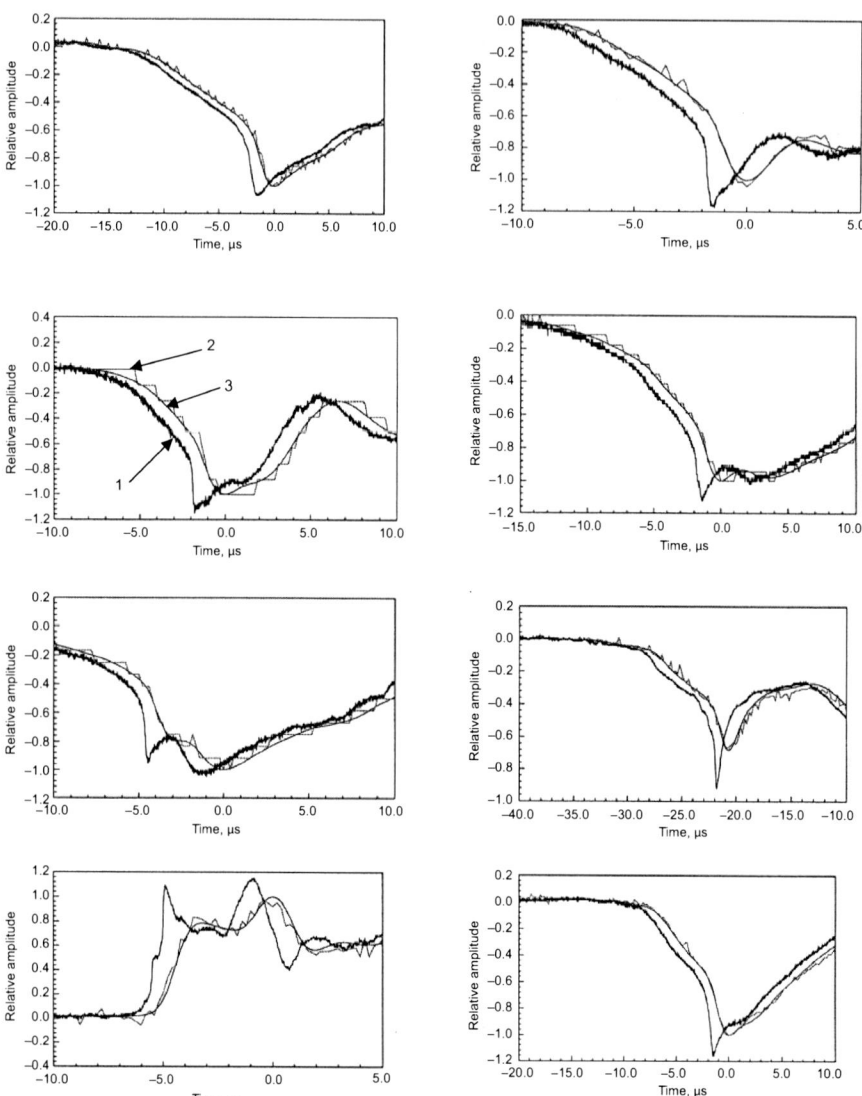

Fig. 12.8 Electric field measured at two stations, one a coastal station (*thick solid line*, marked 1), the other an inland station (*distorted line*, marked 2). The calculation based on theory as outlined in this chapter is shown by the *thin solid line* (marked 3). Note how the predictions based on theory agree with experimental data, indicating that Eq. 12.12 or 12.16 can be used to calculate propagation effects (Adapted from Cooray et al. [9])

References

1. Sommerfeld A (1909) Uber die Ausbreitung der Wellen in der drahtlosen Telegraphie. Ann Phys 28:665
2. Norton KA (1937) Propagation of radio waves over the surface of Earth and in the upper atmosphere, II. Proc IEEE 25:1203–1236
3. Bannister PR (1984) Extension of finitely conducting Earth-image-theory results to any range, NUSC technical report. Naval Underwater Systems Center, New London, Connecticut, USA
4. Cooray V, Lundquist S (1983) Effects of propagation on the risetime and the initial peaks of radiation fields from return strokes. Radio Sci 18:409–415
5. Wait JR (1956) Transient fields of a vertical dipole over homogeneous curved ground. Can J Phys 36:9–17
6. Cooray V (2008) On the accuracy of several approximate theories used in quantifying the propagation effects on lightning generated electromagnetic fields. IEEE Trans Antenna Propag 56(7):1960–1967. doi:10.1109/TAP.2008.924680
7. Cooray V (2009) Propagation effects due to finitely conducting ground on lightning-generated magnetic fields evaluated using Sommerfeld's integrals. IEEE Trans Electromagn Compat 51:526–531
8. Nucci CA, Mazzetti C, Rachidi F, Ianoz M (1988) On lightning return stroke models for LEMP calculations. Paper presented at 19th international conference on lightning protection, Graz, Austria
9. Cooray V, Fernando M, Sörensen T, Götschl T, Pedersen A (2000) Propagation of lightning generated transient electromagnetic fields over finitely conducting ground. J Atmos Sol Terr Phys 62:583–600

Chapter 13
Localization of Lightning Flashes

13.1 Magnetic Direction Finding

As described in Chap. 9, the magnetic field produced by the current flowing in a vertical lightning channel propagates outward with the speed of light along the ground surface and forms circular loops with the center at the point of strike (Fig. 13.1). At any given point the magnetic field is parallel to the tangent to the circle describing the magnetic field. Thus the direction of the magnetic field gives the direction to the point of a lightning strike. Note also that for a given strike point the magnetic field at the point of observation has opposite polarities for negative and positive return strokes. This is so because the directions of positive current are opposite to each other in the two cases.

Assume that the magnetic field is measured at several spatially separated recording stations. Each station provides a direction to the strike point, and by interpolating the directions obtained from several stations, it is possible to obtain the location of the strike point (Fig. 13.2). This method of obtaining the location of lightning strikes is called *magnetic direction finding*. A minimum of two stations are necessary for magnetic direction finding, but in this case, strike points along the line joining the two stations cannot be determined. This problem can be avoided using three magnetic direction finding stations. Another problem that needs to be solved in magnetic direction finding is the ambiguity caused by positive and negative lightning flashes. Since the magnetic field generated by a negative return stroke has a polarity opposite to that of a positive return stroke, magnetic direction finding cannot unambiguously determine whether a signal received, for example, is a negative return stroke, say, from the east or a positive return stroke from the west. This can only be resolved by measuring an electric field that has opposite polarities in the case of positive and negative return strokes. From the measured electric field it is possible to determine whether the magnetic field signal received at a given station is from a positive or a negative flash, and this information, coupled with the data from all the direction finding stations, will give the location of positive or negative ground flashes

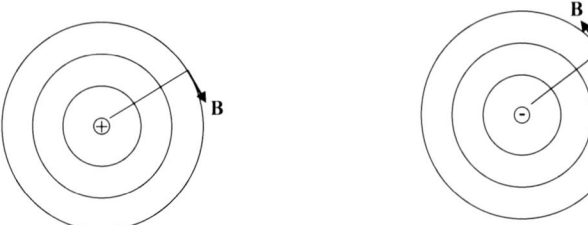

Fig. 13.1 The magnetic field from a lightning flash forms circular loops with the strike point at the center on the surface of the ground. The magnetic field at any point is tangent to the circle at that point. Note that in the case of positive (marked with a + sign) and negative (marked with a − sign) ground flashes, the directions of the magnetic field are opposite to each other (Figure created by author)

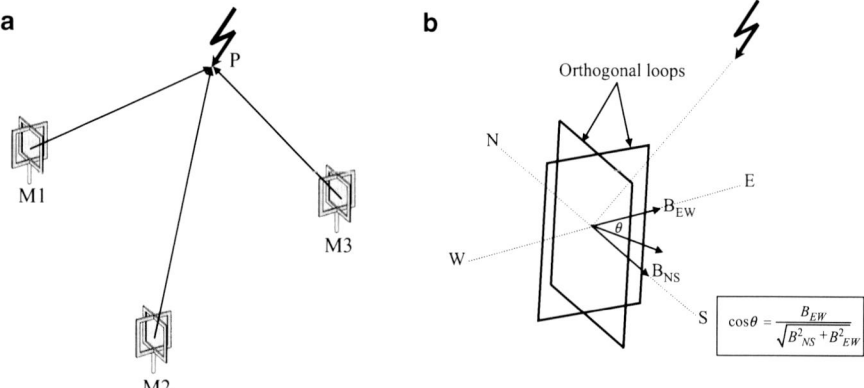

Fig. 13.2 (**a**) In magnetic direction finding three or more stations (M1, M2, and M3) measure the local direction of the magnetic field at their location. The direction of the source is perpendicular to the direction of the magnetic field. Thus, by evaluating the direction of the source from three stations it is possible to obtain the location of the source (P) (Figure created by author). (**b**) The direction and magnitude of the magnetic field at the ground surface can be obtained by measuring the two components of the magnetic field (E–W and N–S) using two loops located perpendicular to each other. From the two components of the magnetic field, i.e., B_{EW} and B_{NS}, it is possible to obtain the direction to the strike point (Figure created by author)

without ambiguity. However, this problem actually arises when only two direction finders are used to locate a lightning flash. With several direction finders the location of a lightning flash can be estimated without ambiguity.

In order for the magnetic direction finding technique to work accurately, the lightning channel must be vertical. However, this is not the case in general. In particular, lightning channels inside a cloud are usually horizontal. However, the channels of lightning flashes that strike the ground are nearly vertical, and this is especially true in the case of the last few hundred meters (i.e., the channel section close to the ground). In magnetic direction finding one can in principle use any process that generates a radiation field to determine the location of a lightning flash, but researchers have decided to use the first few microseconds of the magnetic field

13.1 Magnetic Direction Finding

generated by return strokes to estimate a location [1]. There are several reasons for this choice. The first few microseconds of a return stroke radiation field is generated by the last few hundred meters of a lightning channel (i.e., during the time when the return stroke rises over the last few hundred meters of the leader channel). This channel section is nearly vertical, and therefore the errors in the direction estimation are minimal. The second reason for using the return stroke magnetic field is the fact that the return stroke, being one of the strongest events in a lightning flash, can be detected at large distances. Moreover, it is the return stroke that is of interest in connection with lightning protection. The third reason for using the first few microseconds of a return stroke is to avoid the errors caused by ionospheric reflections (Chap. 9). As the distance to the flash increases, the time separation between the arrival of the ground wave and the reflections from the ionosphere at a given point decreases. One can avoid the influence of ionospheric reflections by measuring the direction to the flash using the first few microseconds of the field. Another advantage of using the return stroke fields in magnetic direction finding is that a rough estimation of the peak current in the return stroke can be obtained from the measured peak return stroke radiation, either magnetic or electric fields (Chap. 10).

Note that the irregularities of the Earth and large metal bodies can also disturb the accuracy of magnetic direction finding. For this and other reasons mentioned earlier, the direction of a lightning strike estimated from a direction finding station may have an error that depends on the particular site where the direction finding station is located. This error may also vary as a function of the azimuthal angle or the direction of the incident magnetic field. Thus, the three directions of the strike point estimated from three direction finding stations may not cross at a given point. Instead, as shown in Fig. 13.3, they give rise to an error triangle. In such cases, the correct location of the lightning stroke must be obtained using optimization procedures [2].

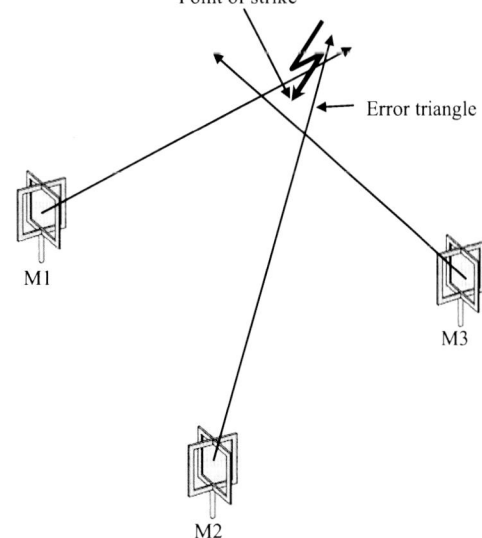

Fig. 13.3 Because of direction finding errors, the three directions given by the three stations may not cross at a single point. This gives rise to an error triangle. In this case, statistical methods must be used to obtain the so-called best location of a flash (Figure created by author based on information available in the literature)

13.2 Time-of-Arrival Method

Assume that the Earth is perfectly conducting and the lightning channel is vertical. As shown in Chap. 9, the electric field in the vicinity of the lightning channel is mainly electrostatic, and as the distance to the point of observation increases, the radiation field separates out. Because of the different rates of attenuation of different field components with distance, the wave shape of the electric field changes with distance. At sufficiently far distances only the radiation field component remains in the electric field, and it maintains its shape with distance. In other words, the temporal features of the radiation field at a distance of 100 km is identical to the temporal features of the radiation field at 200 km. The only difference in the fields measured at 100 km and 200 km is that the amplitude of the radiation field at 100 km is twice the value at 200 km. However, it is important to understand that whether or not the electric field is radiation depends not only on the distance but also on how fast the fields are changing. Let us consider this point in detail.

Recall from the electromagnetic fields of dipoles given in Chap. 3 that the condition necessary for fields to be radiation is that the wavelength λ should be much smaller than the distance d to the point of observation, i.e., $\lambda \ll d$. This translates to $\omega \gg 2\pi c/d$, where c is the speed of light and ω is the angular frequency. Now, consider a time interval Δt in the electric field record. The dominant frequency associated with this time interval is $1/\Delta t$. Thus, the condition for a field within a time Δt to be radiation is that $\Delta t \ll d/(2\pi c)$. Thus, whether or not the electric field at any given time is radiation depends not only on the distance but also on the time scale at which the radiation takes place. In the case of return strokes, the initial peak occurs within approximately 5 μs, and hence the section of the electric field up to the initial peak can be considered mainly radiation, say, at distances greater than approximately 10 km. At these distances a large fraction of the electric field at 100 μs can be electrostatics.

Since the radiation field retains its signature with distance as it propagates, the same feature of the radiation field (e.g., the initial peak) can be identified at stations located at different distances from the lightning flash. Consider three measuring stations located at distances of r_1, r_2, and r_3 from the point of lightning strike. The geometry relevant to the analysis is shown in Fig. 13.4. The distances between the stations are marked d_{12}, d_{13}, and d_{23}. Assume that the same features of the electric (or magnetic) field (say, for example, the initial peak) is felt at these stations at times t_1, t_2, and t_3. Of course, to achieve this result, the clocks at the three stations must be synchronized. For example, Fig. 13.5 shows the location of time-of-arrival recording stations and the electric fields, including the time of arrival, recorded at stations in an experiment conducted in Japan [3]. Since the distance to the strike point is not known, r_1, r_2, and r_3 must be treated as unknowns. To solve for these distances, we need to create three equations from the measured data. The input data available are the time of arrival of the signal (or the same feature of the electromagnetic field, for example, the peak of the signal) at each station and the relative

13.2 Time-of-Arrival Method

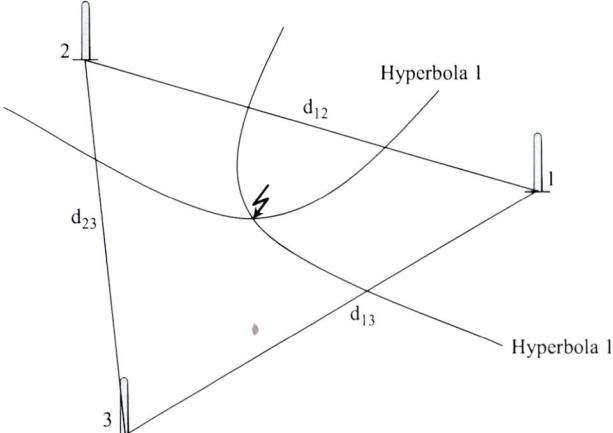

Fig. 13.4 In the time-of-arrival method the possible locations of the source corresponding to a given time of arrival between two stations are located on a hyperbola. Measuring the time intervals of the signal arriving at three stations, marked 1, 2, and 3 in the diagram, one can obtain the location of the source by estimating the point where the two hyperbolas cross each other (Figure created by author based on information available in the literature)

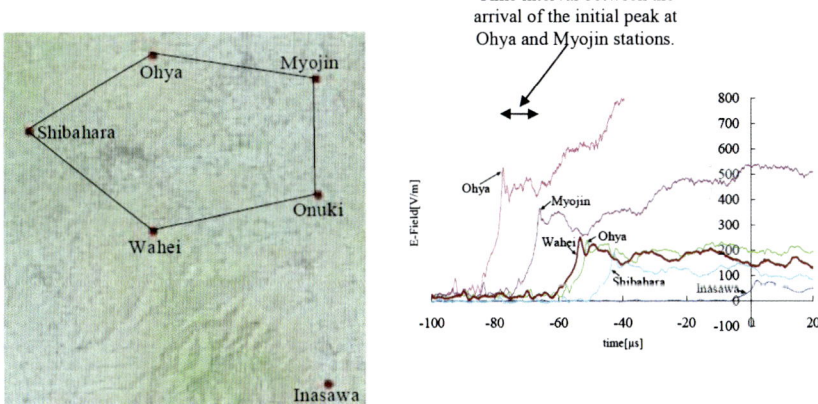

Fig. 13.5 Location of time-of-arrival stations and return stroke field received at five stations in experiment conducted in Japan [3]. Note, for example, that the initial peak of the signal arrives at different times at the five stations

location of the stations. Consider stations 1 and 2. The time interval between the times of arrival is given by

$$\frac{r_1}{c} - \frac{r_2}{c} = t_1 - t_2, \qquad (13.1)$$

where c is the speed of light in free space. Similarly, we can write another equation using the time differences measured between stations 2 and 3. That is,

$$\frac{r_2}{c} - \frac{r_3}{c} = t_2 - t_3. \qquad (13.2)$$

Note that the time difference measured between stations 1 and 3 does not give an independent equation. It can be constructed by the addition of 1 and 2. Equations 13.1 and 13.2 define two hyperbolas, and the point of crossing of these two hyperbolas gives the point of strike of the lightning flash. However, when the lightning flash is located outside the triangle defining the three antennas, the two hyperbolas may cross at two different points, creating an ambiguity in the location of the lightning flash. This location ambiguity can be resolved using four antennas instead of three.

Actually, the time-of-arrival technique can also be applied to locate sources generating electromagnetic fields inside a cloud [4]. Consider two stations, and assume that the distance to the source is much greater than the separation between the two stations d. In this case, the incident electromagnetic field can be treated as a plane wave. The incoming direction of the plane wave can be described by two angles, azimuth θ and elevation φ (Fig. 13.6). The time interval between the arrivals of the same signal at the two stations is given by

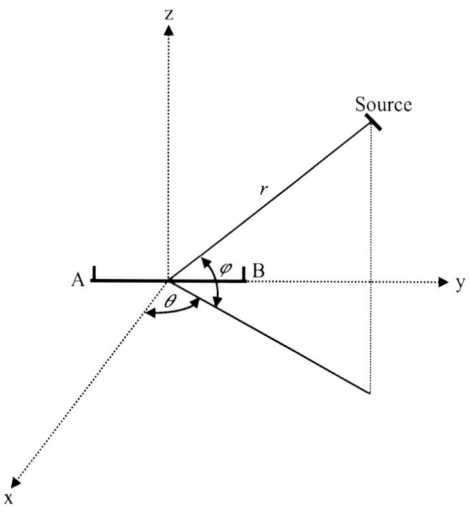

Fig. 13.6 The direction of arrival of a plane wave incident at two stations (A and B) at ground level from a source located at any arbitrary point in space can be described by two angles, the azimuthal angle θ and the angle of elevation φ. Of course, to assume that the incident wave is a plane wave, the distance to the source (marked r) must be much greater than the distance between the two stations (i.e., $r \gg AB$) (Figure created by author based on information available in the literature)

13.2 Time-of-Arrival Method

$$\delta t_1 = \frac{d}{c} \cos \varphi \sin \theta. \tag{13.3}$$

Now consider another pair of stations with the same baseline but located perpendicular to these two stations. The time interval between the arrivals of the same signal at the two new stations is given by

$$\delta t_2 = \frac{d}{c} \cos \varphi \cos \theta. \tag{13.4}$$

These two equations can be solved for θ and φ. This gives the direction to the source of the electromagnetic field. By employing several sets of such station pairs we can estimate the three-dimensional location of the source.

This technique works fine if the ground acts as a perfect conductor. In this case, the signature of the radiation field at any given distance from the source is the same irrespective of the direction of the observation station. However, when the ground conductivity is low and the surface is irregular, the electromagnetic field changes depending on ground conditions (Chap. 12). That is, the signature of the field changes with distance. This is illustrated in Fig. 13.7. In this case, it is difficult to

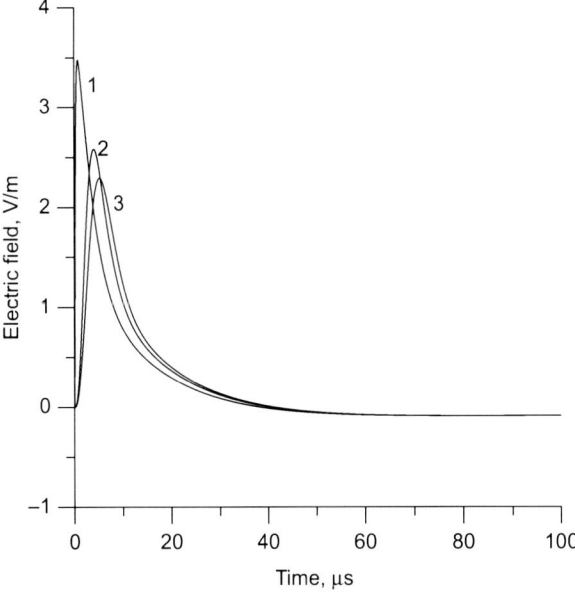

Fig. 13.7 Effect of long-distance propagation of radiation field of a lightning return stroke. (1) Perfect conductor. (2) 100-km propagation over ground of 0.001 S/m conductivity. (3) 200-km propagation over ground of 0.001 S/m conductivity. Note how the rise time of the waveform increases with distance. If the ground were a perfect conductor, the rise time of the signal would have the same value corresponding to the perfect conductor at all distances. This shows that the time of arrival of a signal at two stations depends not only on the relative distance between the stations and the strike point but also on the conductivities of the respective paths of propagation (Figure created by author)

identify the same feature at three stations because the signature of the field reaching different stations will be different. The problem is severe in the case of sources located close to the ground. This indeed is the case in the first few microseconds of a return stroke field where the source is located rather close to the ground. The influence of finitely conducting ground decreases as the height of the source increases. Thus, propagation effects must be taken into consideration when the time-of-arrival method is used to locate lightning flashes. It is important to note that if the distances to the different stations are greater than 100 km or so, then the effect of spherical Earth must also be taken into account in the calculations. The accuracy of the location can also be increased by increasing the number of stations and using optimization procedures to obtain the best location.

The time-of-arrival method can be used in the long-range detection of lightning by time stamping the electromagnetic fields produced by lightning flashes at very large distances in the range of thousands of kilometers. Such a lightning detection system based on the time-of-arrival method to locate lightning is described by Lee [5].

13.3 Interferometric Method

Consider two antennas, separated by a distance d, that can detect electric fields. Assume that far away there is a source that generates an electromagnetic field that varies as a sine or cosine wave with a frequency ω and wavelength λ. The signal spreads like a sphere from the source, but if the separation between the antennas is much smaller than the distance to the source, then the electric field incident on them would be like a plane (part of a very large sphere). At different locations the signal remains a sine wave, but at any given time the phase of the signal is different at different locations (Fig. 13.8). In our case, the phase difference, φ, of the output signals of the two antennas is related to the direction of arrival of the wave (which is described by an azimuthal angle θ and elevation φ; Fig. 13.6) by

$$\phi = 2\pi d \sin\theta \cos\varphi / \lambda. \tag{13.5a}$$

The distance d may have values ranging from a few meters to a few tens of meters depending on the frequency or wavelength of the signal. If the system contains two sets of such antennas with orthogonal baselines, then the phase differences between the outputs of the sets of antennas are given by

$$\phi_1 = 2\pi d \sin\theta \cos\varphi / \lambda, \tag{13.5b}$$

$$\phi_2 = 2\pi d \cos\theta \cos\varphi / \lambda. \tag{13.5c}$$

By measuring the phase difference of the output of two independent pairs one can estimate the azimuth and the elevation of the incoming plane wave. To locate

13.3 Interferometric Method

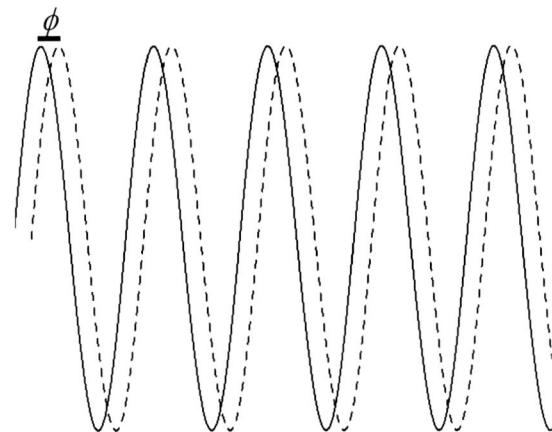

Fig. 13.8 Possible outputs of two tuned antennas separated in space and excited by an incident sinusoidal plane electromagnetic field. Because of the differences in the distance from the source to the two antennas, the output signals are out of phase by ϕ (Figure created by author)

the source, two or more such antenna systems located at distances on the order of 10 km are needed. This method of locating sources that generate electromagnetic fields is called the *interferometric method* [6].

Now, the foregoing concept works with sinusoidal signals, but lightning does not generate such signals. The signals generated by lightning flashes are signatures that vary in time in a complex manner. However, any signal can be separated into a sum of sinusoidal components. This can be done using computers by performing Fourier transformation of the signal or by employing narrowband filters to separate out the desired frequency. In the interferometric systems that are currently being used, narrowband filters are used to extract signals at a given frequency (around hundreds of megahertz). When a transient signal is incident on an antenna, the narrowband filters give rise to outputs that oscillate with the assigned frequency of the narrowband filter. For an input signal that resembles a narrow impulse (a Dirac delta function to be exact) the duration of the output signal depends on the bandwidth of the narrowband filter. As illustrated in Fig. 13.9, the duration of the signal decreases with increasing bandwidth. The duration of the signal is approximately equal to the inverse of bandwidth B, i.e., $1/B$. Of course, if the input signal is a pulse of long duration, then the duration of the output also increases. Indeed, from the known response of a narrowband filter to a Dirac delta function the response of the filter to any other signal can be constructed using convolution theorem. From the output signals of two antennas the phase difference between the two signals can be measured (Fig. 13.10). However, to obtain a phase measurement, the distance between the two antennas or baseline, d, should satisfy the criterion $d/c \leq 1/B$, where c is the speed of light and B is the bandwidth of the antenna system. Furthermore, consider Eq. 13.5a, with $\theta = \pi/2$. If $d = \lambda/2$, then as φ varies from 0 to π, the phase angle ϕ also makes a complete circle and there is a one-to-one correlation between φ and ϕ. However, if the baseline distance is greater than $\lambda/2$, then for a given measurement there are several solutions for the angle φ. This is called *fringe ambiguity*. On the other hand, if the antennas are closely spaced, then the resolution of the measurements is limited.

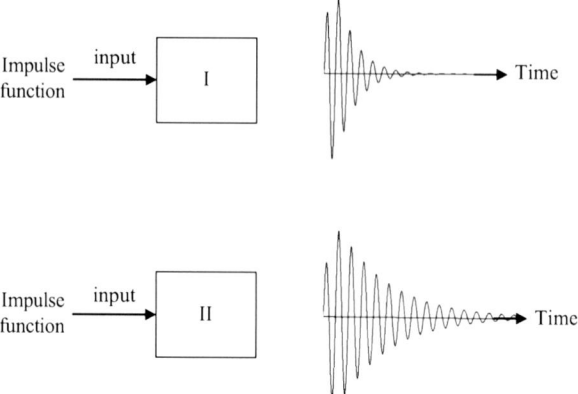

Fig. 13.9 The duration of the response of a tuned circuit to an impulse function (Dirac delta function) depends on the bandwidth of the tuned circuit. As the bandwidth becomes smaller or narrower, the output signal increases in duration. In the example shown in the diagram, circuit I had a broader bandwidth than the II (Figure created by author)

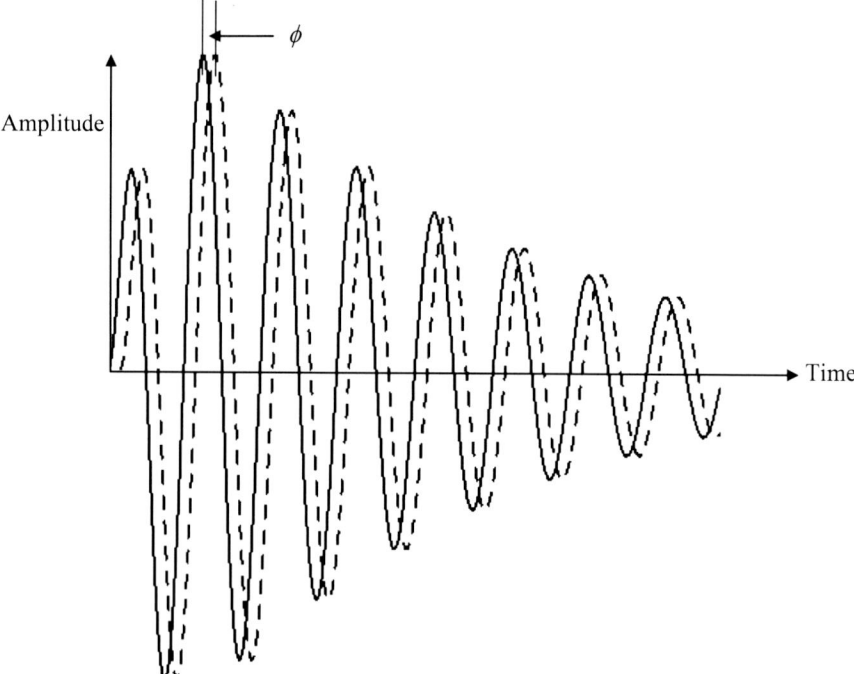

Fig. 13.10 Possible outputs of two tuned circuits located in vicinity of each other to incident electromagnetic field of arbitrary shape. Note the phase difference ϕ between the two signals caused by the differences in the distances from the source to the antennas (Figure created by author)

One can overcome these problems, for example, by employing two pairs of antennas located perpendicular to each other, one having a baseline of $\lambda/2$ and the other a baseline of 4λ [7–9]. Using this method scientists have managed to locate the temporal development of lightning flashes inside clouds [7–9].

13.4 Thunder Ranging

Thunder can be used to determine the distance to a lightning flash, but it cannot determine the actual location or direction. The method is based on the fact that the light signal generated by a lightning flash propagates out at a speed of 3×10^8 m/s, whereas sound waves generated by a flash propagate in air at a speed of approximately 330 m/s. Thus, the light from the flash reaches the observer almost instantaneously, whereas the sound signal is delayed because it covers the same distance at a slower speed. The time interval between the arrival of the light and the arrival of thunder is r/v, where r is the distance to the flash and v is the speed of sound. From the measured time delay it is possible to estimate the distance to a flash. This technique is useful only when the distance to the flash is approximately 10 km or less. The refraction of a sound signal in the atmosphere makes the signal less audible with increasing distance. Actually, only rarely can thunder be heard at distances greater than approximately 20 km.

13.5 Observations from Satellites

Several satellites orbiting the Earth are equipped with light-sensitive detectors to detect light emitted by lightning flashes [10]. Each lightning flash is seen as a dot or a patch of light on the upper surface of a cloud, and the detectors on satellites can record and count these signals to evaluate the global distribution of lightning flashes. The global distribution of lightning flashes as recorded by satellites is shown in Chap. 6.

13.6 Other Methods

Two other techniques different from the ones used routinely and described previously were proposed recently by Lugrin et al. [11] and Rubinstein et al. [12]. They are described in what follows.

13.6.1 Peak Electric Field [11]

This technique utilizes the fact that the radiation field of a return stroke of a lightning flash decreases inversely with distance. Thus, the peak amplitude of the signal measured at several stations can be used to obtain an estimation of the distance to the flash. Consider three stations located at distances of r_1, r_2, and r_3 from a lightning strike (Fig. 13.11). Since the radiation field decreases inversely with distance, the peak of the radiation field measured at the three stations (i.e., P_1, P_2, and P_3) can be written as

$$P_1 = \frac{k}{r_1}; P_2 = \frac{k}{r_2}; P_3 = \frac{k}{r_3}. \qquad (13.6)$$

Now, since the information concerning the location and the distances between P_1, P_2, and P_3 is available, we can write

$$k\left(\frac{1}{r_1} - \frac{1}{r_2}\right) = P_1 - P_2, \qquad (13.7)$$

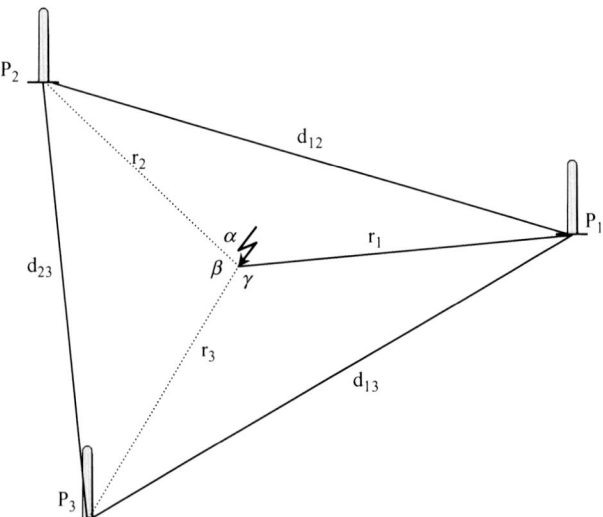

Fig. 13.11 Geometry relevant to calculation of lightning location using peak radiation field method. The three stations where the peak of the radiation field is measured are identified as P_1, P_2, and P_3. Since the peak of the radiation field decreases as $1/r$, the amplitude of the signal received at the three stations depends on the distance from the source to the station. From the measured peak amplitude and the geometry of the location of the stations the strike point of the lightning flash can be estimated (Figure created by author)

$$k\left(\frac{1}{r_2} - \frac{1}{r_3}\right) = P_2 - P_3, \qquad (13.8)$$

$$k\left(\frac{1}{r_1} - \frac{1}{r_3}\right) = P_1 - P_3. \qquad (13.9)$$

The information on the right-hand side of these equations is known. There are four unknowns in the preceding three equations. However, the geometry of the arrangement of stations allows us to write

$$d_{12}^2 = r_1^2 + r_2^2 - r_1 r_2 \cos \alpha, \qquad (13.10)$$

$$d_{13}^2 = r_1^2 + r_3^2 - r_1 r_2 \cos \gamma, \qquad (13.11)$$

$$d_{23}^2 = r_2^2 + r_3^2 - r_1 r_2 \cos \beta, \qquad (13.12)$$

$$\alpha + \beta + \gamma = \pi. \qquad (13.13)$$

Now we have seven equations and seven unknowns, i.e., $k, r_1, r_2, r_3, \alpha, \beta, \gamma$. Thus, solving these equations makes it possible to obtain the location of the point of strike of the lightning flash. For some strike points of lightning flashes these equations may generate two solutions, creating an ambiguity in the location of the point of strike. Such ambiguities can be resolved by utilizing four stations instead of three.

13.6.2 Time-Reversal Method [12]

This technique has been applied to determine the location of sound sources. In the technique, the sound signal is measured at three spatially separated points. The recorded signals are time-reversed and transmitted back (this can be done numerically on a computer once the signals are available). Once this is done, the signals interact with each other and create a peak at a certain spatial location. The location of the peak is the location of the source. The same technique can be utilized in the case of lightning flashes. Instead of sound waves, the electromagnetic radiation reaching several stations can be measured. If the recorded radiation waveforms are time-reversed and transmitted back, they generate a peak at the point of the source, in our case, the point of the lightning strike.

13.7 Lightning Localization Systems in Practice

Several lightning localization systems are currently in operation in different parts of the world. Many of them use magnetic direction finding, time of arrival, the interferometric method, or a combination of these to locate lightning flashes [13].

In lightning protection studies the ground flash density is one of the important parameters necessary in making decisions concerning the vulnerability of structures to lightning. Thus, most lightning detection systems provide information concerning the location of ground flashes. Many systems utilize the fine features of electromagnetic fields to discriminate and identify cloud-to-ground flashes and cloud flashes. Direct electromagnetic field measurements indeed confirm that cloud and ground flashes can be discriminated from the features of the fine structure in electromagnetic fields. However, research based on the LINET lightning location system indicates that some electromagnetic waveforms produced by cloud flashes have features similar to those of return stroke pulses [14]. Of course, it must be recognized that the features of electromagnetic fields change as they propagate over finitely conducting ground (Chap. 12). As they propagate over finitely conducting ground, much of the fine structure disappears, and as a consequence, some of the pulses in cloud flashes after propagating for many tens of kilometers over finitely conducting ground may resemble pulses of return strokes. For example, some of the discharge events that take place inside a cloud, such as preliminary breakdown events, can generate pulses with amplitudes comparable to those of return strokes (Chap. 9). If the fine structure of these pulses disappears as a result of propagation, they may indeed resemble the distant radiation fields of return strokes. The LINET system, for example, utilizes the time of arrival of electromagnetic fields at several stations in special software algorithms to locate and discriminate cloud and ground flashes [14].

Modern-day lightning location systems also provide data on the peak amplitude of first and subsequent stroke currents. According to these data, the mean value of first return stroke peak currents is considerably less than the 30 kA estimated from lightning flashes that strike towers (Chap. 8). There could be several reasons for this discrepancy. First, lightning location systems are capable of detecting a significant fraction of lightning flashes with small currents, which are probably missed in tower measurements because of the small amount of lightning flashes captured by the tower. Second, direction finders might misidentify some subsequent return strokes as first strokes. Since the peak currents of subsequent strokes are less than those of first return strokes, this reduces the mean value of the first return stroke currents. For example, statistics based on return stroke electromagnetic fields measured directly show that the first return stroke peak currents are approximately twice the peak current of subsequent strokes [15]. The mean value of peak currents in subsequent strokes is approximately 12 kA, and therefore, these data suggest that the mean value of first return stroke currents is approximately 24 kA. On the other hand, except for direct measurements all other estimations of first return stroke peak currents are inferred from electromagnetic fields. However, such estimations require information concerning the return stroke speed (Chap. 10). In evaluating the peak currents from the data gathered by direction finding systems, it is assumed that the return stroke speed is the same in both first and subsequent strokes. If the actual return stroke speed in first return strokes is less than that of subsequent strokes, then the mean peak currents of first strokes estimated by direction finding systems must be increased accordingly.

References

1. Krider EP, Noggle RC, Pifer AE, Vance DL (1980) Lightning direction-finding system for forest fire detection. Bull Am Meteorol Soc 61(9):980–986
2. Kenneth L, Cummins KL, Murphy MJ, Tuel JV (2000) Lightning detection methods and meteorological applications. Paper presented at the IV international symposium on military mteorology, Malbork, Poland
3. Koji MK, Hongo Y, Yokoyama S (2009) Estimation of lightning striking points by time of arrival method with small network. Paper presented at the X international symposium on lightning protection Curitiba, Brazil
4. Proctor DE (1971) A hyperbolic system for obtaining VHF radio pictures of lightning. J Geophys Res 76:1478–1489
5. Lee ACL (1989) Ground truth confirmation and theoretical limits of an experimental VLF arrival time difference lightning flash locating system. Q J R Meteorol Soc 115:1147–1166
6. Hayanga CO (1979) Positions and movement of VHF lightning sources determined with microsecond resolution by interferometry. Ph.D. thesis, University of Colo., Boulder
7. Shao XM (1993) The development and structure of lightning discharges observed by VHF radio interferometer. Ph.D. thesis, New Mexico Inst. Mining and Technology, Socorro, New Mexico
8. Shao XM, Krehbiel PR (1996) The spatial and temporal development of intracloud lightning. J Geophys Res 101:26641–26668
9. Shao XM, Krehbiel PR, Thomas RJ, Rison W (1995) Radio interferometric observations of cloud-to-ground lightning phenomena in Florida. J Geophys Res 100:2749–2783
10. Christian HJ, Blakeslee RJ, Goodman SJ (1989) The detection of lightning from geostationary orbit. J Geophys Res 94:13329–13337
11. Lugrin G, Parra NM, Rachidi F, Rubinstein M, Diendorfer G (2013) On the location of lightning discharges using time reversal of electromagnetic fields. Trans IEEE (EMC) 99:1–10. doi:10.1109/TEMC.2013.2266932
12. Rubinstein M, Romero C, Rachidi F, Rubinstein A, Vega F (2010) A two-station lightning location method based on a combination of difference of time of arrival and amplitude attenuation. In: Asia-Pacific symposium on Electromagnetic compatibility, Beijing
13. Cummins KL, Murphy MJ, Tuel JV (2000) Lightning detection methods and meteorological applications. In: Proceedings of the IV international symposium on military meteorology hydro-meteorological support of Allied Forces and PfP Members, Tasks Realization, Malbork, Poland, 26–28 Sept 2000
14. Betz HD, Schmidt K, Oettinger WP, Wirz M (2004) Total VLF/VF-lightning and Pseudo 3D-discriminatuion of intra-cloud and cloud-to ground discharges. In: 18th international lightning detection conference, Helsinki, 7–9 June 2004
15. Sonnadara U, Kathiriarachchi V, Cooray V, Montano R, Götschl T (2014) Performance of lightning locating systems in extracting lightning flash characteristics. J Atmos Sol Terres Phys 112:31–37

Chapter 14
Potential of a Cloud and Its Relationship to Charge Distribution on Stepped Leader and Dart Leader Channels

14.1 The Concept of Cloud Potential

The electric potential of a point in space in the vicinity of Earth can be easily defined by assuming the potential of Earth to be zero. The potential can be calculated by evaluating the line integral of the electrostatic field from a point on the ground to a point where the electrical potential is needed. The path of the integration is immaterial because the result is independent of the path of integration. This is because electrostatic fields are conservative. The potential of a metal object in the vicinity of the ground can be defined in the same way. In finding this potential it is possible to integrate the electric field from the ground to any point of the metal object along any path. This is the case since any point on the metal object has the same potential because, by definition, a metal contains free electrons that move along the surface of the metal and redistribute themselves until the potential at any point of the metal object is the same. But how does one define the potential of a cloud? One problem with the concept of cloud potential is that a cloud is just a collection of charges located on a nonconducting medium, which means that there is no specific definition for the potential of a cloud. The potential of a cloud may vary from one point to another. It may be very high close to the charge centers low at points far from them. Thus, when we speak about cloud potential in connection with a stepped leader, we actually mean the potential of the cloud near the region where the stepped leader was initiated.

In analyzing lightning attachment to grounded structures using the electrogeometrical method (EGM), the EGM striking distance is defined as the distance between the tip of the stepped leader and the grounded structure when the average electric field between them is equal to 500 kV/m (Chap. 17). Now the average electric field between the tip of the stepped leader and any point of the grounded structure is also given by the potential of the tip of the stepped leader divided by the distance between the tip of the stepped leader and the point of interest on the grounded structure. Thus, the striking distance of a stepped leader is given by

$$S = V/5.0 \times 10^5, \quad (14.1)$$

where V is the potential of the tip of the stepped leader in Volts and S is the striking distance in meters. The potential of the tip of the stepped leader is approximately equal to the cloud potential because it is connected to the cloud by a highly conducting channel. Thus, the striking distance is directly related to the potential of the cloud from which the stepped leader emanates. This is why cloud potential is an important parameter in lightning protection. In the following sections a procedure developed by Cooray et al. [1] for obtaining the cloud potential as a function of the first return stroke peak current is described.

14.2 Cloud Potential as a Function of the Return Stroke Peak Current

Now, measurements of the vertical electric field as a function of height below lightning-generating clouds show that initially the field increases over a distance of several hundred meters and that afterward it remains roughly constant with altitude. One such measured example obtained from [2] is shown in Fig. 14.1. The initial

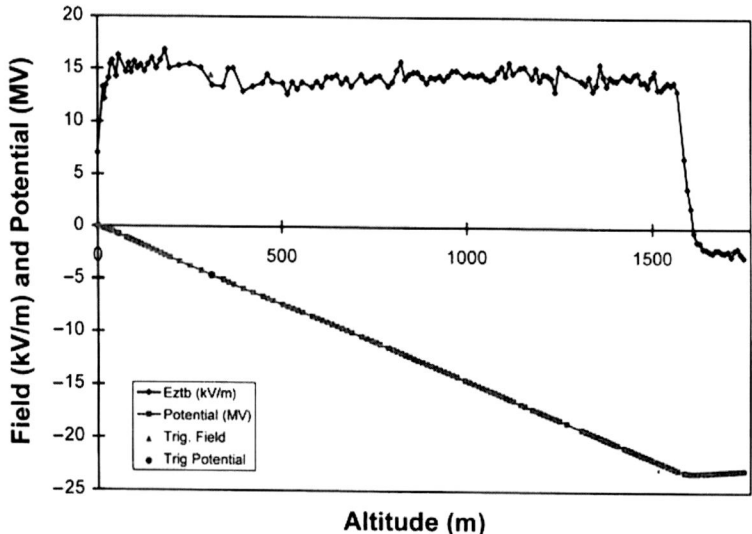

Fig. 14.1 Electric field below thundercloud as function of height (*top diagram*). The diagram at the *bottom* shows the potential as a function of height. Note that, except for the initial rise, the electric field is almost constant with height. Note also that the rocket traveled only to a height that is slightly greater than 1,500 m (Adapted from Willett et al. [2])

14.2 Cloud Potential as a Function of the Return Stroke Peak Current

increase in the field with height is due to the fact that close to ground there is a layer of positive space charge, and this space charge reduces the electric field in the vicinity of the ground. The roughly constant electric field over the first 2 km or so shows that the charged region of the cloud can be represented by a thin flat region giving rise to a roughly constant electric field below the cloud. Let us make a simplifying approximation and represent the negative charged region of the cloud by a flat sheet at a potential $-V$. Thus, the electric field below the cloud, which is uniform under the foregoing approximation, is given by $-V/H$, where H is the height to the charge center. Let us now consider a leader channel (it could be a stepped or a dart leader) extending from this charged region toward the ground (Fig. 14.2). The leader channel is a rather good conductor, and the potential gradient inside the leader channel is approximately 2,000 V/m. Thus, for any given cloud potential the potential at any point of interest on the leader channel is

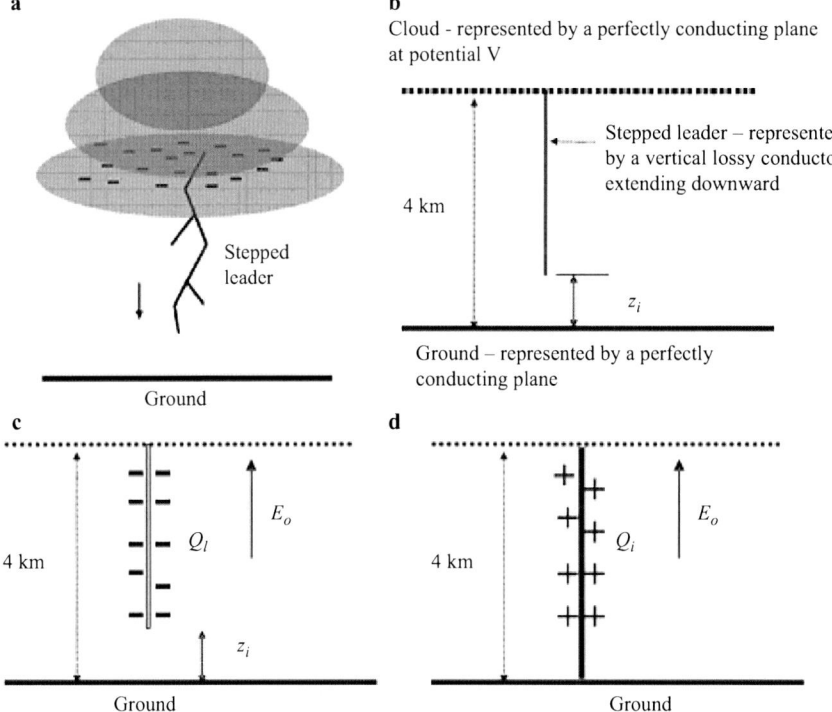

Fig. 14.2 (a) Sketch of a stepped leader approaching ground. (b) Idealization used in computation of charge distribution along leader channel by Cooray et al. [1]. (c) Situation before attachment process and return stroke (leader channel is negatively charged). (d) Situation after return stroke (return stroke channel is positively charged). The total charge that was deposited on the leader channel by the return stroke is $|Q_l|+Q_i$. In other words, the total charge deposited by the return stroke on the leader channel is the sum of the positive charge necessary to neutralize the negative charge on the leader channel and the positive charge induced on the return stroke channel due to the background electric field (Adapted from Cooray et al. [1])

given by $-V + E_g l$, where l is the distance from the point of origin of the stepped leader channel to the point on the leader channel that is of interest and E_g is the potential gradient of the leader channel. The potential at the tip of the stepped leader channel is $-V + E_g L$, where L is the length of the leader channel. To maintain this potential, a negative charge accumulates on the leader channel. The magnitude of this charge Q_l and its distribution along the leader channel can be calculated as a function of the cloud potential using a electromagnetic field simulation program or a charge simulation method (CSM) [3]. If a relationship between the cloud potential and the peak return stroke current is known, these distributions can be described as a function of the return stroke peak current.

Once the leader (stepped or dart) reaches the ground, a return stroke travels along it, changing the potential of the leader channel to zero potential. Two processes that are of interest for us happen during the return stroke. First, it neutralizes the negative charge on the leader channel. Second, since the return stroke channel is also a good conductor and since the background electric field is still the same (assuming, of course, that the removal of the charge from the cloud does not change the background electric field considerably), positive charges are induced on the channel to keep it at ground potential. Thus, the total positive charge that passes into the leader channel from its grounded end is the sum of the charge necessary to neutralize the leader charge and the induced charge. Let us denote this total charge by Q_r. This is equal to the sum of $|Q_l| + Q_i$, where Q_i is the positive charge induced in the return stroke channel. The magnitude of Q_r is approximately equal to $2 Q_l$. For a given cloud potential, Q_r and its distribution along the leader channel can also be estimated using a CSM. If a relationship between the cloud potential and the peak return stroke current is known, these distributions can be described as a function of the return stroke peak current. The results of such an analysis show that the positive charge deposited by the return stroke on the leader channel and the potential of the cloud V are correlated to each other. The best fit for the data points is given by

$$V = 5.86 \times 10^6 + 2.53 \times 10^7 Q_r - 8.53 \times 10^5 Q_r^2, \qquad (14.2)$$

where V is in volts and Q_r is in Coulombs. Recall that in the preceding equation, Q_r is the total charge deposited by the return stroke on the leader channel. This is the positive charge injected into the leader channel from the ground end during the return stroke. Now, the first return stroke duration is approximately 100 µs, and the positive charge deposited on the leader channel by the return stroke can be estimated by integrating the first return stroke current measured at the base of the channel, say, over the first 100 µs. Note that 100 µs as a reasonable time for the return stroke to deposit a charge on the leader channel is estimated by the following consideration. The average speed of the first return stroke over the first 1–2 km of the channel is approximately 10^8 m/s and decreases with increasing channel length. If one assumes that the average speed over the full length of the return stroke channel is approximately 5×10^7 m/s, then it takes approximately

14.2 Cloud Potential as a Function of the Return Stroke Peak Current

100 μs for the return stroke front to traverse the full length of the channel. This is the reasoning behind selecting the value of 100 μs as a representative time for the return stroke to deposit a positive charge on the leader channel. The speed of subsequent return strokes is somewhat greater than the speed of the first return strokes, and therefore 50 μs is a reasonable time for a subsequent return stroke to deposit a charge on the dart leader channel.

Recall that the total duration of the first return stroke currents measured at the channel base is longer than 100 μs. Similarly, the measured subsequent return stroke currents are longer than 50 μs. The charge transferred after these times is probably associated with the transfer of charge after the return stroke has reached the charge center.

Application of this procedure to the first return stroke currents measured by Berger [4] in lightning strikes to towers on Mount San Salvatore shows that Q_r is strongly correlated with the peak of the first return stroke current. To separate first and subsequent return strokes, let us denote this quantity pertinent to first return strokes by Q_{rf}. The relationship found between Q_{rf} (in Coulombs) and the first return stroke peak current I_{pf} (in kiloamperes) can be described by the following equation:

$$Q_{rf} = 0.062 I_{pf}. \tag{14.3}$$

Similarly, the application of this procedure to subsequent return stroke currents measured shows that Q_r is strongly correlated with the peak of the subsequent return stroke current. Let us denote this quantity by Q_{rs} to indicate that it corresponds to subsequent return strokes. The relationship found between Q_{rs} (in Coulombs) and the subsequent return stroke peak current I_{ps} (in kiloamperes) can be described by the following equation:

$$Q_{rs} = 0.028 I_{ps}. \tag{14.4}$$

Equations 14.3 and 14.4 connect the total positive charge deposited by the return stroke to the peak return stroke currents, whereas Eq. 14.2 relates the same quantity to the cloud potential. The combination of Eqs. 14.2 and 14.3 gives the potential of the cloud as a function of the first return stroke current as

$$V = 5.86 \times 10^6 + 1.569 \times 10^6 I_{pf} - 3.279 \times 10^3 I_{pf}^2. \tag{14.5}$$

With a slight reduction in accuracy at low currents, the data can also be represented by the power plot given by

$$V = 3.0 \times 10^6 I_{pf}^{0.813}. \tag{14.6}$$

These equations are plotted in Fig. 14.3. The preceding equation shows that a stepped leader that generates a typical first return stroke current of 30 kA is associated with a cloud potential of 50 MV. This also shows that as the cloud potential decreases, the prospective return stroke peak current generated by the stepped leaders decreases.

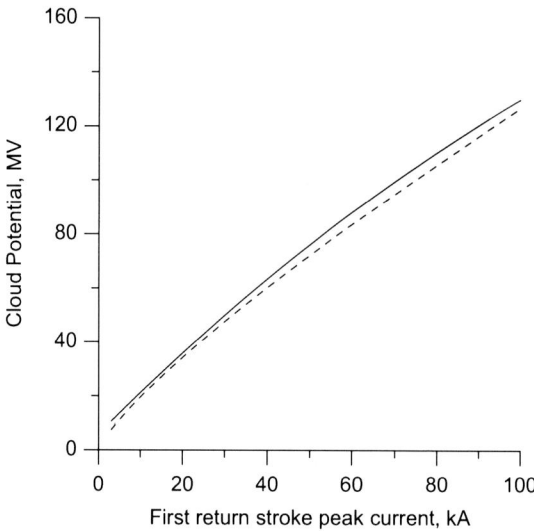

Fig. 14.3 Cloud potential as function of first return stroke peak current, as predicted by Eqs. 14.5 (*solid line*) and 14.6 (*dashed line*). Note that a typical stepped leader that gives rise to a first return stroke current of 30 kA is associated with a cloud potential of approximately 50 MV (Figure created by author)

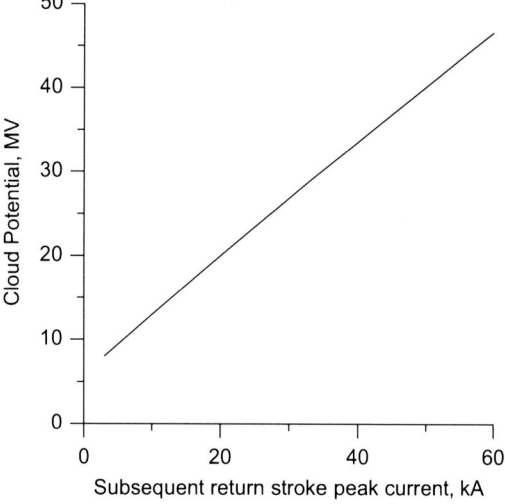

Fig. 14.4 Potential of cloud as function of subsequent return stroke current, as given by Eq. 14.7. Note that a typical dart leader that gives rise to a first return stroke current of 12 kA is associated with a cloud potential of approximately 15 MV (Figure created by author)

Similarly, the combination of Eqs. 14.2 and 14.4 gives the potential of the cloud that exists during subsequent return strokes as a function of subsequent return stroke currents. This potential is given by the following equation:

$$V = 5.86 \times 10^6 + 0.72 \times 10^6 I_{ps} - 0.669 \times 10^3 I_{ps}^2. \tag{14.7}$$

This equation is plotted in Fig. 14.4. The preceding equation shows that a dart leader that generates a typical return stroke current of 12 kA is associated with a cloud potential of 15 MV.

14.3 Analytical Expression for EGM Striking Distance

Equation (14.1) relates the EGM striking distance to the cloud potential. Substituting for the cloud potential as a function of the first return stroke peak current from Eq. 14.5 one obtains the following relationship between the EGM striking distance and the first return stroke peak current:

$$S = 11.7 + 3.14 I_p - 0.656 \times 10^{-2} I_p^2. \tag{14.8}$$

If, instead of Eq. 14.5, Eq. 14.6 is used, then the striking distance becomes

$$S = 6 I_p^{0.813}. \tag{14.9}$$

These two equations are depicted in Fig. 14.5. Note that the striking distance of a 30-kA current is approximately 90 m.

14.4 Distribution of Negative Charge on Stepped Leader and Dart Leader Channel

From the procedure outlined previously it is also possible to estimate the distribution of negative charge along the stepped and dart leader channel and the distribution of the positive charge deposited by the return strokes on the stepped leader and dart leader channels as a function of the return stroke peak current or cloud potential. Results obtained from such an analysis show that the linear charge

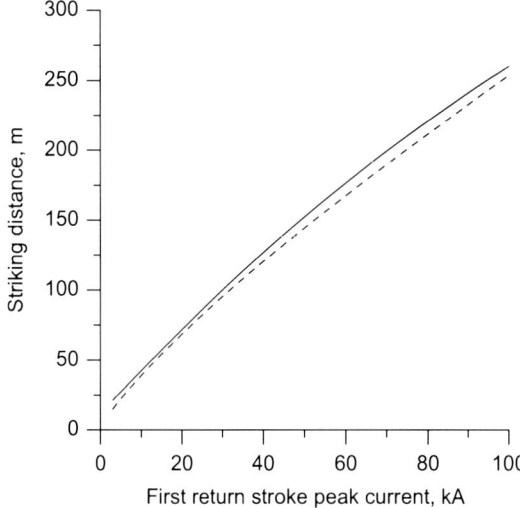

Fig. 14.5 Striking distance as function of prospective first return stroke peak current, as predicted by Eqs. 14.8 (*solid line*) and 14.9 (*dashed line*). Note that the striking distance corresponding to a typical first return stroke current of 30 kA is approximately 100 m (Figure created by author)

distribution (in C/m) on the stepped or dart leader channel when its tip is at a height of z_o above the ground is given as a function of the peak return stroke current by

$$\rho_l(\zeta) = a_o\left(1 - \frac{\zeta}{L-z_o}\right)G(z_o)I_p + \frac{I_p(a+b\zeta)}{1+c\zeta+d\zeta^2}J(z_o), \quad (14.10)$$

$$\zeta = (z - z_o) \quad z > z_o, \quad (14.11)$$

$$G(z_o) = 1 - \frac{z_o}{L}, \quad (14.12)$$

$$J(z_o) = 0.3\alpha + 0.7\beta, \quad (14.13)$$

$$\alpha = \frac{e^{-(z_o-10)}}{75}, \quad (14.14)$$

$$\beta = \left(1 - \frac{z_o}{L}\right), \quad (14.15)$$

where z is the vertical height of the point of observation, L is the total length of the leader channel in meters, and $\rho(\zeta)$ is the charge per unit length of a leader section located at distance ζ from the tip of the leader channel. In the case of stepped leaders, I_p is the first return stroke peak current, $a_o = 1.476 \times 10^{-5}$, $a = 4.857 \times 10^{-5}$, $b = 3.909 \times 10^{-6}$, $c = 0.522$, and $d = 3.73 \times 10^{-3}$. In the case of dart leaders, I_p is the subsequent return stroke peak current, $a_o = 5.09 \times 10^{-6}$, $a = 1.325 \times 10^{-5}$, $b = 7.06 \times 10^{-6}$, $c = 2.089$, and $d = 1.492 \times 10^{-2}$. These charge distributions are depicted in Figs. 14.6 and 14.7 for a 30-kA first return stroke current and for a 12-kA subsequent return stroke current. Note that the preceding equation is valid for $z_o \geq 10$ m. The preceding charge distribution is obtained for $H = 4$ km, but it can also be used for other cloud heights with no significant difference. It is important to point out that, because of the spread in the

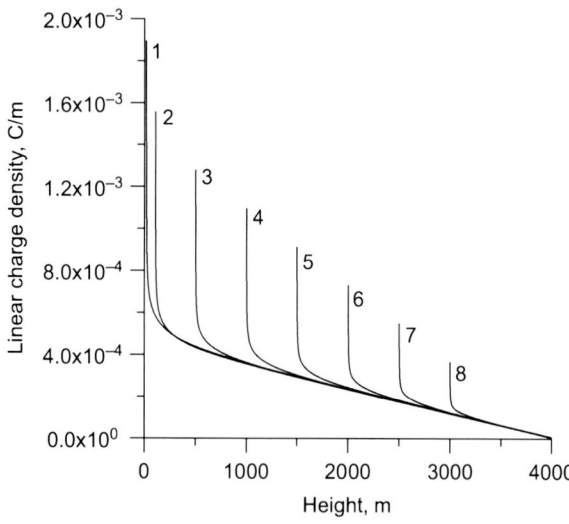

Fig. 14.6 Charge distribution along downward moving stepped leader channel with 30-kA prospective return stroke current when its tip is at different distances from ground: (1) 10 m, (2) 100 m, (3) 500 m, (4) 1,000 m, (5) 1,500 m, (6) 2,000 m, (7) 2,500 m, and (8) 3,000 m. Note that the charge at a given point on the leader channel varies as the leader tip surges ahead of that point (Figure created by author)

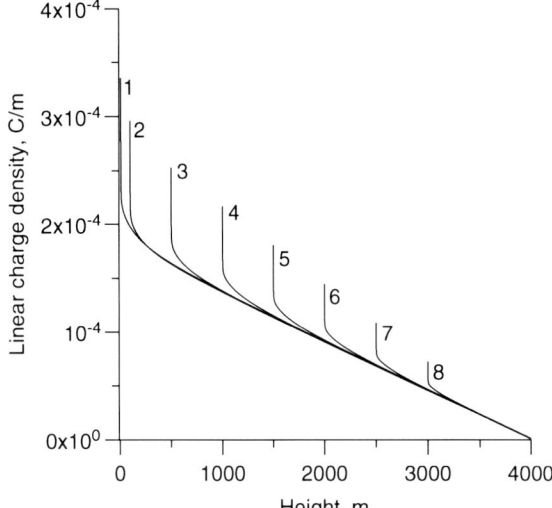

Fig. 14.7 Charge distribution along downward moving dart leader channel with 12-kA prospective return stroke current when its tip is at different distances from ground: (1) 10 m, (2) 100 m, (3) 500 m, (4) 1,000 m, (5) 1,500 m, (6) 2,000 m, (7) 2,500 m, and (8) 3,000 m (Figure created by author)

experimental data used to derive Eqs. 14.3 and 14.4, for a given return stroke current the values of the preceding constants may vary by up to approximately 30 % from one return stroke to another.

Observe that Eq. 14.10 can be written in terms of cloud potential by substituting for I_p in terms of V. Note also that I_p as a function of V can easily be obtained from Eq. 14.5, 14.6, or 14.7 depending on the case of interest. For example, Eq. 14.5 can be numerically inverted to obtain

$$I_{pf} = -3.23 + 5.9 \times 10^{-7} V + 1.46 \times 10^{-15} V^2. \tag{14.16}$$

Substitution of this for I_p in Eq. 14.10 gives the charge distribution of the stepped leader in terms of the cloud potential.

14.5 Distribution of Positive Charge Deposited by Return Stroke on Leader Channel

In the procedure outlined in Sect. 14.2, one can obtain the distribution of the positive charges deposited by the return stroke on the leader channel. Both for first return strokes and subsequent return strokes this distribution is given by

$$\rho_r(z) = a_0 I_p + \frac{I_p(a + bz)}{1 + cz + dz^2}, \tag{14.17}$$

where $\rho(z)$ is the deposited positive charge per unit length at a point located at a height z on the leader channel. The values of the parameters of this equation for first return strokes and subsequent return strokes are given in the previous section. These distributions are plotted in Figs. 14.8 and 14.9 for a 30-kA first return stroke current and for a 12-kA subsequent return stroke current.

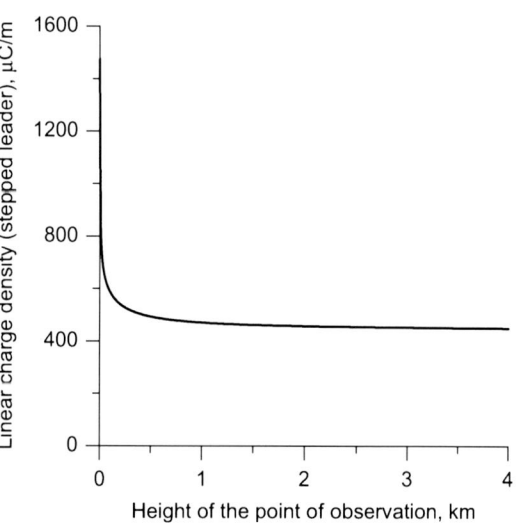

Fig. 14.8 Distribution (linear charge density) of positive charge deposited by 30-kA first return stroke on negative stepped leader channel. Note that, except for the first 500 m or so, the charge deposited by the return stroke on the leader channel is distributed uniformly (Figure created by author)

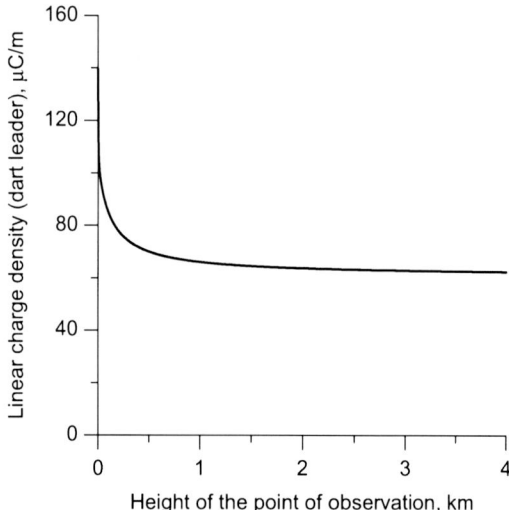

Fig. 14.9 Distribution (linear charge density) of positive charge deposited by 12-kA subsequent return stroke on negative dart leader channel (Figure created by author)

14.6 Energy Dissipation in First Return Strokes

Another interesting exercise that can be done with Eq. 14.10 or 14.11 is to estimate the total energy dissipated by a first return stroke. Now, the charge removed during the first return stroke is given by Eq. 14.3. This charge is transported across a potential difference given by Eq. 14.5 (or 14.6). Thus, the energy dissipated is simply the charge times the potential difference, and the energy dissipated during a return stroke of current I_p (in kA) is given by

$$E = 1.86 \times 10^5 I_p^{1.8}. \tag{14.18}$$

This equation is plotted in Fig. 14.10. According to the results, a typical return stroke of 30 kA dissipates approximately 8.5×10^7 J of energy. Now, since the length of the channel assumed in the analysis is 4 km, a typical return stroke having a current of 30 kA dissipates approximately 2×10^4 J/m of energy. Note that a certain fraction of this energy is dissipated in the leader stage as the leader brings down the charge, neutralized later by the return stroke, closer to the ground.

Recall that in evaluating this energy the charge dissipated by the first return stroke over the first 100 µs is used. Thus, the foregoing result is valid for energy dissipation by first return strokes over the first 100 µs.

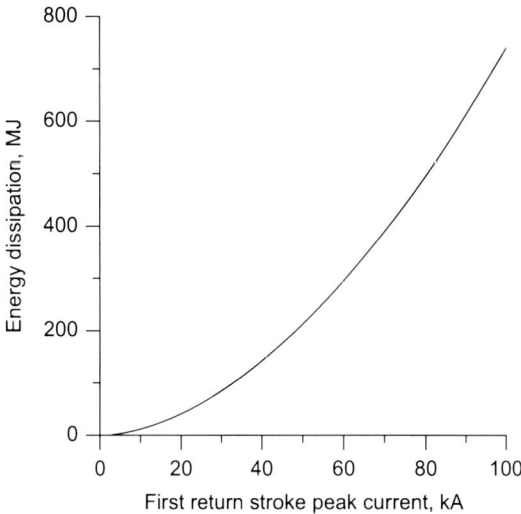

Fig. 14.10 Energy dissipation in first return strokes as function of peak current, as predicted by Eq. 14.18. According to this diagram, a typical first return stroke of a 30-kA current dissipates approximately 85 MJ of energy (Figure created by author)

14.7 Energy Dissipated by Subsequent Return Strokes

The charge dissipated by subsequent return strokes is given by Eq. 14.4, and this charge is transported across a potential as given by Eq. 14.7. Thus, the energy dissipated during a subsequent return stroke having a peak current I_p (in kA) is

$$E = 0.164 \times 10^6 \times I_p + 0.02 \times 10^6 I_p^2 - 0.019 \times 10^3 I_p^3. \qquad (14.19)$$

This equation is plotted in Fig. 14.11. It shows that a typical subsequent return stroke of 12 kA of current will release 4.59 MJ of energy. For a 4-km channel this generates approximately 1 kJ/m of energy. The comment concerning energy dissipation during the leader stage made in the previous section is also applicable here. Recall again that in evaluating this energy the charge dissipated by subsequent return strokes over the first 50 μs is used. Thus the preceding result is valid for energy dissipation by subsequent return strokes over the first 50 μs.

Final Comment It is important to point out that the estimation of charge distributions, cloud potential, and energy dissipation presented in this chapter are valid under two approximations. The first one is that the electric field below the charge center of the cloud is approximately uniform. Of course, the same procedure can be used to obtain charge distributions even in the case where the electric field below the charge center is nonuniform. The second approximation or assumption is that much of the charge deposited by a return stroke on the stepped leader and the dart leader is injected into the channel from its base at ground level in a time on the order of 100 μs and 50 μs, respectively.

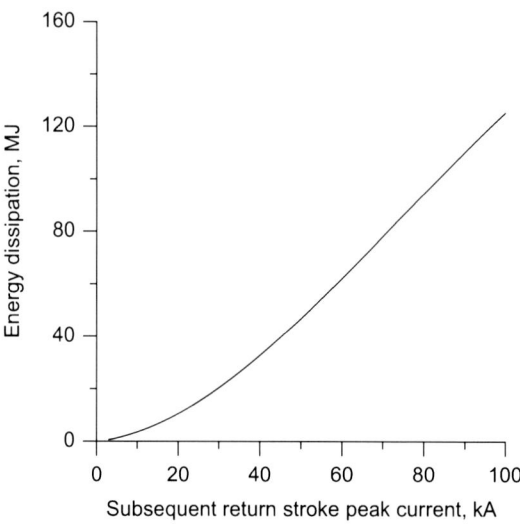

Fig. 14.11 Energy dissipation in subsequent return strokes as function of peak current, as predicted by Eq. 14.19. According to this diagram, a typical subsequent return stroke of a 12-kA current dissipates approximately 4.6 MJ of energy (Figure created by author)

References

1. Cooray V, Rakov V, Theethayi N (2007) The lightning striking distance – revisited. J Electrost 65(5–6):296–306
2. Willett JC, Davies DA, Laroche P (1999) An experimental study of positive leaders initiating rocket triggered lightning. Atmos Res 51:189–219
3. Singer H, Steinbigler H, Weiss P (1968) A charge simulation method for the calculation of high voltage fields. Trans IEEE PAS-93:1160–1168
4. Berger K (1967) Novel observations of lightning discharges: results of research on Mount San Salvatore. J Franklin Inst 283:478–525

Chapter 15
Direct and Indirect Effects of Lightning Flashes

15.1 Introduction

A lightning flash can interact with any object either electrically or mechanically or both. In this chapter, both direct and indirect effects of lightning flashes are described. Following a description of the basic physical phenomena associated with direct and indirect lightning strikes, these effects are illustrated here by considering the effects of lightning strikes on residential houses, wind turbines, trees, airplanes, and power lines.

15.2 Generation of Thunder

During a return stroke, the lightning channel (approximately 1 cm radius; Sect. 15.3) through which the return stroke current propagates is heated very rapidly to approximately 25,000–30,000 K (how this is measured is described in the next section). This heating of the channel causes a rapid increase in the pressure of the channel to approximately ten times above atmospheric pressure. This rapid increase in the pressure creates a shock wave in air, and after traveling a few meters from the channel, it transforms into a sound wave. This sound wave is called *thunder*.

Since there are many return strokes in a lightning flash in general, the process of heating and generation of thunder or the shock wave takes place repeatedly during each return stroke. Now, a typical lightning flash may last for around 200–300 ms and in extreme cases for around 1 s. On the other hand, the generation of thunder from the return stroke channel (or the heating of the channel) takes place almost instantaneously along the whole channel during the return stroke. So then why does the thunder generated by a lightning flash last much longer than 1 s? Because sound travels in air at a speed of approximately 330 m/s. Suppose that an observer is located 1 km from the point of a lightning strike. The entire lightning channel is

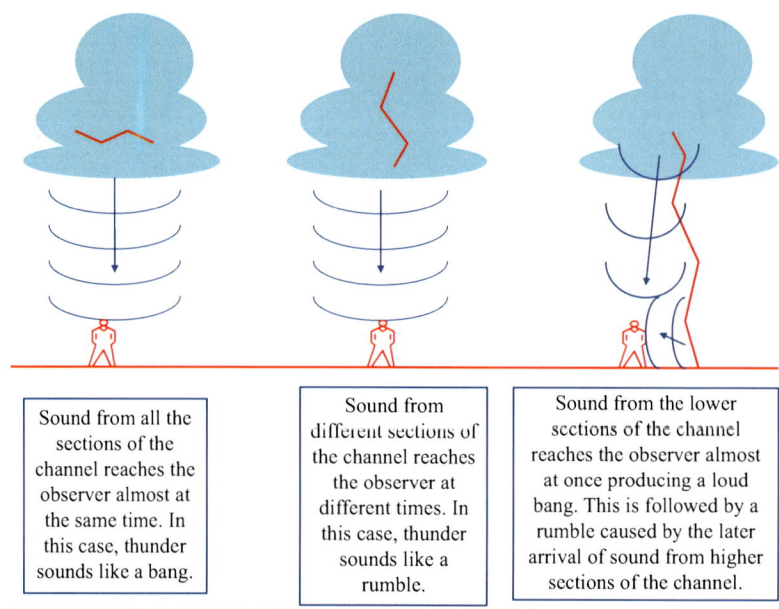

Fig. 15.1 Lightning produces a rich variety of sounds (Figure created by author)

approximately 7,000 m long, and therefore the top of the channel is approximately 7,071 m from the observer. Now, since sound travels at a rate of 330 m/s, the sound from the bottom of the channel reaches the observer in approximately 3 s (i.e., 1,000/330) and the sound from the top of the channel reaches the observer in approximately 23.6 s (i.e. 7071/330) (Fig. 15.1). So thunder lasts for approximately 20 (23 – 3) s. In fact, the duration of the thunder can be used to obtain a rough length of the lightning channel. If the duration of thunder is measured in seconds then dividing that number by 3 will give the minimum length of a lightning channel in kilometers. The actual length of the channel is larger than the length one would estimate by this method. Of course, this approximation is valid if the lightning channel is vertical. In reality, with large horizontal channel sections inside clouds, lightning channel geometry is much more complex. This makes the connection between the duration of thunder and the channel length much more complicated than the simple relationship given previously.

The thunder signal from a lightning flash is not just a monotonic sound but contains peals, claps, and rumble. This rich variety of sound modulations in thunder is caused by the branches and the tortuous nature of the lightning channel. Thunder generated by branches and other channel sections located at different orientations in space reach the observer at different times. Sometimes the thunder generated by all the points along a channel section may reach the observer simultaneously, producing a loud clap. In other cases, sound from different channel sections may reach the observer at different times, producing a rumble. Furthermore, the strength of the sound decreases as it moves in air. If an observer is located close to a lightning strike, the sound from the bottom of the channel, which is located close to the observer, produces a loud clap and the sound from the upper sections of the channel produce a diminishing rumble. If an observer is located very close to a lightning flash, say 100 m or less, thunder sounds like a crack or hissing sound followed by a loud bang. The crack and hissing sound is produced probably by intense corona and connecting leaders generated by objects at ground level. Usually observers can only hear the thunder from a lightning flash when they are located within approximately 20 km away from it.

During a lightning strike to an object, the lightning current may follow the path of least resistance to the ground. This path may contain moisture. An example is when a lightning flash strikes a tree. The lightning current may flow in the inner part of the tree where there is a high moisture content. The rapid heating of the water causes a rapid expansion, leading to an explosion that splits the tree into parts. The same effect can be observed in cases where lightning strike grounded structures. When that happens, the current may travel through building materials such as clay or wet concrete, causing the water vapor to expand explosively, leading to the ejection of material.

15.3 Temperature of a Lightning Channel and How It Is Measured

As mentioned earlier, thunder is generated by the rapid heating of air to approximately 25,000–30,000 K during a lightning flash. But how do we measure this temperature? The procedure used by scientists to measure the temperature of a lightning channel is as follows. First, the light signal generated by the lightning channel is separated into spectral components using a specially designed diffraction grating, and the resulting spectrum is recorded either digitally or on photographic paper. From this spectrum two spectral lines located very close to each other are selected and their intensity measured. Multiplets of doubly ionized nitrogen atoms centered at 399.5, 404.1, 443.3, 463.0, and 567.9 nm had been used for this purpose [1]. If the lightning channel is at thermodynamic equilibrium, then the ratio between the two intensities is related to the temperature of the channel. Thus, by measuring the intensity of two very closely spaced spectral lines (i.e., multiplets of a spectral line), it is possible to estimate the temperature of the lightning channel. In the foregoing statements, *thermodynamic equilibrium* means that all particles in the channel, i.e., neutral atoms, ions, and electrons, are at the same temperature. This is not exactly true in a lightning channel because electrons, which gain energy from electric fields, are usually at a higher temperature than ions and neutral atoms.

On the other hand, the electron densities that occur in a lightning channel can be obtained by measuring the Stark broadening of the spectral lines without making any assumptions concerning thermodynamic equilibrium. Stark broadening of spectral lines is a consequence of the interactions of the electric fields inside a channel with atoms that radiate the spectral line. The electric field experienced by the atoms is related to the electron density in the channel. Thus, measuring the Stark broadening of spectral lines makes it possible to obtain the electron density in a channel. Using this procedure the free electron densities in lightning channels were estimated to be in the range of 10^{17} to 5×10^{18} e/cm^3 [1].

15.4 Thickness of a Lightning Channel

The thickness of a lightning channel can be estimated by photographing it. However, this method always overestimates the thickness of the channel because of the scattering of light on the photographic plate. In the pioneering days of lightning research, researchers estimated the size of lightning channels by placing fiberglass screens across the path of the lightning and measuring the holes created by the lightning in it (Fig. 15.2) [2]. The diameter of a lightning channel can also be calculated using theories on electrical discharges [3]. On the basis of such analyses scientists have estimated that a hot lightning channel is approximately 1 cm in radius.

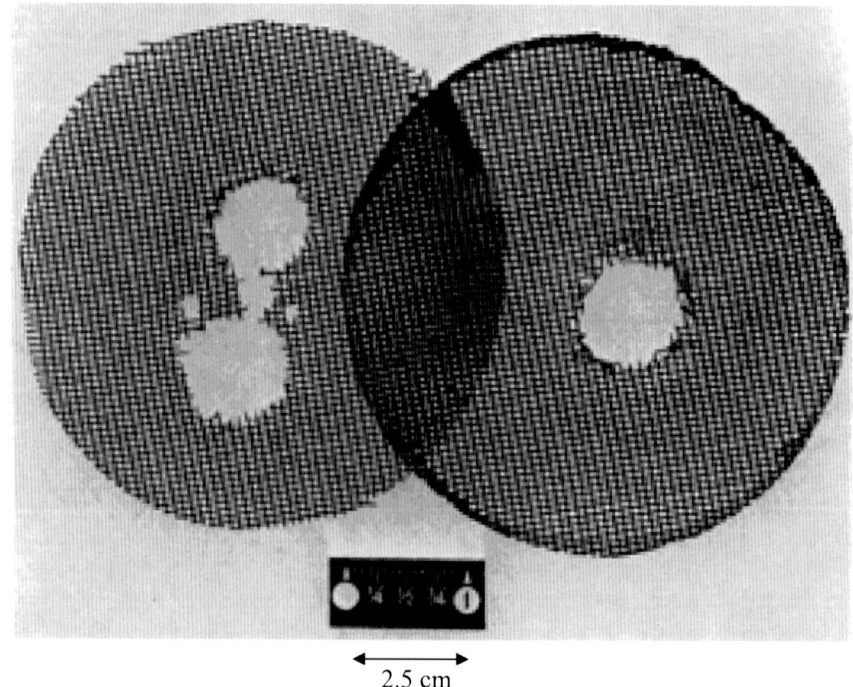

Fig. 15.2 Holes melted in two fiberglass screens by lightning. Four strokes passed through left screen and one through right screen [2]. The size of the holes gives a rough estimation of the thickness of the hot core of the return stroke channel

15.5 Melting of Material

In Chap. 8 was given a description of how to evaluate the energy transferred to a metal in contact with a lightning channel. This energy is directly proportional to the charge transported by the lightning channel and the cathode voltage V_c, which is approximately 15–20 V. Thus, the heat energy, W, that is transferred to the metal across the cathode layer is given by

$$W = QV_c, \quad (15.1)$$

where Q is the charge dissipated in the process under consideration. This heat leads to the melting of metal when the lightning channel is in contact with a metal surface. The amount of metal volume V (in m³) being melted is given by the formula [4]

$$V = (W/\gamma)\frac{1}{c_w\theta_s + c_s}, \quad (15.2)$$

where γ is the density of the material, c_w is the specific heat capacity (in J/kg.K), θ_s is the temperature at which the material melts, and c_s is the specific heat of melting (in Joules per kilogram). These parameters for different types of metal that are used in lightning protection practice are given in Table 15.1. Using the preceding equation and the parameters given in Table 15.1 it is possible to calculate the volume of material melted once the charge Q is given.

Consider a lightning flash transferring 10 C of charge in contact with an aluminum surface. If $V_c = 20$ V, then 200 J of energy is transferred to the surface. For aluminum, $\gamma = 2{,}700$, $c_w = 908$, $\theta_s = 685$, and $c_s = 397 \times 10^3$. Using these parameters we find that the volume of material melted is 73 mm³. That is, each coulomb of charge will melt a volume of 7.3 mm³ of aluminum.

Table 15.1 Various parameters pertinent to calculation of heat dissipation and melting effects during lightning strikes

Parameter	Unit	Aluminum	Copper	Iron
γ	kg/m³	2,700	8,920	7,700
θ_s	°C	685	1,080	1,530
c_s	J/kg	397×10^3	209×10^3	272×10^3
c_w	J/(kg K)	908	385	469
ρ	Ωm	29.0×10^{-9}	17.8×10^{-9}	120×10^{-9}
α	1/K	4×10^{-3}	3.92×10^{-3}	6.5×10^{-3}

15.6 Heating of Material as Lightning Current Passes Through an Object

Consider a lightning strike to an object having a resistance R. The amount of energy dissipated in the object as the lightning current passes through it is given by (Chap. 8)

$$W = R \int_0^\infty i^2(t)\,dt. \qquad (15.3)$$

The release of this energy in the material causes its temperature to rise. The increase in temperature can be calculated easily for an object in the form of a cylinder. Assuming a cylindrical geometry, the increase in the temperature (in degrees Kelvin) of 1 m of material is given by [4]

$$\Delta\theta = \frac{1}{\alpha}\left\{\exp\frac{\alpha\rho\int i^2(t)\,dt}{a^2\gamma c_w} - 1\right\}, \qquad (15.4)$$

where α is the temperature coefficient of the resistance, ρ is the resistivity of the material of the conductor, and a is the cross-sectional area of the conductor. The preceding equation shows that the thinner the conductor, the larger the rise in temperature. If the action integral of the current (i.e., the integral in Eq. 15.3) is 10^6 A^{2}s, then the rise in temperature of a 16 mm^2 copper conductor is 21 K. If the temperature rises above the melting point, it could melt and vaporize explosively. Thus, in selecting the diameter of conductors in lighting protection systems it is important to select the diameter of the conductors such that heating effects can be neglected.

15.7 Attraction of Two Current-Carrying Conductors

During a lightning strike parts of the lightning current might propagate along different conductors. This flow of current along conductors located in the vicinity of each other can cause mechanical stresses on the conductors and, in the worst case, dislodge them from their secured positions. Assume, for example, that the return stroke current is divided into two parts and flows along two parallel conductors (Fig. 15.3a). Denote the separation between them by s, and assume that the radii of the conductors are much smaller than s. Then the force (in Newtons) experienced by per unit length of the two conductors at any given time t is

15.7 Attraction of Two Current-Carrying Conductors

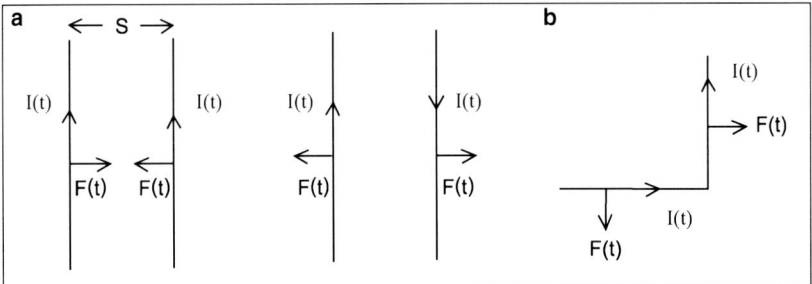

Fig. 15.3 (a) When a lightning current passes through two conductors, there is a force between them. The force is attractive if the current is flowing in the same direction along the two conductors and repulsive if the current is flowing in the opposite direction. (b) When a lightning current flows along a bent conductor, there is a force on the two branches. The force is such that it will act in a direction to straighten the conductor (Figure created by author)

$$F(t) = \frac{\mu_0}{2\pi s} I_1(t) I_2(t). \tag{15.5}$$

If the current is flowing in the same direction along the conductors, then the force is attractive (first diagram of Fig. 15.3a). If the current is flowing in the two conductors in opposite directions, then there will be a repulsive force between them (second diagram of Fig. 15.3a). If the return stroke current $i(t)$ is divided equally among the conductors, then $i_1(t) = i_2(t) = i(t)/2$, and the force between them can be written as

$$F(t) = \frac{\mu_0}{8\pi s} i^2(t). \tag{15.6}$$

Then the impulse I acting between the two conductors because of the flow of the return stroke current through them is

$$I = \frac{\mu_0}{8\pi s} \int_0^{t_d} i^2(t) \, dt, \tag{15.7}$$

where the upper limit of integration t_d is the duration of the return stroke current. This impulse is responsible for the mechanical action or movement of the two conductors. Observe that the impulse is proportional to the action integral of the return stroke current. It is important to remember that the force acts not only on parallel conductors but also when the two current paths form an angle (Fig. 15.3b). In this case, the force acting on the two branches is such that it will try to make them straighter. For this reason, it is necessary to avoid sharp bends in the down conductors used in lightning protection.

15.8 Induction of Voltages Due to Lightning-Generated Magnetic Fields

Consider a lightning strike to a lightning conductor located in a building, as shown in Fig. 15.4a. The current passing through this conductor generates a time-varying magnetic field that can induce voltages in any metallic loop located in its vicinity. To simplify the mathematics, assume that the conductor through which the lightning current is flowing is infinitely long. A conducting loop is located in the vicinity of this conductor, as shown in Fig. 15.4b. The magnetic field generated by the lightning current flow at a radial distance of r from the conductor is

$$B = \frac{\mu_0 i(t)}{2\pi r}. \tag{15.8}$$

The total magnetic flux passing through the loop can be calculated by dividing the loop into infinitesimal strips (dr in width) parallel to the conductor and summing the contribution from all the strips by integration. Thus, the total flux of the magnetic field passing through the square loop of width b is

$$\varphi = \int_{r1}^{r2} \frac{\mu_0 i(t)}{2\pi r} b \, dr. \tag{15.9}$$

After performing the integral we obtain

$$\varphi = \frac{\mu_0 b i(t)}{2\pi} \ln\left(\frac{r_2}{r_1}\right). \tag{15.10}$$

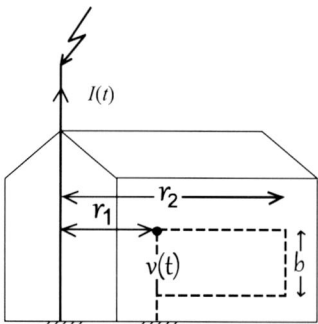

Fig. 15.4 The A lightning current passing through a conductor can induce voltages in conducting loops located nearby. This example is used in the text to evaluate the voltage induced in a conducting loop (*dotted lines*) by the time-varying magnetic field produced by the current flowing in the lightning down conductor (Figure created by author)

15.9 Induction Due to Electric Field

The voltage induced in the loop is then given by

$$V = \frac{\mu_0}{2\pi} \frac{di(t)}{dt} b \ln\left(\frac{r_2}{r_1}\right). \quad (15.11)$$

The peak voltage induced in the loop is given by

$$V_m = \frac{\mu_0 b \ln(r_2/r_1)}{2\pi} \left(\frac{di(t)}{dt}\right)_{max}. \quad (15.12)$$

Thus, the maximum voltage is proportional to the maximum derivative of the current. If the resistance of the loop is equal to R, then the maximum current that will flow through the loop is V_m/R. The maximum voltage induced in the loop can be written as

$$V_m = M \left(\frac{di(t)}{dt}\right)_{max}. \quad (15.13)$$

The parameter M is the mutual inductance between the lightning channel or the conductor through which the current is flowing and the loop. It depends on the geometry of the loop and its location. In the example shown in the figure, M is given by

$$M = \frac{\mu_0}{2\pi} b \ln \frac{r_2}{r_1}. \quad (15.14)$$

Thus, a 1 m² square loop located at a distance of 10 m from a conductor carrying a return stroke current whose maximum rate of change is 100 kA/μs will generate a peak voltage of approximately 2 kV in the loop. It is important to remember that the loops in the vicinity of a lightning path can be formed in various ways. For example, in the case of a person located as shown in Fig. 15.5, a voltage is generated between the person's head and the lightning current path due to induction caused by the magnetic field passing through area A.

15.9 Induction Due to Electric Field

It is not only the magnetic field generated by lightning flashes that can generate voltages and currents because of induction. An electric field can do the same. To illustrate this, consider a small metal sphere located at height h above a perfectly

Fig. 15.5 When a human is located in the vicinity of a conductor struck by lightning, there is a voltage difference between the head and the conductor $U(t)$ because of the magnetic flux passing through the area marked A between the human and the conductor. This voltage is proportional to $A(dI(t)/dt)$, where $dI(t)/dt$ is the rate of change of the current (Figure created by author)

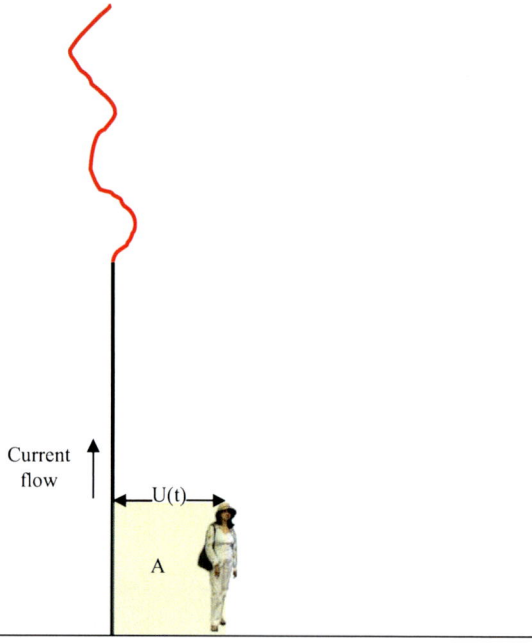

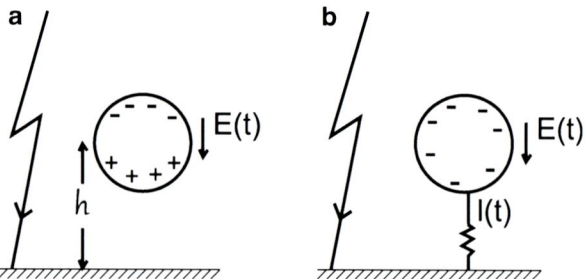

Fig. 15.6 (**a**) Isolated conducting sphere located in an electric field of a lightning flash. Because of action of electric field, charges are displaced on the sphere while keeping the net charge equal to zero. (**b**) If the sphere is connected to the ground, the positive charge flows to the ground (of course, the polarity of this charge depends on the direction of the background electric field), leaving a net negative charge on the sphere. As the electric field varies with time, the net charge on the sphere varies and there is a current flow across the resistor. In this way, an electric field can induce a current in conductors connected to the ground (Figure created by author)

conducting ground at a certain distance from a lightning strike (Fig. 15.6). Let the electric field produced by the lightning flash or the return stroke be $E(t)$. If the sphere is isolated (Fig. 15.6a), this electric field will change the charge distribution on the sphere while keeping the net charge equal to zero. Now, if the sphere is grounded while under the influence of the electric field (Fig. 15.6b), positive charge will go to the ground, leaving behind a net negative charge on the sphere. The charge Q on the sphere is given by

$$Q = hE(t)C, \qquad (15.15)$$

where C is the capacitance of the sphere. This depends on the dimension of the sphere and the height where it is located. In the preceding equation, the influence of the conductor connecting the sphere to the ground is neglected. Moreover, it is assumed that the radius of the sphere is much less than h. As the background electric field changes, the charge on the sphere changes, leading to a current flow along the wire to the ground. This current is given by

$$i(t) = h\frac{dE(t)}{dt}C. \qquad (15.16)$$

Thus, the electric field causes an induction of current in the conductor connecting the sphere to the ground. Of course, in writing down the preceding equation it is assumed that the capacitance C of the sphere and the resistance R of the conductor satisfy the inequality $RC \ll \tau$, where τ is the time scale where significant change in the electric field takes place. Even if this condition is not satisfied, there will still be a current flowing along the conductor, but this current flow will not follow the rate of change of the electric field exactly as written in the preceding equation (Chap. 9). Note that in reality the object need not be a sphere. It could simply be a conductor connected to the ground or any other metal object connected to the ground. Thus, the voltages and currents that are induced during lightning strikes are caused by both the magnetic and electric fields.

15.10 Energy Dissipation During a Lightning Strike

In Chap. 14 it was estimated that the potential of a thundercloud that gives rise to a typical first return stroke with a 30-kA current is approximately 50 MV and its value may decrease to approximately 20 MV during subsequent return strokes. Let us assume therefore that an average potential of approximately 35 MV exists during a typical lightning flash. During a typical lightning flash approximately 7.5 C of charge is transferred from cloud to ground [5]. When this charge is transferred across a potential of 35 MV, the amount of energy released is approximately 2.5×10^8 J. This energy will be released along the lightning channel, and if the length of the lightning channel is approximately 5 km, then the energy released per meter is approximately 5×10^4 J/m. Some part of this energy will be released during leader processes and the rest during the return strokes. In Chap. 14 it was estimated that during the first 100 µs of a typical first return stroke the energy released is approximately 10^4 J/m. This energy is released explosively because it is released within 100 μs. Thus, the average power associated with this energy release is on the order of 10^8 W/m. The peak power dissipation can be approximately ten times greater than this average value. This explosive release of energy during the first return stroke can cause mechanical effects in objects located in the vicinity of the lightning channel.

One interesting question that occupies the lightning research community is whether this energy can be utilized. Let us look for an answer to this question. Each second approximately 100 lightning flashes take place in the atmosphere. If each lightning flash dissipates approximately 2.5×10^8 J of energy on average, then the total energy released per second or the total power generated by lightning flashes is 2.5×10^{10} W. The amount of energy released each year is approximately 200 TWh. This is a considerable amount of energy. To put that in perspective, in Sweden the yearly consumption of electrical energy is approximately 220 TWh. However, if one tries to acquire this energy, several obstacles must be overcome. First, recall that lightning flashes are distributed around the globe and only a small fraction of them is available in a given region. Second, most of the energy from those flashes is dissipated in heating the air in the lightning channel and in creating a shock wave. Only a small fraction of the energy is available at ground level. However, if all that energy were somehow collected, then the energy distributed in each square kilometer of the Earth would be approximately 50 W. Recall that most of this energy is dissipated in the air and only a fraction of it is available at ground level. Of course, this estimate depends on the assumed potential of clouds and the amount of charge dissipated in lightning flashes. However, the estimate is sufficient to illustrate that attempts to generate electric power by storing the energy of lightning flashes is not very viable economically.

15.11 Effects of Direct Lightning Strikes on Structures

The effects of direct strikes on various structures may manifest in various ways depending on the geometrical and electrical characteristics of the given structure. Let us consider several typical structures and try to understand the possible consequences.

15.11.1 *Direct Lightning Strike to a House*

A direct strike to a residential house may cause damage depending on the type of construction and material used. Most of the damage can be avoided by implementing a lightning protection system. Usually, if a house is not protected by a lightning protection system, then a strike will take place to the roof of the house. If the roof is made of flammable material, then the strike can set it on fire, especially if the lightning flash contains a long continuing current. Long continuing currents keep hot channels in contact with roofing materials for a longer time period and therefore can set fire to the material more easily than a lightning strike without a continuing current. If the roof is metallic, then, depending on the thickness of the metallic sheet, the lightning flash may create a hole in the metal, igniting the material located underneath. If the metal roof is not grounded, then there could be sparks jumping directly from the roof to the interior of the house. With tiled roofs, the lightning current may creep through the space between the tiles, again creating a shock wave that could explosively displace them. If the roof is concrete, parts of the concrete could be damaged as a result of water vaporization caused by sparks formed inside cracks in the concrete.

Since modern houses are equipped with electricity, the lightning current may enter into the electrical system. In the worst-case scenario, the current may flash over across phase and ground wires of the electrical system, creating a spark. If the power system is not equipped with a circuit breaker (though circuit breakers can also be fused together by lightning currents, making them ineffective), the spark thus created may not extinguish itself after the cessation of the lightning current because it will be fed with an alternating current by the power line voltage. This long-lasting spark may set fire to any nearby flammable material.

The large currents flowing in the electrical system may destroy most of the electrical devices connected to the power system. These currents may also destroy (sometimes explosively) part of the electrical system itself, for example, switch boxes. They may also cause power surges in other systems not directly connected to the power system through electrical or magnetic induction. Thus, even an electrical apparatus not connected to the power system may still experience an induced voltage due to induction.

A lightning current can cause overvoltages in other systems of conductors such as water pipes, and their voltage can rise to dangerous levels, causing sparks to jump to other objects in their vicinity during the lightning strike, depending on the grounding conditions.

15.11.2 Direct Strike to Trees

In the countryside, trees generally are the highest objects in the area. No wonder, then, that most lightning strikes on land hit trees. Forest fires are the common result. When the lightning stepped leader gets close to the ground, trees, being the highest objects in the vicinity, "reach out" through connecting leaders to meet the downward moving stepped leader. The winning tree of this competition is decided by its height and the conductivity (the ability to transport electric current) of the tree's wood. The conductivity of the wood depends mostly on its moisture content.

The taller the tree and the higher the conductivity of its wood, the greater its chances of becoming attached to the lightning flash. The ability of a tree to attract a lightning flash may, to some extent, also depend on its shape. The reason for this is that the concentration of positive charge that builds up at the top of the tree, and hence the field enhancement when the stepped leader comes down, depends on the shape of the tree. The higher the field enhancement, the easier it is to launch a connecting leader. The ability to launch a connecting leader may also depend on the type of soil on which the tree stands and on its system of roots. The current necessary for the connecting leader is ultimately provided by the soil, and if the soil is moist and conducting, it will facilitate the launching of a connecting leader by the tree. Furthermore, even if the surface soil is not conducting, the roots of some trees may penetrate deep into the ground, where there maybe highly conducting moist layers of soil.

When a lightning flash strikes a tree, a current of the order of 30,000 A on average passes through the tree to the ground. The type of damage sustained by the tree depends on the path taken by this current in its journey toward the ground [6, 7]. The path of least impedance (or resistance) offered by a tree for the path of passage of a lightning current is decided by the way the moisture is distributed within the tree trunk. Usually, the cambium layer (Fig. 15.7) of wood located directly beneath the bark contains a high percentage of moisture, and it therefore provides the path of least resistance for the current to propagate. When the lightning current flows along this path, it heats up this moisture and it vaporizes. As moisture vaporizes to form steam, it expands; in the case of lightning-induced expansion, this happens explosively, blasting the bark from the tree (Fig. 15.8a). Sometimes a striplike furrow can be observed spiraling down the trunk of the tree. In other cases, the bark may be thoroughly soaked with rain water and its surface may provide the least resistive path to the lightning current. Then the current may pass along the surface of the tree, causing little damage other than superficial bark flaking. On the other hand, if the bark is dry and the sapwood (i.e., the layer of wood beneath the

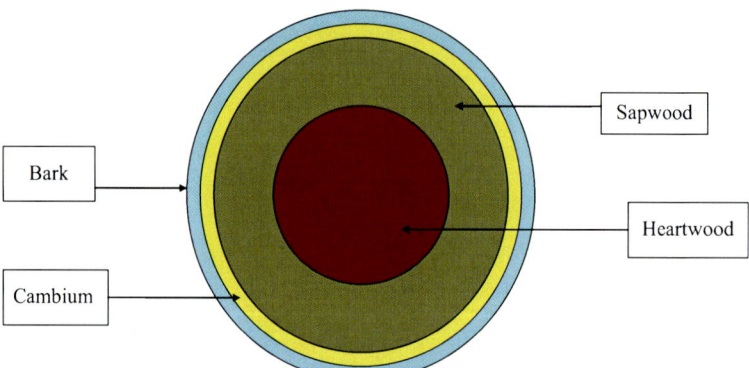

Fig. 15.7 Lightning can damage trees in various ways. A cross section of a tree is shown in the diagram. The damage suffered by a tree during a lightning strike depends on the path taken by the current. If the bark is thoroughly soaked with water, the current may flow along the surface of the tree without causing much damage. If the bark is dry, the current flows through the moist cambium layer. The rapid vaporization of the water as the current flows may cause the bark to split. If the sapwood of the tree contains much moisture, the current may flow deep in the tree instead and the rapid vaporization of water may shatter it (Figure created by author)

15.11 Effects of Direct Lightning Strikes on Structures

Fig. 15.8 (a) A tree where the bark is damaged by a lightning strike. (b) A tree completely shattered by a lightning strike (Adapted from [6])

cambium) contains a lot of moisture, the lightning current may flow deep in the trunk of the tree. In this case, the explosive expansion of moisture may shatter the tree completely (Fig. 15.8b).

There are instances where a single lightning flash can damage several trees [7]. In countries like Sri Lanka, India and Malaysia, lightning flashes cause significant damage in coconut plantations. A coconut tree may take up to 20 years to mature and peak its coconut production. A single lightning flash can destroy many fruit bearing coconut trees causing significant economical losses to the farmers. Figure 15.9 shows the locations of 16 fruit bearing coconut trees (about 10–20 m high) killed by a single lightning flash. The lightning flash struck the tree at the center (marked 0) and killed all the trees located around it up to a distance of 71 feet. The distance to each tree from the tree struck by lightning is given in feet in the diagram. There are three possible ways that several trees can be damaged by a single lightning flash. The first possibility is as follows. As mentioned previously, when the stepped leader approaches the ground, several trees compete to make the connection between the downward moving stepped leader and the ground. These connecting leaders, even if they are unsuccessful in capturing the lightning flash, may carry several tens of Amperes of current, which may be strong enough to cause damage to these trees. Moreover, during the return stroke, the currents in the unsuccessful connecting leaders may rise momentarily to several kiloamperes during back flashover caused by the sudden reduction in the background electric field [8]. In this way, several other trees in the vicinity of a tree that is being struck by a lightning flash could be damaged. The second possibility is as follows. As explained in an earlier chapter, lightning flashes consist of not one but several return strokes. In general, downward moving dart leaders follow the same path taken by the stepped leader and, as a consequence, all return strokes end up at the same point. However,

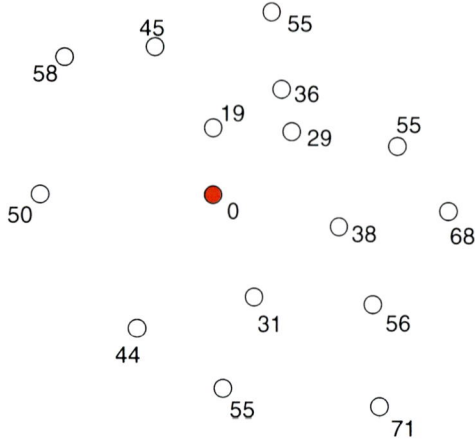

Fig. 15.9 Location of 16 coconut trees killed by a single lightning flash. The distances to the trees from the tree struck by the lightning flash (marked 0) are given in feet in the figure (Diagram courtesy of Mr. Chandra Fonseka)

sometimes the dart leaders deviate from the path taken by the previous leader. In these cases, different return strokes in a single flash may end up on different trees, thereby inducing damage in several of them. The third possibility is as follows. When a lighting flash strikes a tree, the current flows along the tree and, when it reaches the base of the tree, flows radially out into the soil. This ground current can damage the roots of neighboring trees. In the case of the first and third scenarios, i.e., damage caused by connecting leaders and ground currents, the damage may not be visible immediately. But it could weaken the tree's defense system against insects and disease without causing immediate damage, making it more vulnerable to subsequent damage from disease or insects. Its eventual death may be caused by this secondary damage.

A significant fraction of forest fires on Earth are caused by lightning. In the United States approximately 50 % of forest fires are caused by lightning flashes. Figure 15.10 separates the forest fires that occurred in Everglades National Park by cause. Note that approximately 50 % of the fire damage is caused by lightning, and these events are concentrated in the summer months when lightning frequency is high. It is important to note, however, that not all lightning flashes that strike forests cause fires. The probability of a forest fire increases with the duration of the continuing current when other parameters remain the same. In the case of lightning flashes without continuing currents, pieces of wood are heated during return strokes, but the current ceases quickly and the temperature of the lightning channel decreases. To set fire to a piece of wood, the flame (i.e., the lightning channel) must be in contact with the wood for a considerable amount of time. In the case of lightning with long continuing currents, the wood is heated to a high temperature and kept in that state for a long time, thanks to the long duration of the current, so it starts burning. Once the burning process is under way, it can progress without the help of the so-called lightning flame.

15.11 Effects of Direct Lightning Strikes on Structures

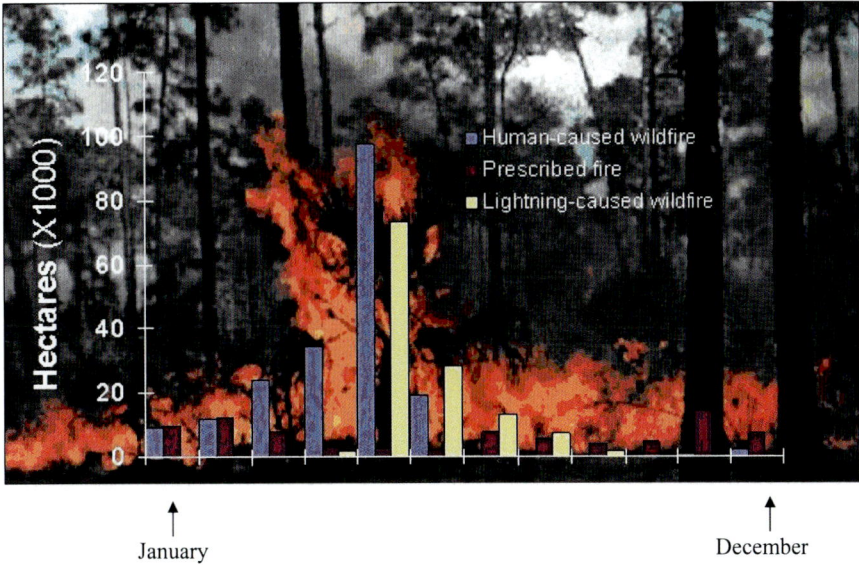

Fig. 15.10 Statistics of area of forest burned in Everglades National Park in the United States as a result of various causes. Note that lightning-caused fires take place during the summer time. http://sofia.usgs.gov/geer/2000/posters/fire/, 2014

15.11.3 Direct Strike to an Airplane

Statistics show that every commercial airplane is struck by lightning once every 3,000 h of flight, or approximately once a year. The infrequency of lightning-caused accidents at present is evidence that present-day airplanes can withstand lightning strikes without experiencing much damage. Of course, from time to time minor electronic problems are experienced during some lightning strikes, and in these cases the planes are diverted to the closest airport for safety reasons.

There are two ways for an airplane to be struck by lightning. The first one is that it intercepts an ongoing lightning strike. For this to happen the airplane must be located in the right place and at the right time to intercept a stepped leader that is moving toward the ground. Such an occurrence is rather rare. For example, at altitudes less than approximately 7 km, only around 10 % of lightning strikes to airplanes are intercepted. At higher altitudes, the percentage is close to zero. What happens frequently is that the airplane itself contributes to the initiation of a lightning flash. Figure 15.11 shows a lightning flash initiated by an airplane during takeoff. Strong electric fields may exist close to the cloud's charge centers. An airplane traveling in the vicinity of thunderclouds is exposed to these high electric fields. As shown in Fig. 15.12, a high electric field may initiate leader discharges of opposite polarity from the extremities of the airplane. One of the leaders may propagate toward one of the charge centers in the cloud and the other leader may travel toward a charge center of opposite polarity or toward the ground.

Fig. 15.11 Video frame of a lightning strike to an aircraft on takeoff from Kamatzu Air Force Base on coast of Sea of Japan during winter. Note that the branches of the upper part of the channel are directed upward, whereas the branches on the lower part of the channel are directed downward. This shows that a bidirectional leader started from the plane with one part propagating upward and the other downward (Courtesy Prof. Z.I. Kawasaki)

If one of the leaders ends on the ground, a ground flash results. If both leaders travel toward charge centers of opposite polarity, then the result is a cloud flash. In either case, the current flowing along the channel (including the return stroke currents in the case of a ground flash) is injected into the airplane at one of the leader initiation points, and this current leaves the airplane from the other leader inception point. The current flows along the metal skin of the airplane. Thus, the metal surface should not contain any gaps that could cause sparks during current flow. One of the most sensitive parts of an aircraft is the fuel system, and a lightning strike should not generate a spark in any part of the fuel system. Moreover, the metal skin of the airplane, especially the metal skin around the fuel tanks, should be thick enough to withstand a lightning current without puncture. In airplane lightning protection, possible lightning strike points are investigated, usually using scale models in the laboratory, and these points are strengthened so that they can take on the full lightning current without causing a puncture. Actually, airplanes are designed in such a way that they should be able to withstand the maximum current from a lightning flash. Figure 15.13 shows the lightning current test waveform that aircrafts should be able to withstand. This current represents one of the most severe lightning currents possible in nature.

15.11 Effects of Direct Lightning Strikes on Structures

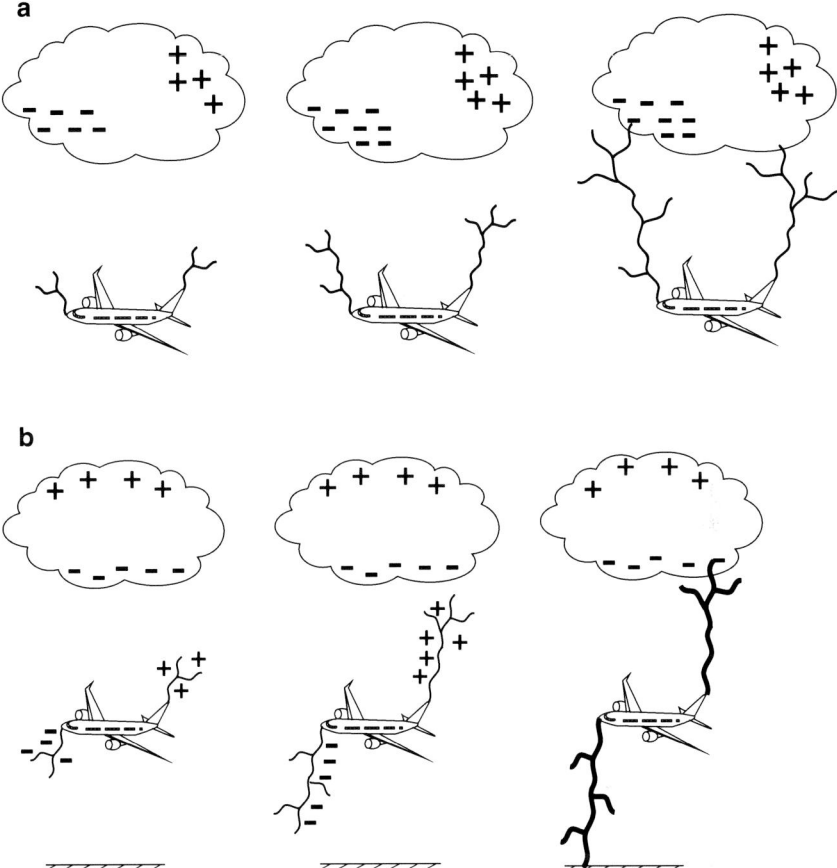

Fig. 15.12 (a) An airplane in the electric field of a thundercloud may give rise to leaders of opposite polarity moving away from the airplane from opposite locations (i.e., a bidirectional leader). If the two leader branches end up in opposite charge centers in the cloud as in (**a**), then a cloud flash results. If one of the leader branches travels to the ground as in (**b**), the result is a ground flash. In either case, the current flowing along the lightning channel passes through the body of the plane (Figure created by author based on information available in the literature)

The attachment process of a lightning flash to a moving airplane is slightly more complicated than in the case of a stationary grounded structure. Once a lightning flash becomes attached to an airplane, physical processes taking place at the points where the current enters the metal skin of the airplane attempt to keep the point of entrance of the current into the airplane the same throughout the duration of the lightning flash. However, as the plane moves forward, the arc channel is continually bent and stretched, and at a certain point in time it becomes favorable energetically for the current entrance point to move to a new location (Fig. 15.14). Thus, during a lightning strike to a moving airplane, the current entry point and exit point move

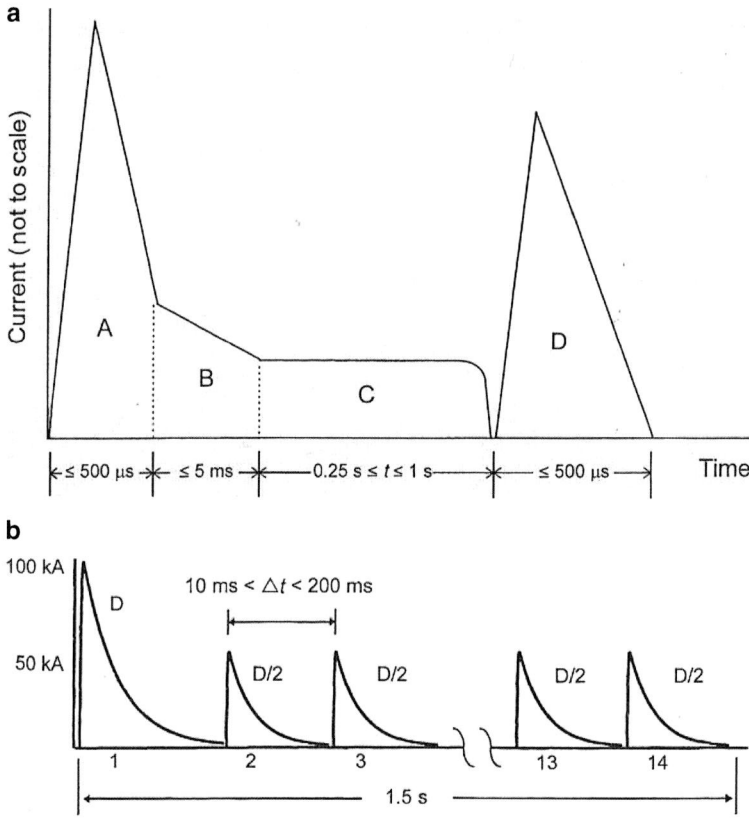

Fig. 15.13 (a) Test current waveform simulating first two strokes of a severe lightning flash to an aircraft. *Component A* (first return stroke): peak amplitude = 200 kA; action integral = 2×10^6 A^{2}s (in 500 μs); duration ≤ 500 μs. *Component B* (intermediate current): maximum charge transfer = 10 C; average amplitude = 2 kA; duration ≤ 5 ms. *Component C* (continuing current): amplitude = 200–800 A; charge transfer = 200 C; duration = 0.25–1 s. *Component D* (subsequent return stroke): peak amplitude = 100 kA; action integral = 0.25×10^6 A^{2}s (in 500 μs); duration ≤ 500 μs (Adapted from aircraft lightning protection standard SAE ARP5412). (**b**) Test current waveform simulating a severe multiple-stroke flash to an aircraft (Adapted from aircraft lightning protection standard SAE ARP5412)

along the body of the airplane (Fig. 15.15). Therefore, a lightning strike could create a series of strike points along the body of the airplane.

The passengers inside the airplane are usually safe from the lightning currents passing along the metal skin of the airplane because the current injected into a metal object flows along the surface of the object without penetrating into the object, i.e., the Faraday cage principle. However, not all parts of the airplane are made of metal. For example, to allow for the passage of radar beams, the front part of the airplane, called the radome, is made of nonmetallic material. This is the most vulnerable part of an airplane. However, protection procedures are usually put in place so that in the case of a lightning strike to the radome, the lightning current passes along the

15.11 Effects of Direct Lightning Strikes on Structures

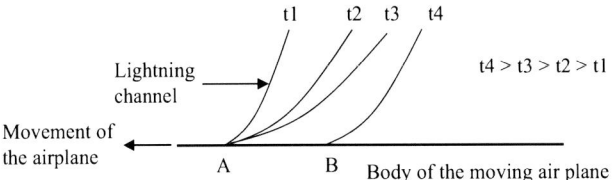

Fig. 15.14 As the airplane moves forward, the lightning channel is stretched, and at a certain time the strike point moves to the right (i.e., from A to B) (Figure created by author)

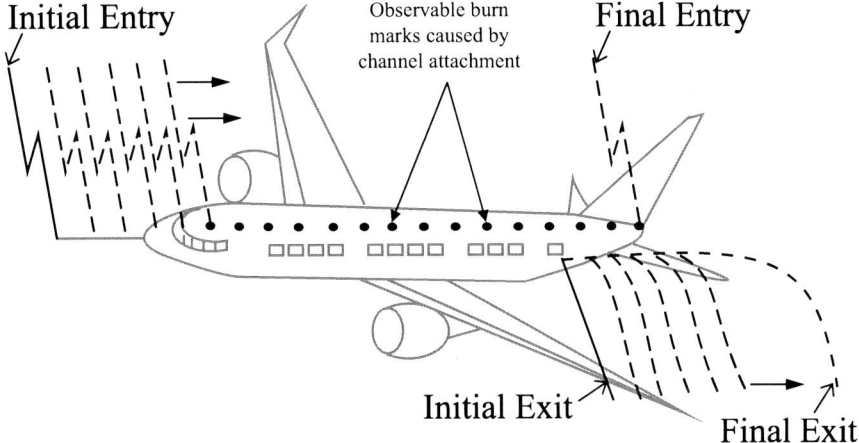

Fig. 15.15 As shown in Fig. 15.14, the movement of the airplane causes the current entry and exit points to move along the body of the airplane. The strike and entry points are observed as a series of spots with burnt paint on the body of the airplane (Figure created by author based on information available in the literature)

radome and enters the skin of the airplane without penetrating into the airplane. This is achieved by locating strips of conducting material on the radome in such a way that they do not obstruct the view of the radar while at the same time providing a safe path for the lightning current across the radome (Fig. 15.16).

During lightning strikes, electromagnetic fields can penetrate through nonconducting regions into the airplane. Moreover, depending on the thickness of the metal skin, even current flowing along the metal skin of the airplane can generate electric and magnetic fields inside the airplane as a result of the skin effect (Chap. 17). These fields, through electric and magnetic induction, can cause induced voltages in the electronic circuitry located inside the airplane. Thus, the sensitive electrical circuits in the airplane are protected using surge diverters (Chap. 17).

Before the 1960s several explosions involving airplanes were caused by lightning strikes. Today, however, accidents are rare, which shows that protection procedures currently in place are effective. In fact, 1967 was the last year a

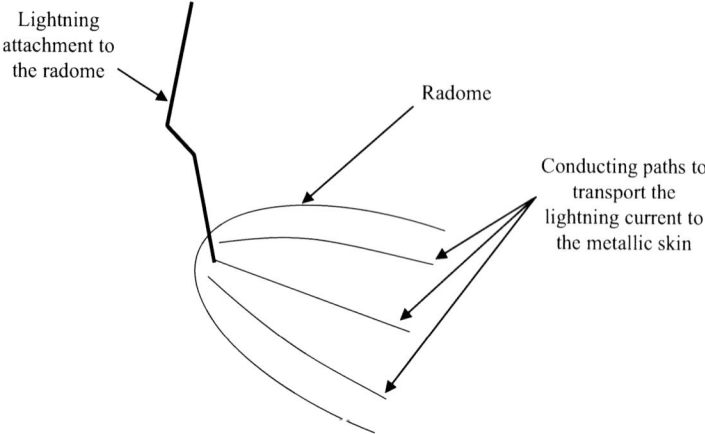

Fig. 15.16 The conductors located on the radome of an airplane to divert lightning currents away from the radome and into the metal skin of the airplane (Figure created by author based on information available in the literature)

lightning strike caused an airliner to crash in the United States. This was caused by an explosion in the fuel tank.

Today the use of composite materials in modern airplanes, like the Boeing 787, with a fuselage made predominantly of carbon fiber, has required additional design features. For example, composite material is made more resistant to lightning by an embedded layer of conducting fibers or a metal screen to conduct lightning currents.

15.11.4 Direct Strike to a Windmill

Alternative energy is the rallying cry of modern society, and wind power is the latest main source of alternative energy around the world. With increasing demand for wind power, the power-generating capacity of wind turbines has also increased over time [9]. However, the higher the power output of the wind turbine, the higher the required wind span of the turbine. A larger wingspan requires greater heights. Actually, the height of a wind turbine increases with increasing power output. Present-day wind turbines can reach heights of more than 100 m. Being rather tall, such wind turbines can initiate lightning flashes when exposed to thundercloud electric fields. Indeed, a wind mill could be a better initiator of lightning flashes than a stationary tower of similar height, for the following reason. In the presence of a background electric field, all pointed structures at ground level go into corona, and the tip of a tall tower is no exception (Fig. 15.17a). Corona discharges create a space charge region around the tip, and this space charge can screen the tip of the tower from an electric field. As the electric field produced by the thundercloud

15.11 Effects of Direct Lightning Strikes on Structures

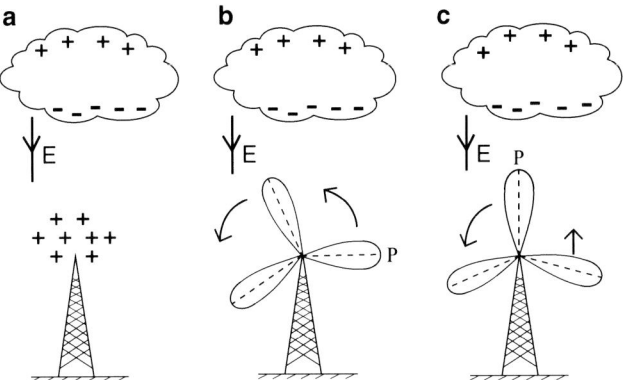

Fig. 15.17 Even though the background electric field remains constant, the electric field at the tip of a blade of a wind turbine changes from a low value to a high value as it rotates in a *vertical plane*. For example, the electric field at the tip of the blade located horizontally [marked P in (**b**)] increases as it rotates and reaches its maximum when the blade is vertical [marked P in (**c**)]. Thus, the electric field at the tip of the blade oscillates as the blade rotates in the background field

slowly increases, the corona discharge intensifies and continues to screen the tip from the increasing strength of the electric field. Thus, to create a connecting leader from the tip of the tower, the field must increase very rapidly to overcome the screening effects of the corona. In a slowly increasing field, the corona generated at the top will inhibit a tower from giving rise to a connecting leader. In the case of wind turbines, corona discharges are generated at the tips of the blades. But the space charge cannot accumulate at the tips because of its displacement resulting from the rotation of the blades. When a blade is located parallel to the ground, the electric field at its tip is rather low, and it increases rapidly as the blade turns up toward the cloud (Fig. 15.17b). The electric field is at its maximum when it is located perpendicular to and away from the ground (Fig. 15.17c). In other words, as the blade rotates, the electric field at its tip increases rapidly, and without a significant amount of corona space charge to screen the tip from the electric field, the conditions necessary for the launching of upward leaders is reached in the background electric field, which would not have generated upward leaders from stationary towers of similar heights.

Lightning can cause severe damage to blades (Fig. 15.18) [10]. Thus, the current in lightning flashes that strike wind turbines must not be allowed to flow along the surface of blades. This is achieved by placing lightning conductors inside the blades and connecting these conductors to the outer surface of the blades through metal receptors located flush with the blades (Fig. 15.19). However, there are cases where a lightning strike misses the receptor and enters the conductor by flashing over through the material of the blade, causing significant damage. Attempts to improve the lightning protection of wind turbines are ongoing in the lightning protection research community.

Fig. 15.18 Damage caused by lightning flash in wind turbine blade (adapted from [10])

Fig. 15.19 Metal receptors located on surface of blade. The metal receptors are connected to the lightning protection down conductors located inside the blade (adapted from [9])

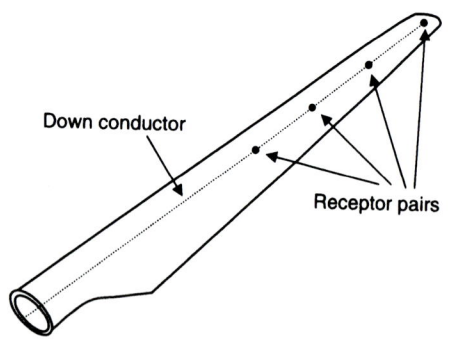

15.11.5 Effects of Lightning on Power Lines

15.11.5.1 Direct Strike to a Power Line

Power lines are protected from direct strikes to phase conductors by a ground wire installed above the phase conductors. Usually, it is this ground wire that receives the lightning strike. However, there are instances where lightning strikes phase conductors directly. Even if lightning strikes a grounded wire, it can still create problems in the phase wire. The reason is as follows. The lightning current injected into the ground wire splits into two equal parts and travels outward along the ground wire (Fig. 15.20). The ground wire is grounded at the tower, and at this point the

Fig. 15.20 When a lightning flash injects a current into a ground conductor, for example, the current splits into two equal parts and travels in the opposite direction at speed c. Once the current pulse reaches the grounded tower, it flows to the ground along the tower

current flows to the ground along the tower. If Z is the grounding impedance at the tower foot, the lightning current flowing through this impedance raises the tower to a potential of $Z\,i(t)$, where $i(t)$ is the current flowing down the tower. This gives rise to a voltage difference between the tower and the phase conductors, and if this voltage is greater than the value necessary to cause an electrical breakdown across the insulators, then a spark results connecting the ground wire and one of the phase conductors (Fig. 15.21). Once this spark is generated, part of the lightning current flows along this path into the phase conductor. Moreover, since an electric spark has a rather low resistance, it is not only the lightning current that passes through the spark channel but also a current driven by the voltage difference between the phase

Fig. 15.21 As the current passes along the tower to the ground, the potential of the tower is raised because of ground impedance. For example, if the ground impedance is Z, then the tower is raised to a potential of $i(t)\ Z$ where $i(t)$ is the current. If this potential is sufficiently high, there could be an electrical breakdown between the tower and the phase conductors

15.11 Effects of Direct Lightning Strikes on Structures

conductor and the ground. This effect can have two consequences. First, part of the lightning current will penetrate into the phase conductors through the spark channel. These phase conductors are usually connected to a distribution transformer from where the electricity is distributed to the consumer. This lightning current, if not properly diverted to the ground before entering the transformer, can cause damage to the transformer windings, which could be very expensive to repair. Moreover, part of the lightning current could enter the consumer side after passing through the transformer, causing problems in the electronics at the consumer's end. The other undesired consequence is the following. Once a spark is created between a ground wire and a phase conductor, the voltage difference between the phase conductor and ground is large enough to maintain the spark. Thus, the current flowing along the phase conductor, which is meant for the consumer, starts to flow into the ground across the spark, cutting off the power supply to the consumer. This is called a *line fault*. The only way to correct the problem is to cut off the electric power to the line until the spark is extinguished. Devices that perform this task are called *circuit breakers*. Usually, after the power is cut off to a line to clear a fault, the power is reconnected to the line after enough time has passed to extinguish the spark by stopping its current flow. However, the subsequent return stroke that originated in the same channel at a later time could reignite the spark and the power to the line would have to be disconnected again. Depending on the number of subsequent strokes and the interval between them, the process may be repeated several times during a single lightning strike.

Insulators in power lines are designed to withstand the high voltage arising in ground wires as a result of lightning strikes. However, sometimes the insulation strength is weakened because of deterioration or growth of organic materials on the insulators, and the sparks that result between the ground wire and the phase conductor along the insulators can further damage the insulators.

15.11.5.2 Indirect Effects of Lightning on Power Lines

In the previous section, we described what happens when a power line is struck by lightning, causing direct current injection into either the ground wire or the phase conductors. However, since both magnetic and electric fields can generate voltages and currents in power lines, even a lightning flash that strikes in the vicinity of power lines can generate large voltages in the line that can cause transformer damage and line faults. Let us consider the basic principles used by scientists to investigate this effect [11, 12].

Transmission lines can transport currents and voltages along it. A transmission line need not be a standard power line to transport power. The cables that transport television signals are also transmission lines. They contain two conductors arranged in a specific manner. The simplest transmission line over ground can be represented as a single conductor located over a ground plane. The ground plane acts as the return conductor through which the current injected from the generator into the power line is returned back to the generator, completing the circuit.

Transmission lines are actually media that support the propagation of voltages and currents as waves. Consider any medium that supports wave propagation. One simple example is a stretched string. If the string is displaced at a given point, the displacement divide into two halves and propagate in opposite directions. Transmission lines behave in exactly the same way. Assume that a current is injected at any given point on a line. The injected current then divides into two equal parts and travels in opposite directions. Now, a current can be injected into a transmission line by connecting a generator to the line. This is what happens during a direct lightning strike. However, a current can also be injected into a transmission line in a more passive way. This is done by applying an electric field in a direction parallel to the conductors of the power line. Since the conductivity of a normal conductor is high, whenever an electric field is applied parallel to the line, free charges start to flow in an attempt to bring the electric field component parallel to the line to zero. Let us consider this problem mathematically. Consider a small section of a transmission line, say, dx. Now let us apply an electric field $E(x,t)$ parallel to the line at that point. This will generate a voltage of $E(x,t)dx$ across the element, resulting in a current $E(x,t)dx/Z$, where Z is the characteristic impedance of the line. The characteristic impedance of a single conductor transmission line located above the Earth is given by

$$Z = \frac{1}{2\pi}\sqrt{\frac{\mu_o}{\varepsilon_o}} \log\left[\frac{2h}{a}\right], \qquad (15.17)$$

where h is the height of the conductor and a is its radius. Now, this current will separate into two equal parts and propagate in opposite directions. With this knowledge, it is possible to make a reasonable estimation of the voltages induced in a power line in many circumstances where the power line configuration is simple, i.e., a single conductor line. It can also be used to evaluate the order of magnitude of induced voltages in more complex configurations. The complexity arises because of the interaction of the electric and magnetic fields generated by the current flow in one conductor with other conductors of the line.

Let us illustrate how to apply the preceding concept using a simple example. Consider a single conductor transmission line of length L where both ends terminate at its characteristic impedance Z. It is important that the terminating resistance be equal to the characteristic impedance because whenever a voltage or a current wave propagating along a transmission line reaches a termination, both the current and the voltage waveforms are reflected. The polarity and the amount of reflected voltage depend on the difference between the characteristic impedance of the line and the impedance connected at the termination. If the impedance connected at the end of the line is equal to the characteristic impedance of the line, the waves are absorbed without any reflection. The preceding configuration, with terminating resistors equal to the characteristic impedance, therefore simplifies the analysis of the problem. On the other hand, if a wave reaches an open end of a transmission line where the impedance is infinite, the reflected voltage wave has the same sign and amplitude as the incident voltage, resulting in a doubling of the voltage. At the

15.11 Effects of Direct Lightning Strikes on Structures

Fig. 15.22 *Top* and *side views* of geometry relevant to calculation of induced voltages in single conductor transmission *line* as a result of a nearby lightning strike. AB is the overhead conductor, E_v is the vertical electric field, and E_h is the horizontal electric field

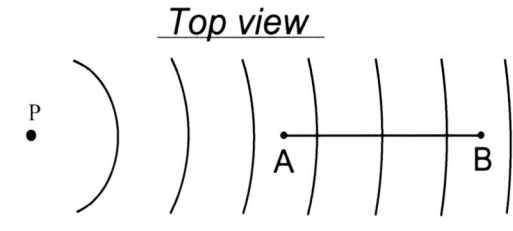

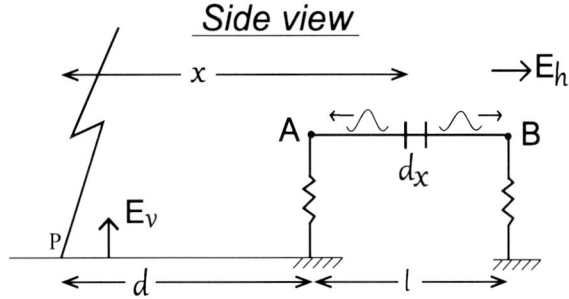

same time, the reflected current waveform has the same amplitude and opposite polarity of the incident current, making the net current at the termination zero. The opposite takes place if the impedance is zero (i.e., short-circuited to the ground). In this case, the net voltage becomes zero at the termination and the current is doubled. To simplify the analysis, we will calculate the voltage induced in the terminations of a single conductor transmission line for the special case where the propagation direction of the electromagnetic field is along the line.

The side and top views of the geometry relevant to the problem under consideration are shown in Fig. 15.22. Points A and B are the locations of terminations of the single-conductor transmission line, and P is the strike point of the lightning flash. The impedances at the terminations are assumed to be equal to the characteristic impedance of the line, Z. The transmission line is assumed to be straight and horizontal, and the lightning channel is assumed to be straight and vertical. During the lightning strike the electromagnetic field propagates in a radial direction along the ground plane, and since the power line is located parallel to the radial direction, the direction of the propagation of the electromagnetic field is parallel to the direction of the line. In other words, the wave front of the electromagnetic field is perpendicular to the line (Fig. 15.22).

Let $t = 0$ be the time of occurrence of the lightning flash. Consider a small line element dx located at a distance x from the lightning channel. The electric field associated with the electromagnetic field generated by the lightning flash can be resolved into two parts, one parallel to the ground plane, $E_h(x,t)$ (i.e., in the positive x-direction), and the other perpendicular to the ground plane, $E_v(x,t)$ (i.e., in the positive z-direction). The polarity of the voltage at ground terminations is considered positive if it drives a positive current into the ground through the termination impedance. If the ground plane is perfectly conducting, the horizontal electric component is zero at ground level but it is not zero at points above the ground plane. The horizontal electric field induces a current of magnitude $E_h(x, t-x/c)dx/Z$ in the line element dx, and half of it propagates at the speed of light toward termination A and the other half propagates toward termination B. Let dI_B be the current generated by element dx reaching termination B. This is given by

$$dI_B = \frac{1}{2Z} E_h[x, t - x/c - (d + l - x)/c] dx. \tag{15.18}$$

Note that the field is incident on the line element at time x/c (its value is zero at times before that), and the current signal takes a time $(d+l-x)/c$ to reach the termination. Note that c is the speed of light, and this is the speed of propagation of current and voltage pulses along the line. Equation 15.18 can be written as

$$dI_B = \frac{1}{2Z} E_h(x, t - d/c - l/c) dx. \tag{15.19}$$

The total current reaching termination B due to the entire line is given by

$$I_B = \frac{1}{2Z} \int_d^{d+l} E_h(x, t - d/c - l/c) dx. \tag{15.20}$$

If the way in which the field changes with distance is known, this can be evaluated numerically. On the other hand, assume that the distance to line d from the lightning flash is much greater than the length of the line. In this case, the electric fields can be assumed to be of the same magnitude along the entire line (i.e., E_h is independent of x but is still a function of time), and in this case the current injected into termination B becomes

$$I_B = \frac{1}{2Z} E_h(t - d/c - l/c) l. \tag{15.21}$$

Since all this current is absorbed into the termination without any reflection, the voltage V_B induced in termination B as a result of the horizontal electric field associated with the lightning flash (note that this voltage is positive because the horizontal electric field is considered positive along the positive x-direction and it will drive positive charges through the termination into the ground) is

$$V_B = \frac{1}{2} \int_0^l E_h(x, t - d/c - l/c) dx. \tag{15.22}$$

15.11 Effects of Direct Lightning Strikes on Structures

Now, let us consider termination A. Let dI_A be the current generated by element dx reaching termination A. This is given by

$$dI_A = -\frac{1}{2Z} E_h(x, t + d/c - 2x/c) dx. \qquad (15.23)$$

The total current reaching termination A due to the entire line is given by

$$I_A = -\frac{1}{2Z} \int_0^l E_h(x, t + d/c - 2x/c) dx. \qquad (15.24)$$

The voltage V_A induced in termination A because of the horizontal electric field associated with the lightning flash is

$$V_A = -\frac{1}{2} \int_0^l E_h(x, t + d/c - 2x/c) dx. \qquad (15.25)$$

If the temporal variation of the horizontal electric field as a function of x is known, this can be calculated numerically.

So far, we have considered only the voltage induced in the terminations of the line due to the horizontal electric field. The vertical field can induce voltages and currents in vertical conductors, and, as in the case of voltages induced by a horizontal electric field, they propagate in both directions along the line. Let us first consider termination A. The current induced by the vertical field on the vertical conductor at the termination is equal to $E_v(d, t - d/c)h/Z$. The polarity of this current is such that it transports positive charge in the positive direction of E_v. In writing this equation the whole vertical conductor was taken as one element. Half of this current is absorbed into the termination and the other half propagates along the line toward termination B (Fig. 15.23). The voltage generated by the current passing through termination A is $-E_v(d, t - d/c)h/2$. The current reaching termination B because of the vertical conductor at termination A is $E_v(d, t - d/c - l/c)h/2Z$. The voltage generated by this current across the impedance at termination B is $E_v(d, t - d/c - l/c)h/2$. Now, let us consider the vertical conductor at termination B. The current induced by the vertical electric field on the vertical conductor at termination B is equal to $E_v(d + l, t - d/c - l/c)h/Z$. Half of this current is absorbed into termination B and the other half propagates along the line toward termination A. The voltage generated across the termination impedance at B by this current is $-E_v(d + l, t - d/c - l/c)h/2$. The current reaching termination A because of the vertical conductor at termination B is $E_v(d + l, t - d/c - 2l/c)h/2Z$. The voltage resulting from this current at termination A is $E_v(d + l, t - d/c - 2l/c)h/2$. Now we are in a position to write down the expression for the total voltage appearing across terminations at A and B.

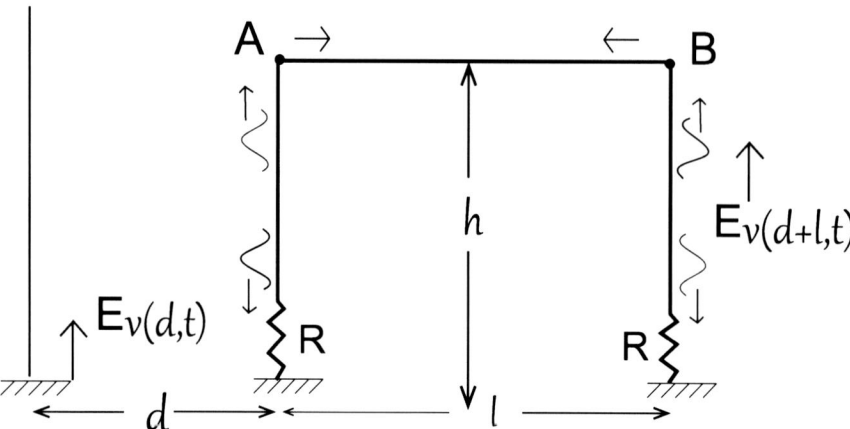

Fig. 15.23 Currents and voltages are induced in vertical conductors by a vertical electric field. Half of the current induced in the vertical conductor travels to the ground through the grounding impedance R and the other half enters the horizontal part of the conductor (i.e., AB)

The total voltage induced in termination B is given by

$$V_{B,\text{total}} = \frac{1}{2}\int_0^l E_h(x, t - d/c - l/c)dx - E_v(d + l, t - d/c - l/c)h/2$$
$$+ E_v(d, t - d/c - l/c)h/2. \quad (15.26)$$

The total voltage induced in termination A is given by

$$V_{A,\text{total}} = -\frac{1}{2}\int_0^l E_h(x, t + d/c - 2x/c)dx + E_v(d + l, t - d/c - 2l/c)h/2$$
$$- E_v(d, t - d/c)h/2. \quad (15.27)$$

If both the vertical electric field and the horizontal electric field can be assumed to have the same amplitude and temporal wave shape along the line (i.e., when the distance from the lightning flash to the line is much greater than the length of the line), then the total induced voltage reduces to

$$V_{B,\text{total}} = \frac{1}{2}lE_h(t - d/c - l/c) - E_v(t - d/c - l/c)h/2$$
$$+ E_v(t - d/c - l/c)h/2 \quad (15.28)$$

and

$$V_{A,\text{total}} = -\frac{1}{2}\int_0^l E_h(t+d/c-2x/c)dx + E_v(t-d/c-2l/c)h/2$$
$$- E_v(d,t-d/c)h/2. \qquad (15.29)$$

The same technique can be used to calculate the voltages induced in terminations of a power line when the position of the strike point is located in any arbitrary direction. In this case, it is necessary to keep track of the time of incidence of the electromagnetic field at different elements of the line. Moreover, the fact that the horizontal electric field is always radial to the strike point (and hence may make different angles with different line elements) should also be taken into account. Since currents are generated by the electric field component parallel to the line, it is necessary to consider only the component of the horizontal electric field parallel to the line elements in the analysis.

The foregoing analysis shows that knowledge of the horizontal electric field and the vertical electric field generated by a lightning flash makes it possible to estimate the voltages and currents induced in power lines using simple procedures. Of course, when the line configuration is complex, it is much more convenient to follow numerical procedures. But the foregoing analysis illustrates the basic principles involved in evaluating the voltages induced in power lines resulting from lightning flashes striking in their vicinity.

References

1. Orville RE (1977) Lightning spectroscopy. In: Golde RH (ed) Lightning, vol 1. Academic, London
2. Uman MA (1964) The diameter of lightning. J Geophys Res 69:583–585
3. Braginskiii SI (1958) Theory of the development of spark channel. Sov Phys JETP (English Trans.) 34:1068–1074
4. Lightning protection for electrical and electronic equipment, A publication of The Institute of Electrical Installation Engineers of Japan, Japan, 2011. ISBN: 978-4-9902110-7-3
5. Berger K (1967) Novel observations on lightning discharges: results of research on mount San Salvatore. J Franklin Inst 283:478–525
6. Fernando F, Mäkelä J, Cooray V (2010) Lightning and trees. In: Cooray V (ed) Lighting protection. IET Publishers, London
7. Taylor AR (1977) Lightning and trees. In: Golde RH (ed) Lightning, vol 2. Academic, London
8. Becerra M, Cooray V (2009) On the interaction of lightning upward connecting positive leaders with humans. Trans IEEE (EMC) 51:1001–1008
9. Sorensen T (2010) Lightning protection of wind turbines. In: Cooray V (ed) Lightning protection. IET Publishers, London
10. Yokoyama S (2013) Lightning protection of wind turbine blades. Electr Power Syst Res 94:3–9

11. Cooray V, De La Rosa F (1986) Shapes and amplitudes of the initial peaks of lightning induced voltage in power lines over finitely conducting earth: theory and comparison with experiment. IEEE Trans Antenna Propag AP-34:88–92
12. Nucci CA, Rachidi F (2010) Lightning protection of medium voltage lines. In: Cooray V (ed) Lighting protection. Institution of Engineering and Technology, London

Chapter 16
Interaction of Lightning Flashes with Humans

16.1 Lightning Causes Human Casualties

Every year around three billion lightning flashes occur around the world. In tropical regions, approximately 10–20 % of lightning flashes strike the ground while the rest take place inside a cloud. In temperate regions, the corresponding figure is approximately 50 %. From time to time lightning flashes striking the ground interact with humans, causing injuries and sometimes death. Statistics concerning the number of deaths caused by lightning are available only for a few countries. Statistics from many countries in tropical regions are not available, but the number of deaths caused by lightning is likely to be higher in those countries as a result of the higher number of lightning strikes and the amount of time spent outside and in unprotected buildings. The global mortality rate from lightning could be around 1,000 per year. Table 16.1 tabulates the lightning-caused fatalities in several countries.

As mentioned earlier, lightning fatalities increase with an increasing number of lightning ground flash density or an increasing number of thunderstorm days. Table 16.2 shows how the lightning-caused death rate varies with the number of thunderstorm days.

16.2 Different Ways in Which Lightning Can Interact with Humans

There are six different ways in which humans can be affected by a lightning strike: direct strike, side flash, touch voltage, step voltage, connecting leaders, and shock waves [3–5]. In the case of a direct strike, the lightning channel terminates on the

Table 16.1 Annual lightning death rates per million people [1]

Country	Year(s)	Fatalities	Death rate
Austria	10	48	0.6
	5	Not stated	103
	2005	133	0.9
Bavaria	1843–1873	1,355	8.7
China	1997–2005	3,776	0.3
	2004	(875)	0.7
	2006	(410)	0.3
	2007	(564)	0.4
European Russia	1870–1874	2,131	6.2
German Federal Republic	1952–1960	Not stated	0.8
Hungary	10	6	0.5
India	8	473	0.1
Ireland	20	7	0.1
Lithuania	Recent	0.5	0.1
Malaysia	Recent	100–150	3.4
Nepal	2004	(100)	2.7
Prussia	1869–1877	1,004	4.3
South Africa	1963–1969	Not stated	1.5
	4	Not stated	8.8 rural
			1.5 urban
Sri Lanka	2003	49	2.4
Switzerland	1876–1877	33	6.0
Vietnam	Recent	Dozens	1.2
Zimbabwe	Recent	150	21.3
		100	14.2

Table 16.2 Correlation between death rate from lightning and thunder days per year [2]

Typical range in thunder days per year	Approximate mean annual lightning death rate per 10^6 population
5–20	0.2
20–40	0.4
40–80	1.0
80–200	1.7

body of a person, exposing it to the full lightning current. The channel usually terminates on the head or the upper part of the body. This mechanism probably accounts for the largest number of lightning deaths. Let us consider these possibilities one by one.

16.2.1 Direct Strike

During a direct strike to a human being, the current that appears at the base of the lightning channel is injected into the body from the point of strike. The point of strike is usually either the head or the upper body. Let us consider what happens during the passage of the return stroke current, the most energetic component of the lightning flash. As described in Chap. 8, the return stroke current starts from a small value and increases very rapidly. In a typical case, it increases from a small value to approximately 30 kA in approximately 5 µs. After reaching this peak value the current decays gradually to zero in approximately 100–200 µs. When a person is struck by lightning, the current passing through the body also starts at a small value and continues to increase very rapidly. The body acts as a resistive object, and therefore there will be a voltage difference across the body (i.e., between the entry point of the current, probably the head, and the current's exit point, probably the feet). As the lightning current increases, this voltage difference across the body also increases. When this voltage difference reaches a certain critical value, it leads to the creation of a discharge path along the outer skin of the body. The resistance of this discharge path is much less than that of the body, and as a consequence, a very large fraction of the lightning current will pass along this path. This reduces the current passing through the body to a very small value. This event is called a *surface flashover*. These events are depicted pictorially in Fig. 16.1.

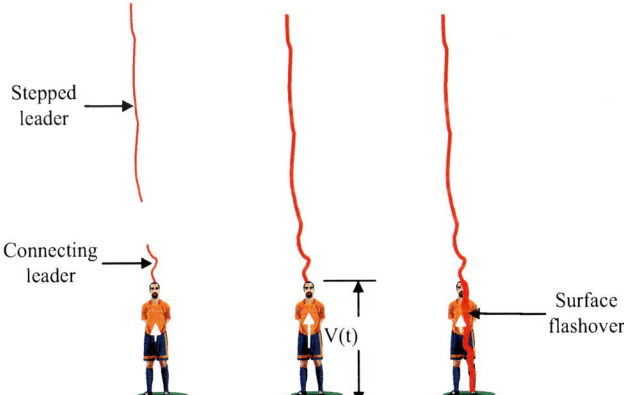

Fig. 16.1 Events that might take place when lightning strikes a human being. As the stepped leader approaches the ground, a connecting leader is generated, probably from the person's head. After the connection is made between the connecting leader and the downward moving stepped leader, the return stroke current starts to flow through the body. At the same time, because of the resistance of the body, a voltage difference $V(t)$ is created between the person's head and the ground. This voltage increases as the return stroke current increases, and when this voltage reaches a critical value, a discharge path is created along the surface (or skin) of the body. A main portion of the lightning current takes this new path to the ground, and as a result the current flowing through the body decreases (Figure created by author)

Let us now enter in some numerical values in the previously described problem to understand the consequences better. A 2-m-tall person requires a voltage difference of approximately 1 million V to create a spark from head to toe along the skin. However, if the skin is wet and salty, the required voltage may be as low as 0.4 million V. The resistance of the body of an average human being is approximately 1,000 Ω [4]. Let us consider the situation of a 2-m-tall human struck on the head by an average 30-kA lightning flash. As the current in the body increases, the voltage difference across the body increases along with it, and when the current reaches 1,000 A, the voltage difference across the body reaches 1 million V (i.e., 1,000 Ω × 1,000 A). At this stage, a spark channel is created along the outer surface of the body, and most of the current starts to flow along the surface of the body. At this time, the current flowing through the body decreases to approximately 5 A. If the current rises to its peak value in 5 μs, the surface flashover takes place in approximately 150 ns if the peak current of the lightning flash is 30 kA. If the rise time of the current is 1 μs, the *surface flashover* may take place within approximately 30 ns from the time of strike. Therefore, the current passing through the body increases to approximately 1,000 A in about 30–150 ns and then decreases suddenly to approximately 5 A, and this low current continues with diminishing amplitude to flow for the duration of the first return stroke. If there are several return strokes in the flash, this process is repeated several times. On the other hand, if there is no surface flashover, the full lightning current (i.e., approximately 30 kA in the case of a typical return stroke) passes through the body. Since it is the current passing through the body and not the current flowing along the surface of the skin that causes internal damage, the injuries the body experiences are much more severe in the absence of a surface flashover. As mentioned previously, the voltage required across the body to create a surface flashover decreases if the skin is wet and salty. In other words, the wetness of the skin caused by rain may reduce the voltage needed to cause a surface flashover, and as a consequence, the current passing through the body will decrease. Thus, a person caught in a thunderstorm should give priority to finding a suitable shelter to protect herself from lightning and not from rain. As we will see later, sometimes a shelter used as protection from rain will increase the chances of being struck by lightning.

The voltage required across a body to create a surface flashover may also depend on the type of clothes and the jewelry the victim is wearing. Metal objects such as metal necklaces may decrease the voltage required to produce a surface flashover. Thus, a necklace may save a victim's life by diverting the current that otherwise would pass through the heart or other sensitive organs to outside the body.

16.2.2 Side Flash

When lightning strikes a tree, for example, the current injected by the lightning flash into the tree flows along the trunk of the tree to the ground. If a person is standing close to a tree struck by lightning, a potential difference is created between

16.2 Different Ways in Which Lightning Can Interact with Humans

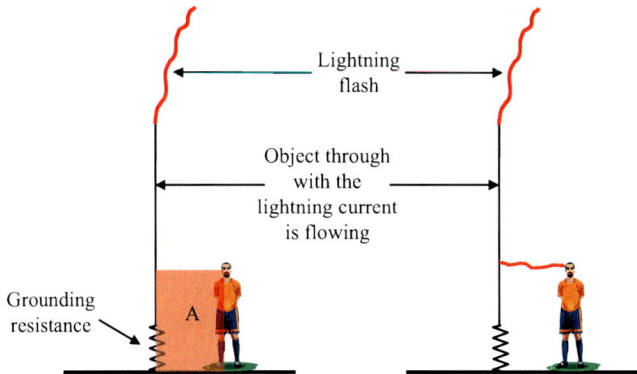

Fig. 16.2 Mechanism of side flash to person from a conductor struck by lightning. Because of the grounding impedance and the inductance of the conductor, it is at a higher potential than ground when the lightning current is flowing through it. This creates a potential difference between the person, who is at ground potential, and the conductor. Moreover, the time-varying magnetic field passing through the area marked A also induces a voltage between the person and the conductor. This voltage difference can give rise to a spark between the conductor and the person. Part of the lightning current passes through this spark channel and passes to the ground through the person (Figure created by author)

the trunk of the tree and the person. This potential difference is caused by the resistive voltage drop along the tree and the magnetic induction between the tree and the human, or both these effects. For example, refer to the situation depicted in Fig. 16.2a. Using the theory developed in Chap. 15, the maximum voltage that appears between the head and the path of the lightning channel is given by

$$V_m = \frac{\mu_0 h \ln(d/a)}{2\pi} \left(\frac{di(t)}{dt}\right)_{max}, \quad (16.1)$$

where a is the radius of the current path, h is the height of the person, d is the distance from the lightning path to the person, and $di(t)/dt$ is the derivative of the current flowing in the lightning channel. Note that the maximum voltage is proportional to the maximum derivative of the current. Consider a lightning flash with a return stroke having a peak current of 30 kA and maximum derivative of 100 kA/μs, and assume that the radius of the current path (i.e., a) is 1 cm. For a 2-m-high human being located at a distance of 1 m, the maximum voltage induced between the head and the current path due to magnetic induction is approximately 2×10^5 V. If the resistance at the ground end of the current path is 10 Ω, then the peak resistive voltage drop between the head and the lightning channel is approximately 3×10^5 V/m (assuming the person is at ground potential). Even though the peak values of these voltages occur at slightly different times, their combination could be large enough to establish a discharge path between the tree and the person. Moreover, a discharge between the head and the tree could also be established already at the connecting leader stage of the lightning flash. A portion of the lightning current passing along the tree may flow along this discharge path and

Fig. 16.3 Side flashes can be experienced from the branches of a tree. As the lightning current flows along the tree, different parts of the tree can be raised to a high potential. This may lead to an electric spark to a human being located in the vicinity of the tree from a region of the tree that is raised to a high potential and located close to the person. Part of the lightning current flows through this spark and to the ground through the person. This may happen, for example, from a branch of a tree (Figure created by author)

then through the body to the ground (Fig. 16.2b). Such an event is called a *side flash*. It is important to note here that a large fraction of the lightning injuries that take place outdoors are caused by side flashes from trees while the trees are being used as a shelter from rain. This highlights the danger of this sort of practice.

One can experience side flashes not only from the trunk of a tree but also from branches of a tree struck by lightning (Fig. 16.3). In fact, one can also experience side flashes inside buildings not equipped with proper lightning protection systems. The lightning current, in its search of the least resistive path (or, more correctly, the path of least impedance) to the ground, may find the path across the body of a victim to be less resistive. In such situations, a person may experience a side flash from an object located inside the building through which the lightning current is flowing. For example, a person can receive side flashes inside houses from water pipes, electrical apparatus, or land-based telephones. Most lightning-related injuries that happen inside buildings are indeed caused by such side flashes.

16.2.3 Touch Voltage

When lightning current flows along an object, i.e., a tree or along a wire fence struck by lightning, a potential difference is created between the ground and any other point on the object. Consider the situation shown in Fig. 16.4. The object

16.2 Different Ways in Which Lightning Can Interact with Humans

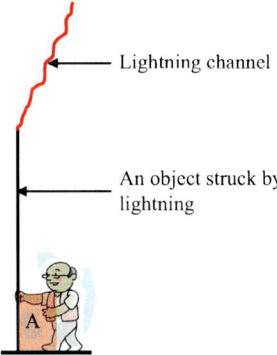

Fig. 16.4 Touch voltage. Because of inductance and grounding resistance, a conductor through which lightning current is flowing could be at a higher potential than that of a human being at ground potential. This may cause a part of the lightning current to flow to the ground through the person. Moreover, a voltage difference is also caused by the changing magnetic field passing through the area marked A. This voltage also contributes to the current passing through the person. The voltage that drives a current through the person in this manner is called a *touch voltage* (Figure created by author)

through which the current is flowing is depicted by the cylindrical conductor. The resistance of part of the conductor between the point of touch and the ground and the grounding resistance at the ground termination is lumped together in resistance R. If $R = 10\,\Omega$ and the peak current is 30 kA, then the peak potential difference that develops across the body is 3×10^5 V. Moreover, as in the previous case, the inductive potential difference that develops as a result of the magnetic flux passing through the loop shown in Fig. 16.4 also contributes to the potential difference. This potential difference causes a current to flow through the body of the victim from the contact point to the ground, causing injuries. This is called an *injury due to touch voltage*. Touch voltages can also result during climbing. A nearby lightning strike generates a potential difference between the point of attachment of hands and the feet, causing a current to flow from the hand to the feet or vice versa, inflicting injuries (Fig. 16.5).

16.2.4 Step Voltage

There is another way that a person can be injured by being in the vicinity of a lightning strike. Such injuries are caused by the lightning current flowing in the ground and are called *injuries due to step voltages*.

When a lightning flash strikes the ground, the current flows radially outward in all directions from the strike point. As the current disperses out from the strike point, and if the ground is resistive, the current flow creates a voltage with a very high value at the strike point that decreases as one moves away from the strike

Fig. 16.5 During a lightning strike to the ground, the current injected by the lightning flash into the ground spreads outward from the point of strike into the soil. As the current passes through the soil, different parts of the ground may be raised to a different potential. This could create a potential difference, for example, between the hand and the feet of a person located as in the figure. This voltage can cause a current to flow (*arrows*) through the body of the person. This situation may arise, for example, in the case of a lightning strike in the vicinity of a person climbing a mountain (Figure created by author)

point. If a person happened to be standing in the vicinity of a strike point, with one foot directed toward the strike point and the other away from it, there would be a potential difference between one foot and the other. This would cause a current to flow through the body. Depending on the polarity of the lightning flash, the current enters the body through the foot located close to the point of strike and exits through the other, or vice versa. This is illustrated in Fig. 16.6. The strength of this current passing through the body, which is called the *step current*, depends on many parameters. First, the greater the strength of the current in the lightning flash, the larger the step current. Second, the closer the location of the victim to the strike point, the larger the step current. Third, the lower the conductivity of the ground (for example, wet soil has a high conductivity, whereas dry, stony ground has a low conductivity), the higher the step current. This is because the lower the conductivity of the ground, the higher the resistance with which it opposes the flow of current. Thus, with decreasing conductivity of the ground a current flowing along it finds the path across the body to be increasingly attractive, and as a consequence, more and more current will take the path across the body. Finally, the magnitude of the step current also depends on the separation between the feet. The larger the separation between the feet, the larger the current. Thus, the worst-case situation is one in which a victim happens to be standing with feet far apart, with one foot directed toward the strike point and the other directed away, on stony ground in the vicinity of the strike point of a strong lightning flash. Of course, if the victim happens to be

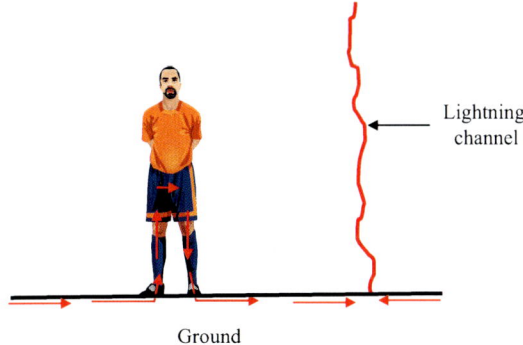

Fig. 16.6 Mechanism of step voltage. As the lightning current flows through the ground, different parts of the ground are raised to different potentials. When a person is standing as in the figure, one foot of the person will be at a different potential with respect to the other foot. This potential difference is called a *step voltage*. This voltage may drive a current through the body of the person as indicated in the figure. Since this current does not flow through the vital parts of the body, such as the brain or heart, the step voltage may not injure the person critically (Figure created by author)

standing with both feet at equal distance from the strike point (this could be the case if the victim is facing the point of strike), he will not be exposed to a step current. If the victim happens to be running during the time of strike, only one of his feet will be on the ground, and again no step current will flow through her body. The step current causes less severe injuries than a direct strike because the current does not flow across vital organs such as the heart and brain. However, it can cause paralysis of the lower limbs, and the person may collapse to the ground immediately.

With horses and cows, the injuries caused by step currents are more severe. There are two reasons for this. First, the separation between the hind and front feet of cattle is larger than that of a human stride. Thus the step current flowing through the body of horses or cattle is larger than that of a human. Second, the step current flowing through the body of a horse or a cow passes through the heart of the animal, causing severe injuries.

Of course, in some circumstances, ground currents can pass through a person's heart. This can happen if the person is performing some manual task or holding a metal object that is connected to the ground. In this case, the current enters the body through the feet and exits it through the hands into the metal object. An example of this is illustrated in Fig. 16.7. The effect can be severe compared to a normal step voltage because the current now passes through the heart.

If the person happens to be sitting or lying down close to the point of strike, the magnitude and path of the current passing through the body may depend on the way in which the body contacts the ground. In the case of a person lying on the ground, the current may enter the body through the head and part of the current may pass through vital organs such as the brain or heart, causing severe injuries (Fig. 16.8).

In general, a lightning flash consists of several strokes, and the point of termination of different strokes may not be the same. That is, the first stroke of the flash

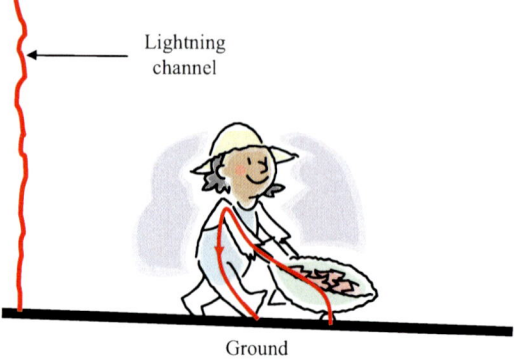

Fig. 16.7 There are situations where the step voltage may critically injure a person. Consider, for example, the situation shown in the figure. There is a potential difference between the feet of the person and the wheel of the wheelbarrow because of the lightning current flowing in the ground. This potential difference may cause a current to pass through the person as shown in the figure. Since this current passes through the person's heart, the injuries received could be severe (Figure created by author)

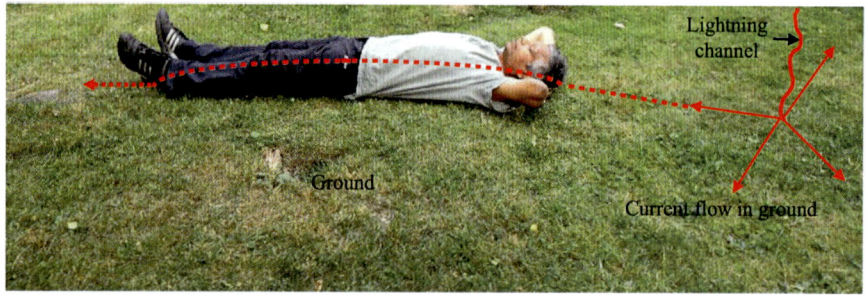

Fig. 16.8 A person lying down on the ground can be exposed to ground currents that pass through the heart. In this case, since different parts of the ground are at different potentials, a current could enter the body from the person's head and pass to the ground from different parts of the body that are in contact with ground. In this situation, the current may flow both through the brain and the heart and could inflict severe injuries (Figure created by author)

may strike the ground or some other object near a person, and a subsequent flash may strike the person directly. In this case, the person is exposed to the step voltage of the first stroke, and the subsequent stroke will strike him directly.

16.2.4.1 Mathematical Expression for Step Voltage

Consider a person located at a distance r from a lightning strike. Let s be the separation between the person's feet (Fig. 16.9). If it is assumed that the lightning current spreads out uniformly around a hemisphere (which is a reasonable assumption if the soil is uniform and isotropic), then the total current passing through any

16.2 Different Ways in Which Lightning Can Interact with Humans

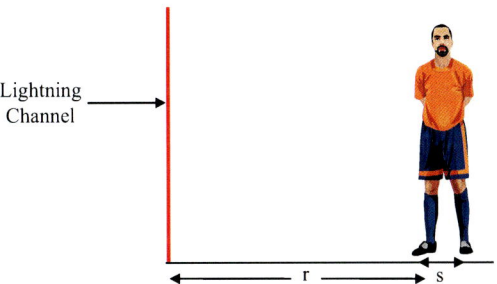

Fig. 16.9 Geometry relevant to calculation of magnitude of step voltage (Figure created by author)

hemisphere at any given radial distance r will be equal to the current injected by the lightning strike. This is the case since we assume that no charges are accumulated at any place in the soil. Thus, the current density passing through any hemisphere at any given time t at any distance r is given by

$$J(t) = \frac{I(t)}{2\pi r^2}. \qquad (16.2)$$

Therefore, the electric field at a distance r on the surface (or at any point on a hemisphere of radius r) is given by

$$E(t) = \frac{I(t)}{2\sigma \pi r^2}. \qquad (16.3)$$

Thus, the potential difference V between the feet is given by

$$V = \frac{I(t)}{2\sigma \pi} \left(\frac{1}{r} - \frac{1}{r+s} \right). \qquad (16.4)$$

The maximum potential between the feet is

$$V_p = \frac{I_p}{2\sigma \pi} \left(\frac{1}{r} - \frac{1}{r+s} \right), \qquad (16.5)$$

where I_p is the peak current. Assume that the peak current is 30 kA, $r = 10$ m, $s = 0.5$ m, and $\sigma = 0.001$ S/m. Then the voltage difference between the feet is approximately 20 kV. Now, the resistance of the body for a current path entering from one foot and moving out from the other, including the skin resistance, is approximately $20 \times 10^3 \, \Omega$ [4]. Thus, the peak current flowing through the body is approximately 1 A. Of course, if the person is wearing shoes, then the current must also flow through the soles of the shoes, and this increase the resistance of the current path. Thus, wearing shoes can reduce a current passing through the body due to step voltages.

16.2.5 Connecting Leader

Another way a person can receive injuries from a lightning flash, although only recently identified in the literature, is through a connecting leader current [5, 6]. As described in previous chapters, as a stepped leader comes to within approximately a few hundred meters of the ground, several connecting leaders may rise from several grounded objects toward the downward moving stepped leader. Only one of these connecting leaders makes the connection between the stepped leader and the ground.

A person situated near a downward moving stepped leader also sends up a connecting leader toward the downward moving stepped leader. These so-called unsuccessful connecting leaders may harbor currents as high as 10 A that last for a few tens of microseconds. However, these currents are much smaller in amplitude and shorter in duration compared to the currents experienced by a person receiving a direct strike (i.e., one who has launched a so-called successful connecting leader). Thus, the injuries inflicted on humans by currents in connecting leaders may not be as severe as those from a direct lightning strike. However, depending on their duration, these currents can cause temporary paralysis of the limbs, and the victim may collapse to the ground. The victim might be oblivious to the event or be confused and disoriented after it.

This injury from a connecting leader can happen only if the person is located in a specified range of distances from the lightning channel, and this distance range depends on the strength of the stepped leader or the return stroke current [6]. This range of distances is marked 'aborted leader zone' in Fig. 16.10. If the person is located closer than the lower limit of this distance range to a downward moving stepped leader, he or she will experience a direct strike. In other words, at distances closer than the lower limit of this range, the person is located within the striking distance of the stepped leader (Chap. 17). If the person is located at distances greater than the maximum limit of this range, then the electric field generated by the stepped leader would not be able to generate a connecting leader. Figure 16.10 shows the regions, as a function of peak current of the return stroke, where a person would either receive a direct strike or only launch an unsuccessful connecting leader.

16.2.6 Shock Waves

Injuries can also be caused by shock waves created by a lightning channel. During a lightning strike, the channel temperature rises to approximately 25,000–30,000 K in a few microseconds, and as a result, the pressure in the channel may increase to several atmospheres. The resulting rapid expansion of air creates a shock wave. This shock wave can injure a human being located in the vicinity of the lightning flash. The pressure associated with the shock wave decreases with the distance rapidly, so that the shock wave can injure a human being only when located very close to the strike point of the lightning flash.

16.3 Different Types of Injury

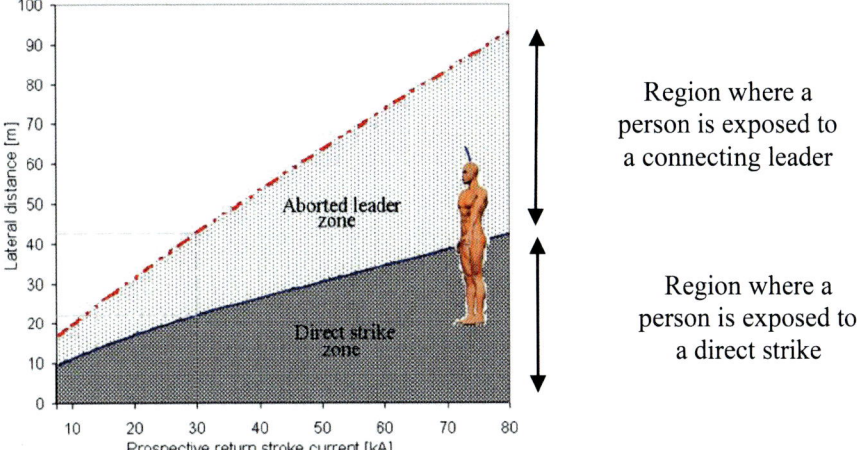

Fig. 16.10 Region or lateral distance from lightning strike where a person is exposed to a connecting leader. This region is shown as a function of the peak current of the first return stroke of the lightning flash. Note that for a return stroke with a 30-kA peak current, any person standing within approximately 20 m from the downward moving stepped leader could be struck directly by lightning. A person located within 20–30 m of the stepped leader (or strike point) will be exposed to a connecting leader (Adapted from [6])

16.2.7 A Single Lightning Strike Can Cause Injuries to Many People

It is also important to note that a single lightning flash can injure several people at the same time. If a lightning strike takes place in the vicinity of a group of people, one or several people can be struck directly by lightning (recall that different leaders and, hence, return strokes in the same flash can terminate on different people in a group) while the others can experience injuries from connecting leaders and step currents. If a group of people are holding hands, for example, the lightning current may pass through each member of the chain of people holding hands, causing severe injuries to the whole group (Fig. 16.11).

16.3 Different Types of Injury

The lightning current flowing inside a body, though small thanks to surface flashover, can cause various types of injury by the heating of tissue, electrolysis, and by upsetting the electrical state of excitable tissue (i.e., depolarization) [3, 4, 7, 8]. These effects are controlled by the way in which the lightning current is distributed inside the body. This in turn depends on the conductivity of bodily fluids and different types of tissue in the body. The current flowing outside can also cause injuries from heat and shock waves. Let us consider these effects one at a time.

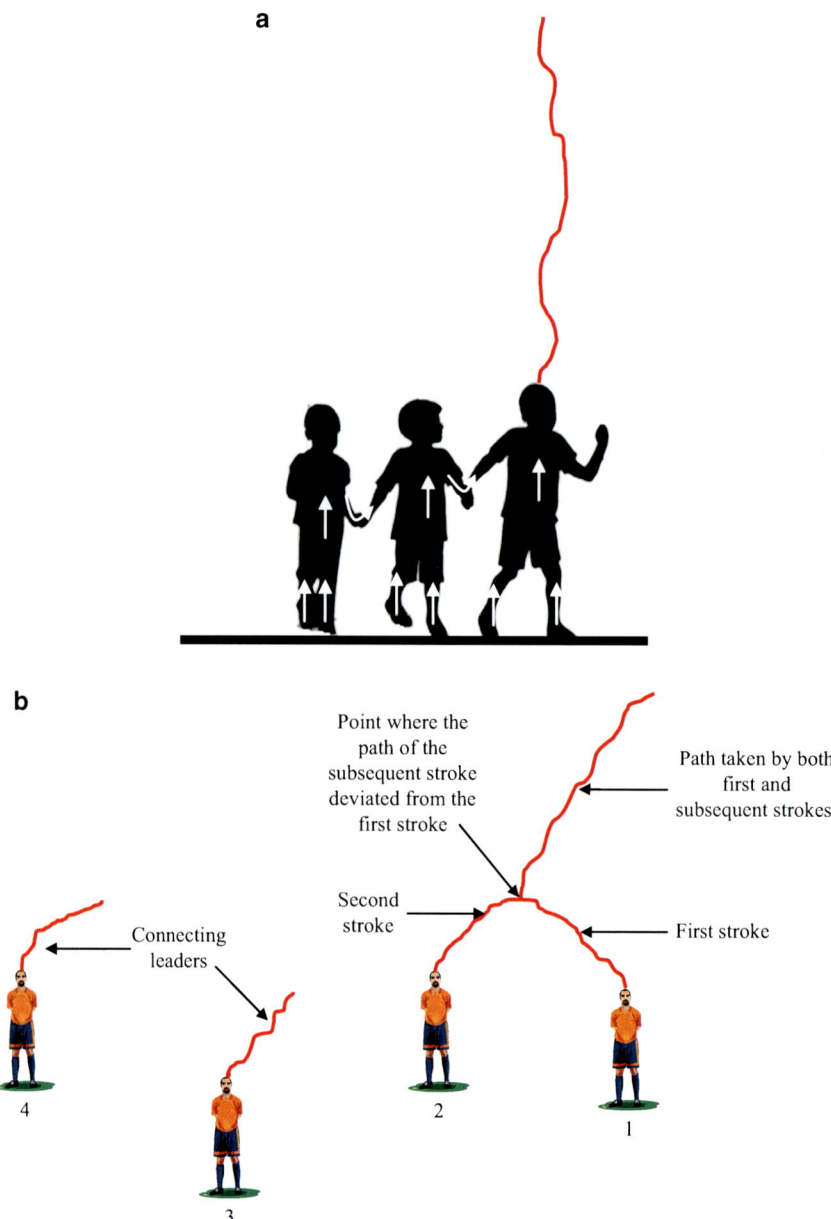

Fig. 16.11 (a) Example where a single lightning flash can injure several people: a lightning strike to a group of children holding hands. In this case the current passes from the body of the directly struck child to the other children. The *arrows* denote the direction of current flow (Figure created by author; children's images courtesy roseannapiter.com). (b) Second example of a single lightning flash injuring several people: a situation like this may occur in the case of a group of people located in a field (during a football game, for example) during a thunderstorm. A lightning strike in this case can cause injuries by step currents, connecting leaders (3 and 4) and a direct strike (1 and 2). Note that different strokes of the same flash can end up on different people, causing several people to receive direct strikes (Figure created by author)

16.3 Different Types of Injury

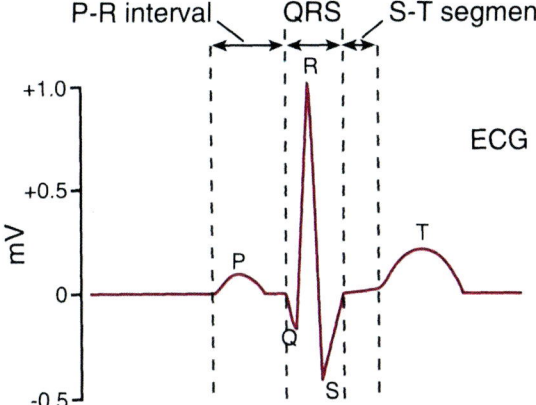

Fig. 16.12 Example of an electrocardiogram. The heart is more sensitive to electrical currents when it is in phase T, where the repolarization of the ventricles takes place. Any external electrical current that transgresses this portion of the cycle may produce the most deleterious effects. Thus, the injury received by a person during a lightning strike may depend on the phase of the heart in its cardiac cycle at the time of strike (Adapted from [9])

16.3.1 Cardiac and Respiratory Arrest

Cardiac and respiratory arrest is the major cause of death following a lightning strike. With appropriate first aid it is reversible in some cases. However, the mortality from lightning strikes remains approximately 20 %. A small number of patients can be resuscitated with external cardiac massage and expired air ventilation after cardiac arrest due to lightning injury, demonstrating the importance of this as primary first aid. The probability of cardiac arrest may also depend on the location of the heart's cycle at the time of the lightning strike.

The heart is controlled by a cycle of electrical impulses, and the sensitivity of the heart to external electric currents varies depending on where it is in this cycle at the time of lightning strike. The cumulative effects of the electrical impulses that control the heart are called an *electrocardiogram* (Fig. 16.12). The electrical cycle of the heart starts with the P wave that triggers the QRS impulse, which is caused by the contraction of the ventricles (heart chambers) and is responsible for the main pumping action. When this is completed, the T phase restores the electrical state of the heart to its initial conditions and prepares it for the next beat. It is during this T phase that the heart is more sensitive to electric currents. Therefore, whether or not a person suffers a cardiac arrest may depend not only on the magnitude of current flowing inside the body but also on the state of the heart during the lightning strike with respect to its electrical cycle. In many cases, victims can suffer both cardiac arrest and respiratory arrest. The respiratory arrest may last longer than the cardiac arrest, and as a consequence, the heart, even if it starts beating again after the first arrest, may not receive enough oxygen and suffer a second cardiac arrest. In addition to these effects, the current passing through the body can elevate the temperature of the fluids and muscles on its path, and as a result, internal organs may suffer fatal injuries.

The lack of oxygen to the heart may lead to permanent damage of the myocardium, but more importantly, the lack of oxygenated blood reaching the brain quickly leads to the death of brain tissue.

16.3.2 Injuries to the Eye

In the case of a lightning strike, both the current passing through the head and the strong radiation produced by the channel may cause a series of medical problems in the eye. The cataract is the most common long-term injury reported in lightning strikes. During lightning strikes a small break in the macula (highly pigmented yellow spot near the center of the retina of the human eye) can occur, causing blurred and distorted central vision. Such an injury is called a *macular hole*. A lightning injury may lead to a pulling or shifting of the retina from its normal position. Such damage is called *retinal detachment*. In addition to retinal detachment, lightning can induce wrinkles in the retinal tissue in one or more areas. These wrinkles cause small blind spots and are called *retinal folds*.

16.3.3 Injuries to the Ear

The most common injury is the rupture of the eardrum. The cause of this is likely the shock wave produced by the lightning flash. During a direct lightning strike to the upper part of the body, the ears can be located within a few centimeters of the lightning channel. The overpressure within a few centimeters of the lightning channel can reach values as large as 10 atm. This overpressure is the cause of the rapture of the eardrum. In some cases, even if the eardrum remains intact, the victim may still suffer from varying degrees of permanent hearing loss and so-called ringing in the ear (tinnitus). This is probably caused by damage to the hair cells and nerves in the cochlea, either from the shock wave or by the flow of current through it. The blast can also cause damage to ossicles inside the ear that can result in conductive deafness, especially at high frequency.

16.3.4 Injuries to the Nervous System

The flow of lightning current inside a body can cause intracranial hemorrhage, swelling of tissues (edema), and neuronal injury. These in turn can cause prolonged or even permanent neurological symptoms. The nervous system can also be affected by the lack of oxygen resulting from cardiorespiratory arrest. Lightning can also cause intense vasospasms and constriction of blood vessels and, thus, restriction of blood flow (and thus oxygen) to parts of the body. A lack of oxygen to a tissue is termed *tissue ischemia* and can cause further injuries to individual parts of the nervous system. A large current flowing through the brain can also lead to neuronal damage, which can lead to permanent brain damage.

16.3.5 Paralysis During Lightning Strikes

Lightning victims can be affected by temporary paralysis of the arms or legs through which lightning current has passed. This is called *keraunoparalysis* [4]. Usually after some time, limbs regain their normal functioning without intervention by a medical practitioner.

Lightning victims may experience a loss of consciousness for varying periods. Many victims have no recollection of the event, and in some cases memory of events a few days to a few weeks before and after the lightning event could be affected. Lightning can cause other specific items of brain dysfunction, for example, aphasia, an impairment of language expression. This may affect the production or comprehension of speech. The ability to read or write may also be affected.

In addition to keraunoparalysis, lightning victims may experience weakness, numbness, and tingling in muscles and tissues that may last for several weeks to years.

16.3.6 Burns Caused by Lightning

The burns caused by lightning are usually minor and require little treatment. Lightning can cause burn injuries ranging from superficial burns to full-thickness burns. Burns can occur anywhere – on the head, neck, trunk, upper extremities, hands, lower extremities, or legs. There are several ways in which lightning can cause burn injuries. When an electric discharge in air terminates on a solid body, a voltage difference of approximately 10–20 V is created across a thin layer of gas and vaporized solid matter (Chap. 15). In the case of metal objects, this is called a *cathode fall* and has a thickness of less than 1 mm. A similar *electrode layer* may arise at the gas–solid interface of the entrance and exit points of the lightning current into and out of the body. The heat generated in this gas layer is proportional to the total charge passing through the layer. This heat can cause full-thickness burns in the body tissue in contact with it. In lightning-burn victims it is often observed a characteristic burn pattern in the form of small, circular, full-thickness burns involving the sides of the soles of the feet and the tips of the toes. These are probably caused as the lightning current exits from the body by creating an electric discharge between the feet and the ground.

Fig. 16.13 Skin discolorations that resemble feathery marks are common in lightning-strike victims. These *feathery marks* could be a reaction of the skin to the surface electrical discharges flowing on the skin during the lightning strike (Adapted from [10])

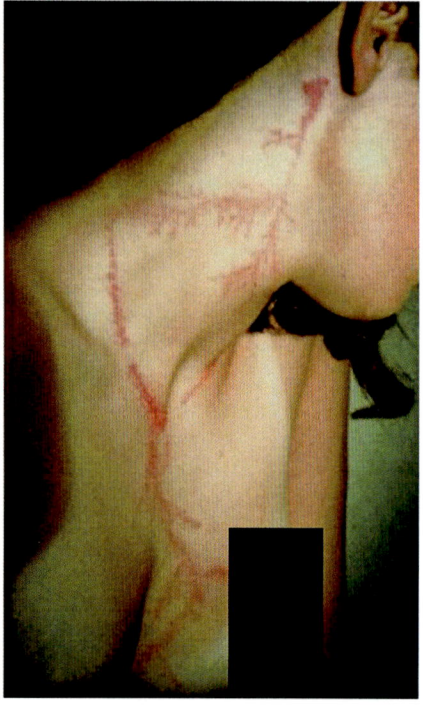

Many lightning-strike victims also develop a skin discoloration that resembles reddish brown feathery skin markings (Fig. 16.13). These markings, sometimes known as *keraunographic marks* or *arborizations*, are probably caused by the streamerlike electrical discharges connected to the main discharge channel propagating over the surface of the skin. The pattern on the skin may be an inflammatory reaction to the effects of these discharges. These patterns usually disappear within a day or two. Indeed, the pattern of discharge observed on the skin of lightning victims is very similar to those that result when electrical discharges are directed onto insulating photographic paper, i.e., Lichtenberg figures.

16.3.7 Psychological Injuries

In addition to physical damage, lightning victims may experience a range of psychological problems. These include the fear of thunderstorms, anxiety, depression, disturbances in sleep rhythms, panic attacks (a sudden rush of uncomfortable physical symptoms such as increased heart rate, dizziness or light-headedness, shortness of breath, inability to concentrate, and confusion), disorders of memory, learning, concentration, and higher mental facility. Some lightning victims repeatedly reexperience the ordeal in the form of flashback episodes, memories, nightmares, or frightening thoughts, especially when they are exposed to events or objects reminiscent of the trauma, for example thunderstorms or sudden bright lights. These problems may lead to altered bowel habits, constipation, or gastric dilation, in which the stomach becomes excessively dilated with gas, causing it to expand.

16.3.8 Injuries Caused by Shock Waves

As mentioned earlier, during a lightning flash the channel temperature may increase to approximately 30,000 K within a few microseconds. This rapid heating leads to the creation of a shock wave in the vicinity of the channel. As mentioned previously, the shock wave associated with the lightning flash may reach overpressures of approximately 10 atm in the vicinity of the channel. In addition to causing damage in the ear and eyes, this shock wave can also damage other internal organs such as the spleen, liver, lungs, and gastrointestinal tract. Moreover, it may displace the victim suddenly from one place to another, causing head and other traumatic injuries.

Blunt injuries can also be received from material ejected from a struck object. For example, when lightning strikes trees, the trunk of the tree can explode and the splinters can cause injuries in those standing nearby (Chap. 15). Blunt injuries can also be caused by flying objects inside buildings. During a lightning strike to an unprotected building, central power distribution switches, television sets, and antenna cables may explode causing injuries. Injuries may also be associated with falls from a location, like a cliff, in which a victim finds himself during a lightning strike.

16.3.9 Disability Caused by Lightning

Even though the risk of being killed by lightning is very small and a significant number of lightning victims survive the lightning strike, medical scientists working in the field believe that disabilities resulting from a lightning strike constitute a serious problem [11]. The primary cause of such disabilities is injuries to the nervous system coupled with psychological trauma. According to the available data, many survivors of lightning strikes are unable to return to their previous occupations, which may have devastating consequences both to the individuals concerned and their families. Primary areas of disability caused by lightning involve neurocognitive functions such as deficits in short-term memory and processing new information, personality changes, easy fatigability, decreased work capacity, chronic pain syndromes, sleep difficulties, dizziness, and severe headache.

References

1. Holle RA (2008) Annual rates of lightning fatalities by country. In: The proceedings of the 20th international lightning detection conference, Tucson, AZ, USA
2. Mackerras D (1992) Occurrence of lightning death and injury. In: Andrews CJ et al (eds) Lightning injuries: electrical, medical, and legal aspects. CRC Press, Boca Raton
3. Cooray V, Cooray C, Andrews C (2010) Lightning caused injuries in humans. In: Cooray V (ed) Lightning protection. IET Publishers, London
4. Andrews C (2003) Electrical aspects of lightning strike to humans. In: Cooray V (ed) The lightning flash. IET Publishers, London
5. Mackerras D (1992) Protection from lightning. In: Andrews CJ et al (eds) Lightning injuries: electrical, medical, and legal aspects. CRC Press, Boca Raton
6. Becerra M, Cooray V (2009) On the interaction of lightning upward connecting positive leaders with humans. Trans IEEE (EMC) 51:1001–1008
7. Andrews CJ, Cooper MA (1992) Clinical presentation of the lightning victim. In: Andrews CJ et al (eds) Lightning injuries: electrical, medical, and legal aspects. CRC Press, Boca Raton
8. Andrews CJ, Eadie M, ten Duis HJ, Raphael B, Cash R, Fraunfelder F, Meyer M, La Bergstrom V (1992) Pathophysiology of lightning injury. In: Andrews CJ et al (eds) Lightning injuries: electrical, medical, and legal aspects. CRC Press, Boca Raton
9. Rhoades RA, Tanner GA (2003) Medical physiology. Lippincott Williams & Wilkins, New York
10. Huss F, Erlandsson U, Cooray V, Kratz G, Sjöberg F (2004) Blixtolyckor – mix av elektriskt, termiskt och multipelt trauma. Lakartidningen 101(28–29):2328–2331
11. Cooper MA, Andrews CJ (2003) Disability, not death is the issue. In: International conference on lightning and static electricity, Blackpool, UK

Chapter 17
Basic Principles of Lightning Protection

17.1 Function of a Lightning Conductor

A lightning conductor is a device invented by Benjamin Franklin to protect buildings from lightning strikes. It provides a low-resistance path for the lightning current to flow to the ground, preventing any damage that would have resulted if the lightning current had passed through the building to the ground. Let us now consider how a lightning conductor acts during a lightning strike.

As described in previous chapters, when the lower end of a stepped leader is roughly 100–200 m from the Earth, connecting leaders are likely to initiate at protruding earthed objects and to propagate toward the tip of the stepped leader channel. Several connecting leaders may start from different objects or from different protrusions on the same object. Usually only one of these connecting leaders reaches the stepped leader and the protrusion or the object that sent up the successful connecting leader receives the current in the lightning flash. However, the lightning conductor, being located at the highest point of the structure, will be the first object on the structure to be affected by the electric field of the stepped leader and, hence, the first object on the structure to launch a connecting leader. Because of this lead, it is usually the lightning conductor that succeeds in making a connection with the downward moving stepped leader and receives the current in the lightning flash. Thus, the function of a lightning conductor is to divert to itself a lightning discharge that might otherwise strike a vulnerable part of the structure requiring protection and to safely conduct the current in the lightning flash to the ground.

Benjamin Franklin assumed that a lightning conductor functioned by generating a positive corona charge from its tip that will flow into the cloud and neutralizes the negative charge. For this reason, the tips of lightning conductors were made sharp to promote the generation of corona currents. However, today we know that this concept does not work because the charge generated by such corona currents is very small in comparison to the charge in the cloud. But as we will discuss later, this same discarded concept is being used by some lightning protection system manufacturers to create and commercialize lightning protection systems.

Recent research has shown that when allowed to compete with each other, lightning rods with moderately blunt tips perform better than rods with sharp tips.

17.2 Attractive Range of a Lightning Conductor – Electro-Geometrical Method

The electro-geometrical method (EGM) is a simple procedure used by lightning protection engineers to evaluate the attractive range of a lightning conductor [1]. Let us first examine the basic assumptions of this method.

Two basic assumptions are made in EGM. The first assumption is that in evaluating a lightning attachment to structures one can neglect the effect of connecting leaders. Since connecting leaders are an integral part of lightning attachment, this assumption is not justified. However, the length of the connecting leader depends on the height of the structure. It increases with increasing structure height. This assumption is therefore justified only in the case of short structures where the influence of the connecting leader on the lightning attachment is slight. With increasing structure height the effect of connecting leaders on lightning attachment increases, and one would expect the results obtained from EGM to deviate from reality. The second assumption is that a stepped leader is attracted to a grounded structure when it reaches a critical distance from the grounded structure. This critical distance is called the *striking distance*. It can be described as the distance of the field of vision of the stepped leader. If the distance from a grounded structure to the stepped leader is larger than the striking distance, then the stepped leader is not aware of the grounded structure. The reason for this assumption in EGM is as follows. As the distance between a grounded structure and the tip of the stepped leader decreases, the average electric field in the space between the grounded structure and the tip of the stepped leader increases. When this average electric field reaches a critical value, the positive streamers (in the case of a negative stepped leader, of course) generated from the extremity of the structure will be able to propagate in the space between the structure and the stepped leader connecting the stepped leader to the grounded structure. When the electric field that exists between the stepped leader and the grounded structure reaches this critical value, the attachment process is said to have reached the final jump condition. Once the final jump condition is established beween a point on the structure and the stepped leader the attachment of the lightning flash to that point is imminent. As we saw in Chap. 2, the background electric field necessary for the propagation of positive streamers is approximately 5×10^5 V/m. Therefore, when the final jump condition is reached, the average electric field between the grounded structure and the stepped leader is equal to the aforementioned value. From another point of view,

17.2 Attractive Range of a Lightning Conductor – Electro-Geometrical Method

the average electric field between the stepped leader and the ground is given by V/d, where V is the potential of the tip of the stepped leader (in volts) and d (in meters) is the separation between the structure and the tip of the stepped leader. Thus, the striking distance S is given by

$$S = V/5 \times 10^5. \tag{17.1}$$

Observe that the electric field in the space between the stepped leader and the grounded structure is determined by the charge on the stepped leader. Of course with increasing V the charge on the stepped leader increases. Thus, the greater the charge, the larger the striking distance (i.e., the distance over which the final jump condition is reached). The charge on the leader channel is related to the peak current of the prospective return stroke: as the charge on the leader channel increases, the peak return stroke current increases. Thus, a larger charge corresponds to a larger return stroke current. Consequently, the striking distance of the stepped leader increases with increasing prospective return stroke peak current. The same conclusion was arrived at by analyzing the potential of the stepped leader in Chap. 14. In lightning protection standards it is assumed that the striking distance, S (given in meters), is related to the prospective peak return stroke current (i.e., the return stroke generated by the given stepped leader), I_p (given in kA), by the equation [1]

$$S = 10 I_p^{0.65}. \tag{17.2}$$

This equation is based on the voltages necessary for the breakdown of long gaps in the laboratory. Recall that in Chap. 14, the striking distance was estimated as a function of the first return stroke peak current by plugging in an expression for the potential of the stepped leader as a function of return stroke peak current. The expression arrived at in Chap. 14 is

$$S = 11.7 + 3.14 I_p - 0.656 \times 10^{-2} I_p^2. \tag{17.3}$$

Figure 17.1 depicts these two functions. Note that from 3 to 30 kA the difference between the two curves is less than 10 %. The difference increases to approximately 25 % at 100 kA.

Once the striking distance is known as a function of return stroke peak current, it is possible to evaluate how a lightning conductor will function. Let us consider a lightning conductor of height h. We will also consider the approach of a stepped leader, with a prospective return stroke current of I_p having a striking distance of S, toward the ground in the vicinity of this lightning conductor. For simplicity, we assume that the path of the stepped leader is straight and vertical. In reality, the direction of approach of the stepped leader can have any angle. Now, the downward moving leader has two options. It can strike either the lightning conductor or the ground. At any given moment in time the tip of the stepped leader is at a certain distance from the rod, say S_r, and a

Fig. 17.1 Striking distance as given by Eqs. 17.2 and 17.3 (Figure created by author)

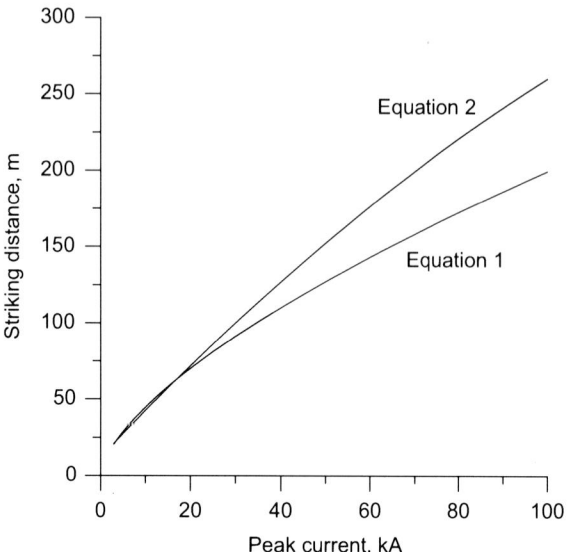

certain distance from the ground, say S_g. If S_r becomes equal to S before S_g, then the stepped leader will strike the rod. If S_g becomes equal to S before S_r, then the stepped leader will go to ground. With this information in mind, let us consider what happens as the lateral distance from the lightning conductor to the stepped leader, say d, changes. The possible outcomes of this exercise are shown in Fig. 17.2. Two situations are depicted in this figure, namely, for $S < h$ and $S > h$. Recall that what is given in Fig. 17.2 is just a cross section of three-dimensional space. The flat line shows the final location of the tip of the stepped leader where it will be attracted to the ground, and the curved region shows the location of the stepped leader tip where it will be attracted to the conductor. In the case where $S < h$, the stepped leader can end on the side of the conductor, whereas when $S > h$ it is always attracted to the tip of the conductor. The maximum lateral distance from which a stepped leader is attracted to a lightning conductor is called the *attractive radius*. From the preceding analysis the attractive radius R of a lightning conductor can be derived as

$$R = \sqrt{S^2 - (S - h)^2} \quad \text{for} \quad S > h, \tag{17.4}$$

$$R = S \quad \text{for} \quad S \leq h. \tag{17.5}$$

By combining these equations with Eq. 17.2 or 17.3, one can estimate the attractive range of lightning conductors of different heights as a function of peak return stroke current.

17.2 Attractive Range of a Lightning Conductor – Electro-Geometrical Method

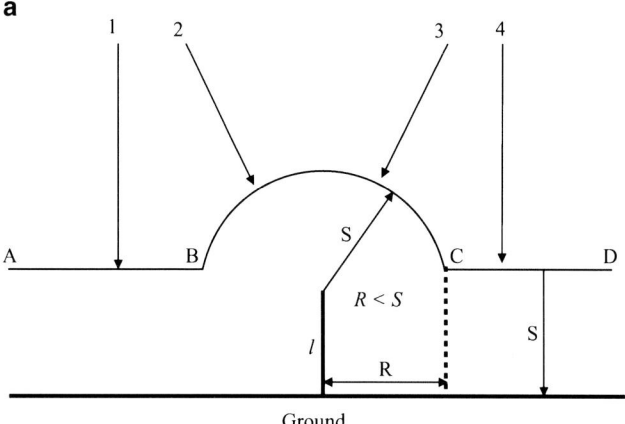

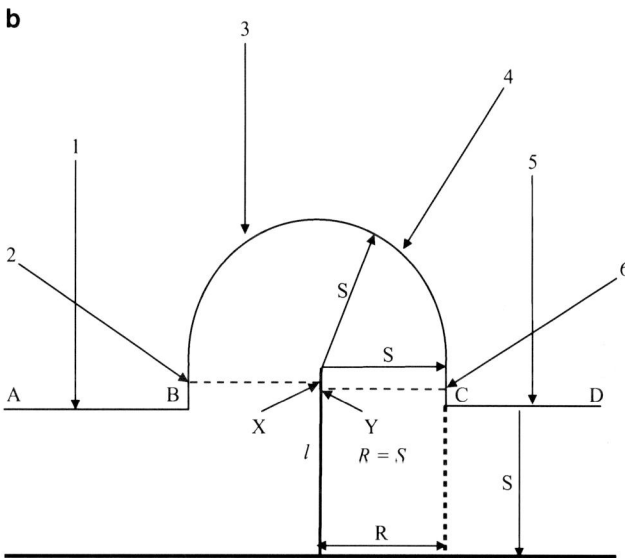

Fig. 17.2 (**a**) Cross section of three-dimensional surface describing possible outcomes in case of a stepped leader approaching ground in vicinity of a lightning conductor for the case $S > l$ where S is the striking distance and l the height of the structure or the conductor. The *straight line* sections (i.e., AB and CD) show cases where the stepped leader goes to ground, and the *curved region* shows cases where the stepped leader is attracted to the lightning conductor. If the downward moving stepped leader ends on the *straight line section*, it goes to ground. Otherwise, it ends on the conductor. The downward moving stepped leaders 1 and 4 go to ground, while stepped leaders 2 and 3 end on the conductor. In the figure, R is the attractive radius of the lightning conductor, and in this case, $R < S$ (Figure created by author). (**b**) Cross section of three-dimensional surface describing possible outcomes in case of a stepped leader approaching ground in vicinity of lightning conductor for the case $S < l$ where S is the striking distance and l the height of the structure or the conductor. The *straight line sections* (i.e., AB and CD) show cases where the stepped leader goes to ground and the *curved region* shows cases where the stepped leader is attracted to the lightning conductor. If the downward moving stepped leader ends on the *straight line section*, it goes to ground. Otherwise, it ends on the conductor. In this case, the conductor can

17.3 Estimating the Location of Lightning Conductors Using EGM – Rolling Sphere Method

The assumption that the striking distance of a downward moving stepped leader depends only on the prospective return stroke current made it possible to utilize the concept very conveniently in lightning protection. Since the EGM striking distance depends only on the return stroke current, it can be assumed that whatever the structure under consideration, the part of the structure that comes within the striking distance is the part that receives the lightning strike. Since S is constant (or assumed to be constant) for a given stepped leader (or a given prospective return stroke current), it can be assumed that there is a spherical region of radius S with the tip of the stepped leader at the center having the following property: the first structure or part of the structure that enters this spherical volume is the one to receive the lightning strike. This concept can be utilized in locating lightning conductors on a grounded structure. Since the goal of lightning protection is for a stepped leader approaching a structure from any direction to be attracted to the lightning conductors, the lightning conductors should be placed on the structure in such a way that when a sphere of radius S is rolled over the protected structure, the sphere touches only the conductors of the lightning protection system (Fig. 17.3). Once this is done, the first object that enters the sphere of vision of the stepped leader is a lightning conductor, and as a consequence, the leaders coming down in the vicinity of and over the structure will be attracted to the lightning conductor and spare the structure. This method of locating lightning conductors on a structure is called the *rolling sphere method* [1].

As is apparent from Eq. 17.1, the striking distance is a function of the prospective return stroke peak current, as is the radius of the sphere that should be rolled around the structure to determine the location of lightning conductors. So what is the radius of the sphere to be selected in lightning protection? To solve this problem, lightning protection engineers have categorized lightning protection of structures into four levels. These levels are presented in Table 17.1. If a structure is protected at Level I, then any return stroke peak current larger than 3 kA will be stopped by the lightning protection system. That is, the radius of the sphere used in designing the lightning protection system will be 20 m. If a structure is protected at level II, then any peak current larger than 5 kA will be intercepted by the lightning protection system. At level III the safety level is 10 kA and at level IV 16 kA.

Consider a structure that is protected at level IV. This does not mean that all leaders that come down over the structure having prospective return stroke currents smaller than 16 kA will strike it. Many of these stepped leaders will also be captured

---

Fig. 17.2 (continued) be struck below its top by a stepped leader that approaches the conductor at an angle. The stepped leaders marked 1 and 5 strike the ground, while stepped leaders 3 and 4 strike the tip of the conductor. Stepped leaders 2 and 6 strike the side of the conductor at points X and Y, respectively. In the figure, R is the attractive radius of the lightning conductor, and in this case, $R = S$ (Figure created by author)

17.3 Estimating the Location of Lightning Conductors Using EGM – Rolling...

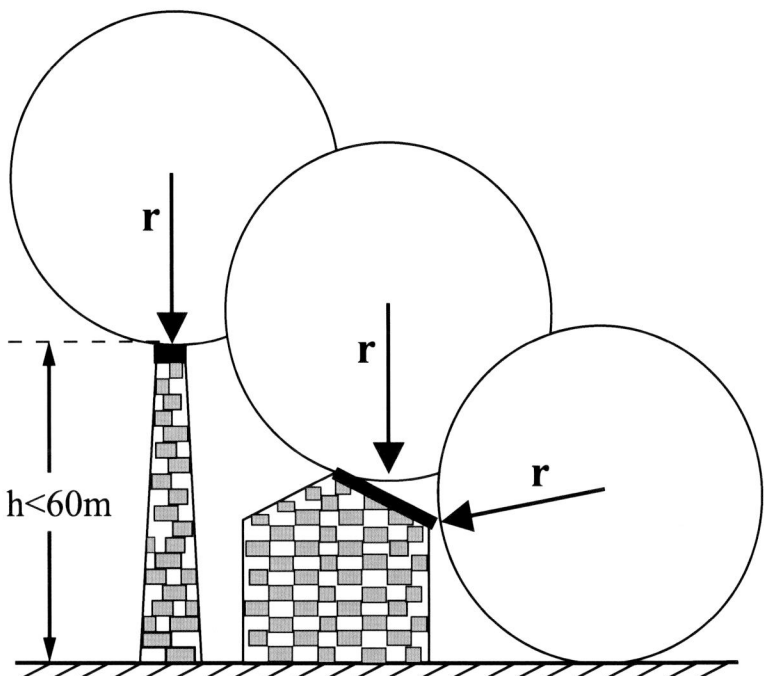

Fig. 17.3 Rolling sphere method of lightning protection. The radius of the rolling sphere that should be selected in the analysis depends on the level of protection or risk level selected. In the case of risk levels I–IV, the corresponding radii are 20, 30, 45, and 60 m, respectively. The corresponding critical currents are 3, 5, 10, and 16 kA. The lightning protection system is designed in such a way that when the sphere is rolled over the structure, it makes contact only with the lightning protection system (Adapted from [1])

Table 17.1 Minimum current allowed to penetrate the structure corresponding to different classes of the lightning protection system

Class of lightning protection system	Minimum associated current (kA)
I	3
II	5
III	10
IV	16

by the lightning protection system. For example, any stepped leader having a prospective return stroke current less than 16 kA approaching the structure within its striking distance from a lightning conductor will be intercepted by the lightning conductor (Fig. 17.4).

Observe that having a lightning protection system does not mean that the structure will not be struck by lightning. It will be struck by lightning having current peaks smaller than one associated with the level of protection. Thus, before

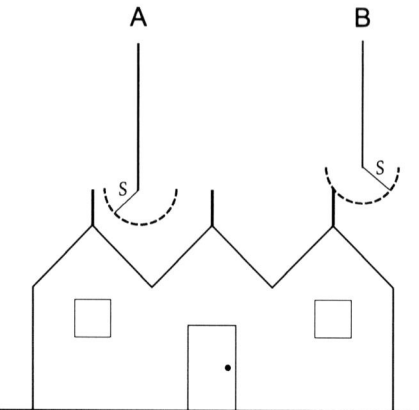

Fig. 17.4 If a structure is protected based on a certain risk level (i.e., level I, II, III, or IV) with an optimum current (e.g., level I corresponds to 3 kA), this does not mean that all stepped leaders with prospective currents less than that optimum value will strike the ground. Some of them are still captured by the lightning protection system. For example, stepped leader B is captured by the conductor, whereas stepped leader A, though it has the same prospective return stroke current (or the same rolling sphere radius), does strike the structure (Figure created by author)

installing a lightning protection system, one must have some idea of the magnitude of the peak return stroke current that the structure can safely withstand. Once this decision is made, the level of protection necessary can be selected. Observe that the higher the sensitivity of the structure (or its contents) to lightning flashes, the higher the level of protection that should be selected. Accordingly, for sensitive structures level I of lightning protection must be selected.

The lowest level of protection at which a structure is protected according to lightning protection standards is level IV. This corresponds to a striking distance of 60 m and, hence, a rolling sphere radius of 60 m. What happens when the structure is taller than 60 m? In this case the rolling sphere method indicates that the sides of the structure can be struck by lightning (Fig. 17.5). This fact was also illustrated in Fig. 17.2b. This makes it necessary to provide lightning protection at the side of the structure if the structure is taller than 60 m.

17.4 Angle of Protection

Consider a tall lightning conductor. Applying the rolling sphere method to this conductor we observe that there is a region around the conductor where stepped leaders do not penetrate. This region is identified in Fig. 17.6. The smaller the sphere radius, the smaller the protected volume. In other words, the volume of protection offered by a lightning conductor increases with increasing return stroke current. The protected region can be approximately described by a conical region,

17.4 Angle of Protection

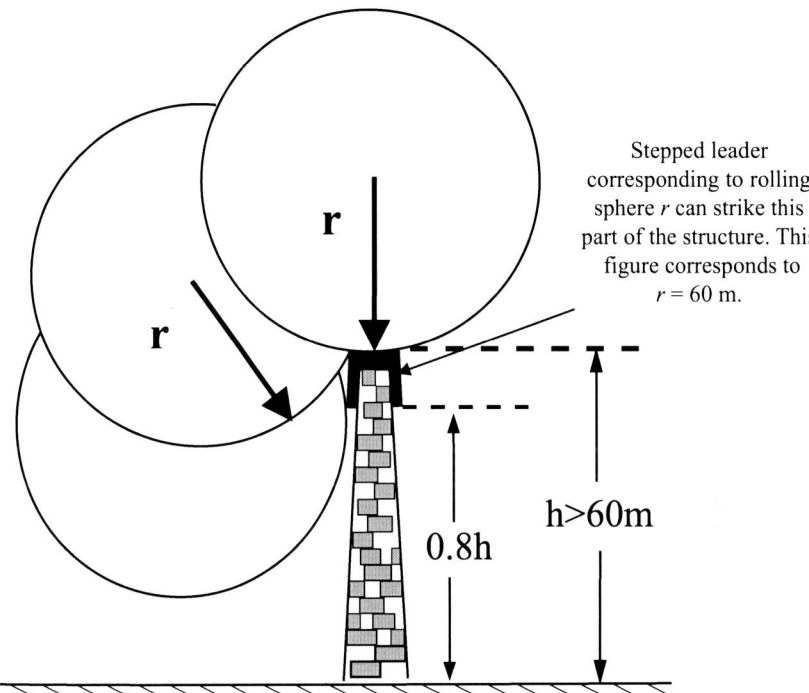

Fig. 17.5 If the structure is higher than the radius of the rolling sphere used in the design, then lightning flashes can strike the sides of the structure. The side of the structure where a lightning flash could terminate is marked by a *thick line* (Adapted from [1])

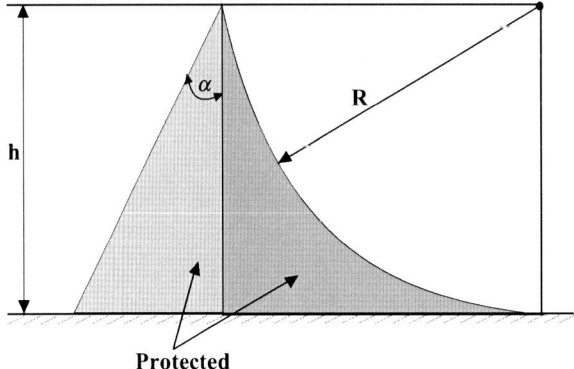

Fig. 17.6 Volume of space protected by lightning conductor. *Right side*: volume protected by conductor according to rolling sphere method; *left side*: this volume is approximated by a conical region (note that the figure is a cross section of a three-dimensional volume). The definition of the angle of protection, marked α, is based on this approximation (Figure created by author)

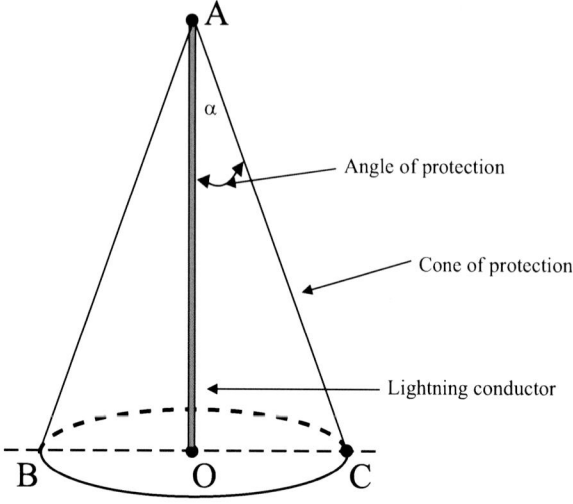

Fig. 17.7 Definition of angle of protection. The volume protected by a lightning conductor according to the angle of protection method (Adapted from [1])

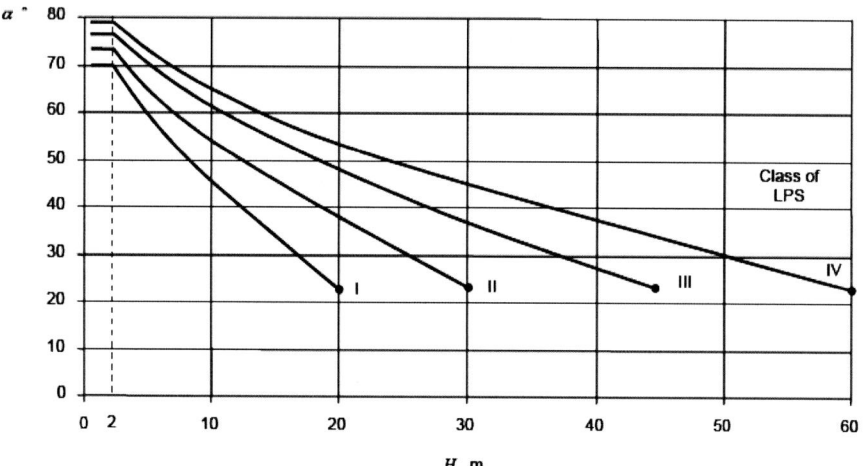

Fig. 17.8 Angle of protection corresponding to conductors of different heights and for different protection levels according to IEC lightning protection standards [1]. Note that for protection level I, the corresponding rolling sphere radius is 20 m and the angle of protection is defined only for conductors of length up to 20 m. Similar consideration applies to protection levels II–IV (Adapted from [1])

and this has come into practice as the cone of protection of the protection angle method. That is, each conductor protects a conical region around it, and the angle at the apex of the cone is called the *angle of protection* (Fig. 17.7). Naturally, the angle of protection decreases with decreasing current amplitude. Figure 17.8 shows a cone of protection corresponding to conductors of different heights.

17.5 Volume of Protection Provided by Horizontal and Vertical Conductors

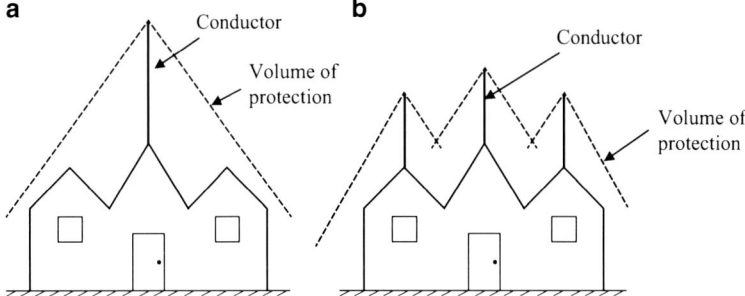

Fig. 17.9 A structure can be protected either by a tall single conductor as in (**a**) or by a large number of short conductors as in (**b**). The main idea is that the structure lies in the protection zone of the conductor(s) (Figure created by author)

The cone of protection can be used in estimating the location of lightning conductors on a structure. Thus, an extensive structure can be protected by a small number of tall conductors or a larger number of small conductors (Fig. 17.9). The idea is to set up lightning conductors in such a way that the cumulative effect of the cone of protection of all the conductors covers the structure completely, providing it with protection from lightning flashes.

17.5 Volume of Protection Provided by Horizontal and Vertical Conductors

The protection volume offered by a horizontal conductor located at a height h from the ground can be calculated using either the rolling sphere method or the protection angle method. The shape of this volume is shown in Fig. 17.10a. The volume of protection offered by a horizontal conductor terminated at both sides by vertical conductors is shown in Fig. 17.10b. Note again that the volume of protection depends on the peak amplitude of the return stroke current. The volume of protection offered by vertical and horizontal conductors can be used to protect grounded structures employing simple procedures. For example, as shown in Fig. 17.11a, two conductors can be erected in such a way that the structure is within the cone of protection of the conductors. As shown in Fig. 17.11b, a combination of two vertical conductors and a horizontal conductor can be applied to create a protective volume within the location of conductors. Figure 17.11c shows how three vertical conductors or towers are used to protect rockets on a launching pad.

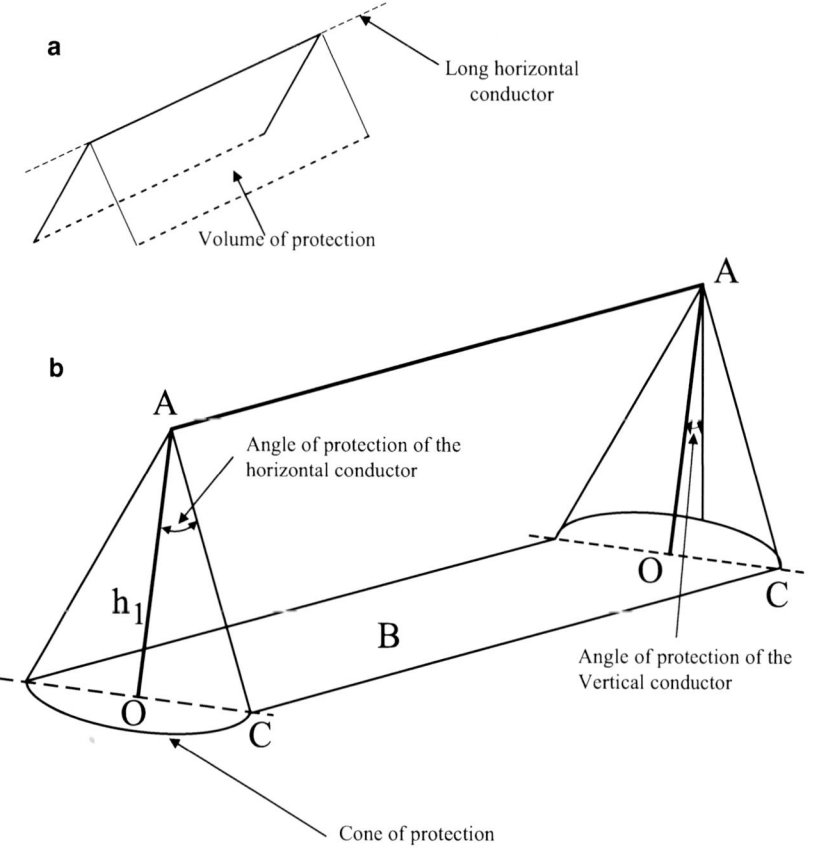

Fig. 17.10 (a) Volume of protection provided by a long horizontal conductor. (b) Volume of protection provided by a horizontal conductor terminated at both sides by a vertical conductor. This is actually the sum of the volumes protected by vertical and horizontal conductors (Adapted from [1])

17.6 Estimating the Number of Lightning Flashes Striking a Structure Over a Given Period of Time

The principles outlined previously can be used to estimate the number of lightning flashes striking a given structure over a given time interval. To estimate that, first we must know the ground flash density, the number of lightning flashes striking a unit area (usually 1 km^2) over a given period of time (usually 1 year), in the region under consideration. Let us denote this by N_g. Since the first return stroke current amplitude changes from one lightning flash to another, we should know the probability distribution $f(i_p)$ of the first return stroke currents in lightning flashes in the region under consideration. The probability distribution $f(i_p)$ is defined in such a way that $f(i_p)di_p$

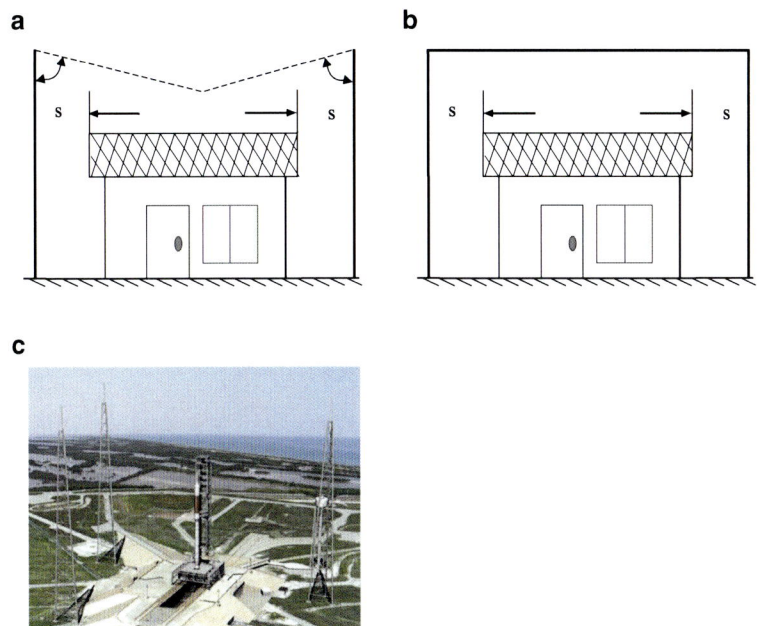

Fig. 17.11 Different methods of protecting a structure using vertical and horizontal conductors. (**a**) Placing two vertical conductors by the side of the structure so that the structure is within the cone of protection of the two conductors. Note that to avoid side flashes from the conductor, there should be a minimum critical distance (marked *s*) between the conductor and the structure. (**b**) By introducing a horizontal conductor that connects the two vertical conductors, it is possible to increase the level of protection. (**c**) A rocket in a launching pad protected by three tall masts arranged in a triangle around the launching pad (panels a and b are adapted from [1] and panel c is from http://www.nasa.gov/mission_pages/constellation/multimedia/LPS_concept.html, 2014)

gives the fraction of lightning flashes having peak current amplitudes in a range of i_p to $i_p + di_p$. Now, let the attractive area of the structure for a peak current i_p be $R(i_p)$. Thus, all lightning flashes with stepped leaders having prospective return stroke currents i_p and $i_p + di_p$ approaching the structure within a radius of $R(i_p)$ will strike the structure. Because N_g is the number of lightning flashes per unit area, $N_g f(i_p) di_p$ gives the total number of lightning flashes having current amplitudes in the range i_p and $i_p + di_p$ that occur in a unit area. Thus, in an area of $\pi R^2(i_p)$ we have $\pi R^2(i_p) N_g f(i_p) di_p$ number of lightning flashes with currents in the previously given range. They all will strike the structure because their attractive radius is equal to $R(i_p)$. Now, to estimate the total number of lightning flashes striking the structure, we must sum up the contributions from all the ranges of currents, and this can be done by integration. Thus, the total number of lightning flashes that strike the structure over a given period of time (specified in the definition of ground flash density) is

$$N = \int_0^\infty \pi R^2(i_p) N_g f(i_p) di_p. \tag{17.6}$$

Thus, knowing the ground flash density, the attractive radius of the structure as a function of return stroke peak current and the lightning current distribution makes it possible to estimate the total number of lightning flashes striking a structure over a given time.

17.7 Basic Features of External Lightning Protection System

So far we have discussed the attractive range of lightning conductors and the method used by scientists to estimate the location of the conductors. As mentioned earlier, the function of lightning conductors is to intercept a lightning flash and safely transport the lightning current to the ground so that the structure will not experience a lightning strike. The part of the lightning protection system that intercepts the lightning flashes is called the *air terminals*, and they form one unit of the external lightning protection system (Fig. 17.12). The dimensions of the air terminals should be such that they will not explode or vaporize even in the case of very large lightning currents. The correct dimensions are specified in the international lightning protection standard [1].

A lightning current intercepted by air terminals must be safely transported to the ground by the lightning protection system. The conductors that do this job are called *down conductors*. In placing down conductors, one must consider several facts. As the currents flow along these conductors, they will be raised to a high potential (depending on the self-inductance and the impedance at the point of grounding), and there should be no metal structures in the vicinity of these down conductors on which a side flash could be generated. If this is not possible in practice, then these

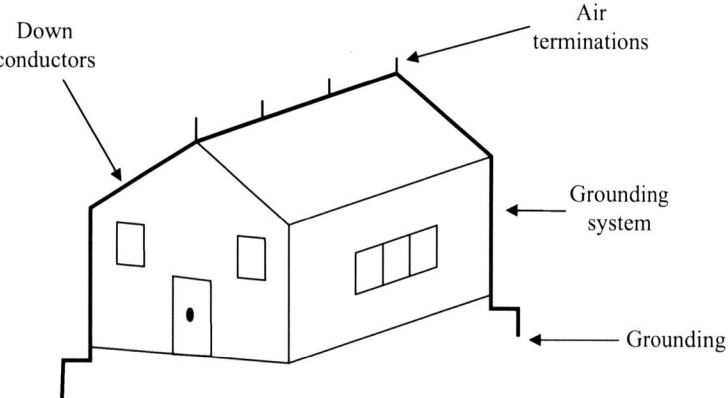

Fig. 17.12 Different parts of external lightning protection system. It consists of air termination, down conductors, and a grounding system (Figure created by author)

17.7 Basic Features of External Lightning Protection System

Fig. 17.13 During a lightning strike to a lightning conductor, even if the grounding resistance is zero, different parts of the conductor could be at different potentials because of the inductance of the conductor. For this reason, any conductors in the vicinity of down conductors should be bonded to the down conductor or the separation between them should be such that no side flashes can occur from the down conductor to the conductor connected to ground (Figure created by author)

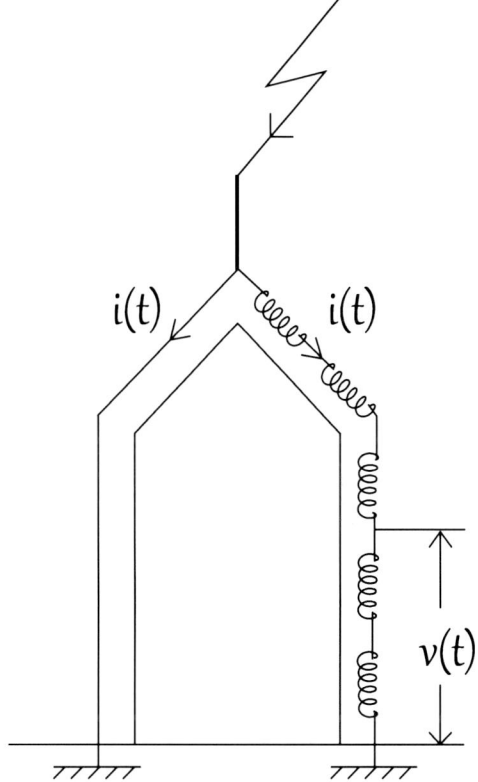

additional conductors or metal elements must be bonded to the down conductors to make their potentials equal. This is called *potential equalization*. If the down conductors are located along reinforced concrete walls, they should be connected to the reinforcing steel at the top and the bottom of the structure so as to avoid any voltage difference between the down conductors and the reinforced steel structure. The probability of occurrence of side flashes can also be reduced by reducing the resistance or impedance at the ground end of the down conductors. But even when the ground resistance is zero, there is still a voltage difference between the top of the down conductor and its lower end because of the self-inductance of the down conductor (Fig. 17.13). Second, the current flowing along the down conductor generates a magnetic field that varies with time. The interaction of this magnetic field with any other conducting loops in the vicinity of the down conductors can generate induced voltages in these loops, disturbing or damaging any electronic equipment connected to them. Moreover, these induced voltages can generate sparks between small gaps in conductors. The down conductors should be arranged in such a way so as to minimize these unwanted voltages. This can be done by symmetrically placing the down conductors (at least two of them) around the structure to share the current and to minimize the magnetic field inside the structure. Of course, all such down conductors and the air terminals should be connected to each other. Placing several down conductors around the building also makes it possible to approximate the topology of the lightning protection system to a Faraday cage, which is an effective method of protecting the structure.

The goal of lightning protection systems is to transport lightning current safely to ground. This means the current transported down by the down conductors should dissipate safely into the ground without increasing the potential of the down conductor to high values capable of causing sparks. When grounding down conductors, attention must be paid to several important facts. First, at the point of grounding the resistance must be very small. For example, if the grounding resistance is R, then the peak voltage that develops at the grounding point is $I_p R$, where I_p is the peak current. For a 10-Ω grounding impedance the voltage generated by a 30-kA peak current is 300 kV. This will be the voltage to which the down conductor is raised. Different arrangements of ground conductors to reduce the grounding resistance are shown in Fig. 17.14. Another fact to take into account in grounding is the corrosion of metal parts. Ground conductors should not be made of aluminum, and if the soil is corrosive, then a corrosive protection cover in the ground rods, including part of the down conductor, must be provided.

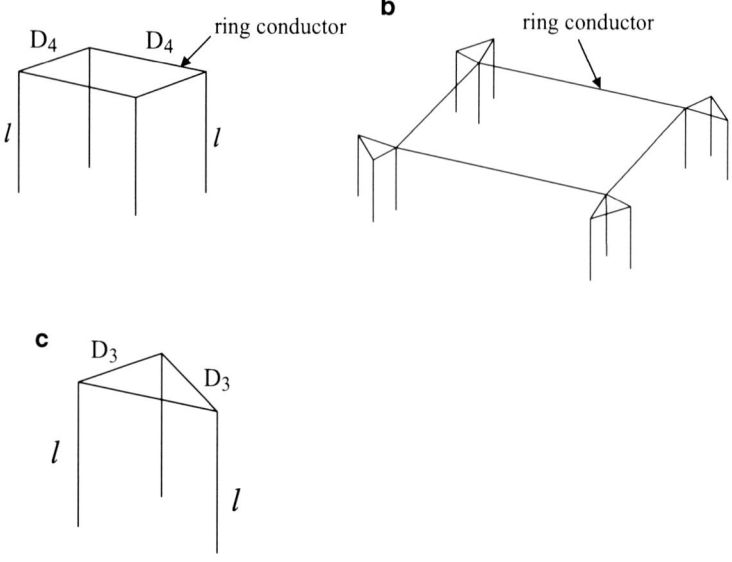

Fig. 17.14 To reduce the ground resistance, grounding rods of the grounding system should extend vertically to a depth of 3 m. Different arrangements of grounding rods as recommended by IEC are shown in the diagram [1]. (**a**) This arrangement could be used as a grounding system of a small metallic shelter (D4: sides of shelter). (**b**) Grounding system including ring conductor and triangular prismatic arrangement of vertical ground conductors. (**c**) Prismatic arrangement of vertical ground conductors

17.8 Mesh Protection Method

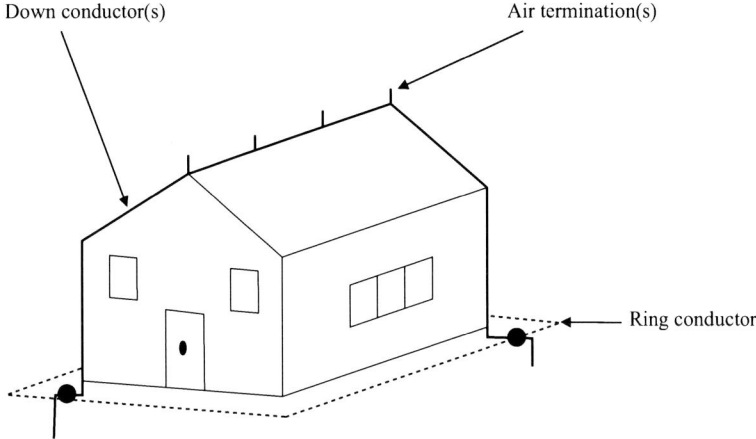

Fig. 17.15 The ring conductor can reduce the large potential differences that may occur at different points on the ground located inside it during lightning strikes. The ring conductor should be bonded to the down conductors (Figure created by author)

As a current enters the ground along the ground conductors, the current flow into the ground creates a potential difference at different points on the ground, and this can cause injuries because of step potentials during lightning strikes (Chap. 16). To reduce such voltages, a ring conductor that connects to all the down conductors is placed around the structure (Fig. 17.15). Such a conductor reduces the differences in potential at different points on the ground and reduce the strength of the step potentials.

17.8 Mesh Protection Method

As the preceding discussion makes clear, a lightning protection system cannot completely remove the lightning strikes to a structure. Lightning flashes with small currents could penetrate the structure by bypassing the lightning protection system. However, there might be structures where even a small lightning current could cause severe consequences, for example, structures storing ammunition and flammable liquids. On the other hand, according to the Faraday cage principle, if a current is injected into a closed metal container, the current will flow only along the outer surface of the container and no current will penetrate into the inner space. Thus, in principle, by enclosing a structure with a metal screen one can completely remove the lightning strikes to the structure. Obviously, this cannot be done in practice, but a derivative of this concept exists as a lightning protection method. It is called the *mesh method of lightning protection*. The idea is to cover part of the structure by a metal mesh. Usually, this method is used to protect large flat areas of a roof. The idea is to place a metal mesh at the top of the structure, instead of lightning conductors, and connect the mesh to ground using down conductors. In this method, the selection of the mesh size is done using a principle identical to the one used in evaluating the separation between lightning terminations. The current that is allowed to creep into the structure across the mesh is given by the same level of protection as was given earlier. The lower the amplitude of the currents allowed to strike the structure, the smaller the dimension of the cells of the mesh (Fig. 17.16). Reducing the size of the cells of the mesh makes it possible in principle to reduce the number of lightning flashes striking the structure to insignificant levels. The mesh can also be used to protect the sides of a tall structure where lightning flashes are expected to terminate (Fig. 17.17).

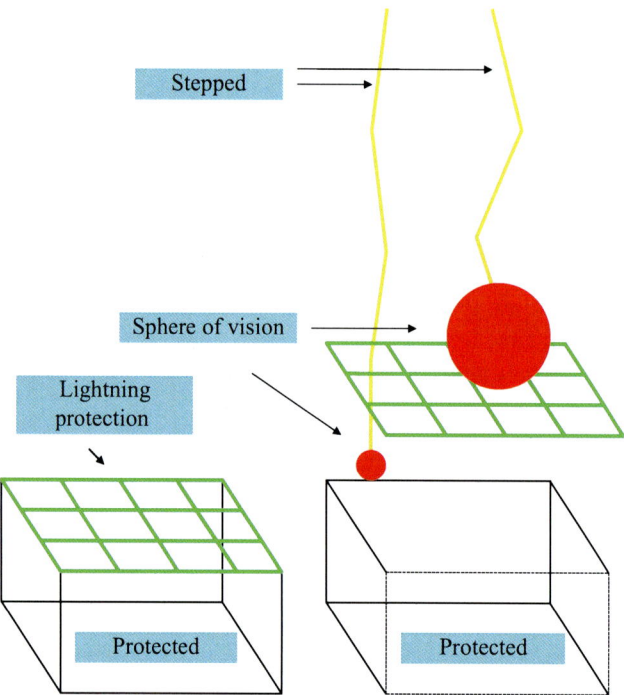

Fig. 17.16 Mesh method of lightning protection. One can visualize that at the tip of each stepped leader is a sphere of vision. The radius of the sphere of vision is equal to the radius of the corresponding rolling sphere. The stepped leader becomes attached to the first object that comes into the sphere of vision. The idea of the mesh method is to remove or filter out stepped leaders having a sphere of vision (or rolling sphere) larger than a certain radius. Since this radius is related to the peak current of the prospective return stroke, the idea is to filter out return strokes having peak currents larger than a certain critical value. For example, a mesh designed for lightning protection level (risk level) I filters out peak return stroke currents larger than 3 kA. Stepped leaders with prospective peak currents smaller than this critical value can penetrate through the mesh and strike the structure (Figure created by author)

17.9 Summary of External Lightning Protection System 319

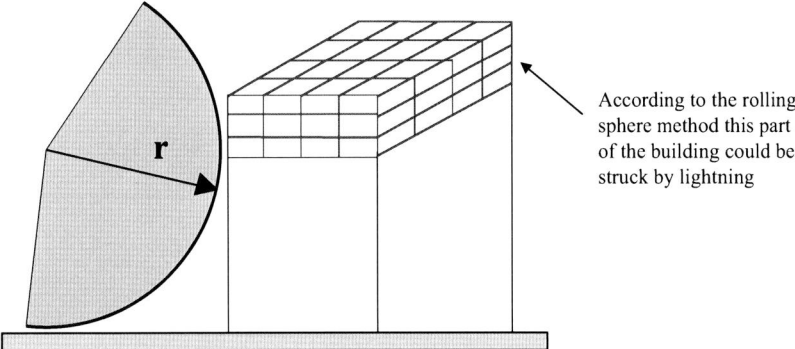

Fig. 17.17 The mesh method can be used to protect the sides of a structure. The mesh can be used not only to protect the roof but also sides of the tall building that might be struck by lightning. The height of the building is greater than the radius of the rolling sphere (Adapted from [1])

Table 17.2 Mesh sizes corresponding to different classes of the lightning protection system

Class of lightning protection system	Mesh size (m × m)
I	5 × 5
II	10 × 10
III	15 × 15
IV	20 × 20

Like the lightning protection using the rolling sphere method, the protection offered by the mesh method is divided into several levels. The mesh sizes corresponding to different levels of protection are given in Table 17.2. In practice, small air terminations are provided on the mesh to intercept lightning flashes.

Consider a stepped leader associated with a given rolling sphere radius. Figure 17.18 illustrates the location of the sphere at the interception of the stepped leader by the conductors of the mesh. Note that the sphere can penetrate through the mesh to a certain distance. To avoid any lightning strikes to the structure resulting from this penetration of part of the sphere through the mesh, in practice, the mesh is located slightly above the ground plane.

17.9 Summary of External Lightning Protection System

The aforementioned methods of protection – rolling sphere method, protection angle method, and mesh method – are applied in the protection of houses, tents, boats, power lines, and other structures. Basically, each system consists of lightning receptors, down conductors, and a grounding system. As mentioned previously, the goal is to safely conduct the lightning current to ground.

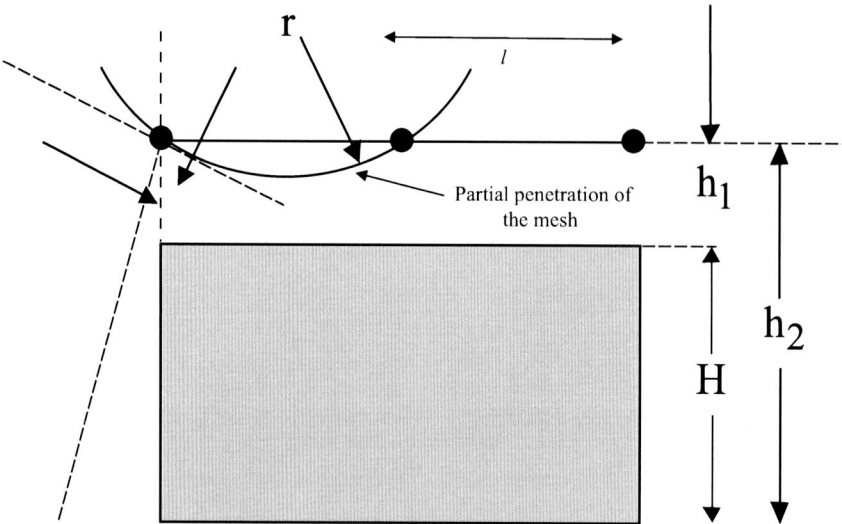

Fig. 17.18 To take into account the partial penetration of the sphere through the mesh, in practice the mesh is located slightly above the roof that is to be protected. In the diagram, the mesh is located at a height of h_1 from the roof. Note also that in the diagram, the parameter l is the size of the sides of the squares in the mesh (Adapted from [1])

17.10 Internal Lightning Protection System

It is important to understand that in any structure those belonging to the lightning protection system are not the only metal conductors crisscrossing different parts of the structure. A typical structure contains conductors belonging to a power system, telecommunications system, gas supply system, and water and drainage system. In the case of a lightning strike, all conductors belonging to these systems will be raised to a high potential, and if they are not properly bonded together, then large voltage differences can arise between conductors of different systems, causing flashover from one system to another. Thus, protecting a building from the effects of lightning requires not only an external lightning protection system but also procedures to reduce the occurrence of overvoltages in different conductors inside the building and mitigate their effects. The set of procedures developed to reduce the effects of lightning inside a building is collectively called the *internal lightning protection system*.

The first step to developing such a system is to bond all the conductors entering the building to a common ground bar, which is also bonded to the external lightning protection system. This is called *potential equalization* (Fig. 17.19). In some cases, the conductors entering a building cannot be bonded to ground at the entrance to the building. Conductors bringing in power to the building are an example. Thus, during a lightning strike, large voltages may appear in these conductors with respect to the other conductors that are bonded together to ground (including the ground wire of the incoming power). Since these conductors cannot be directly

17.10 Internal Lightning Protection System

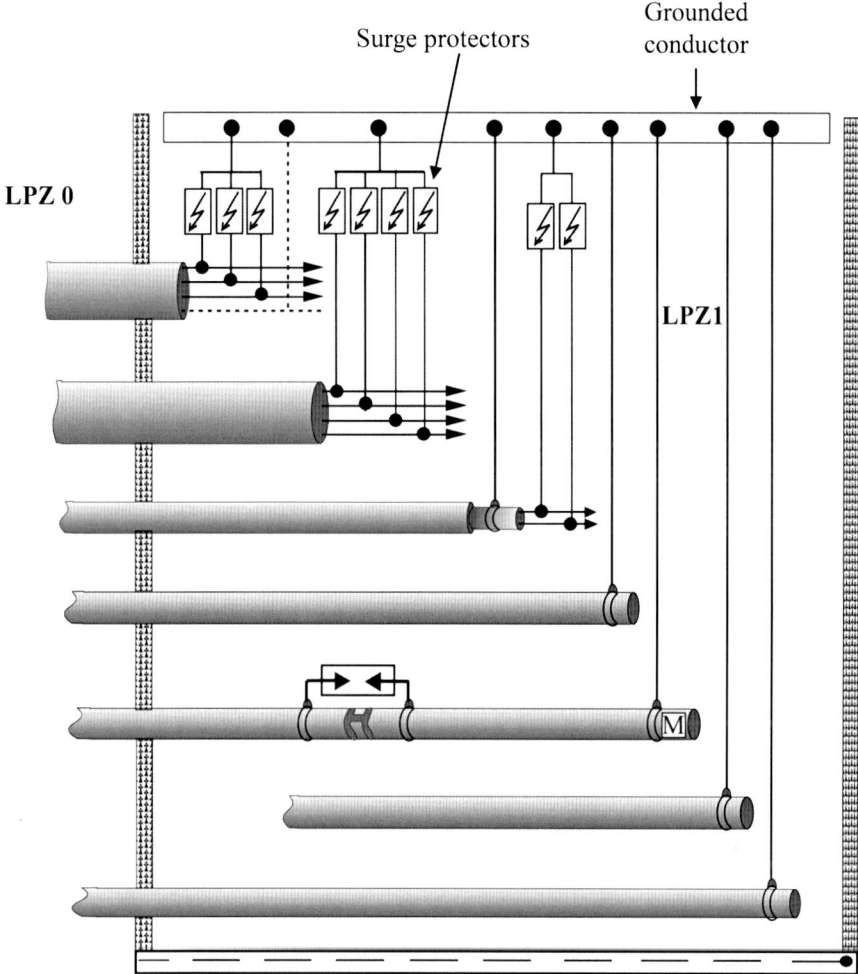

Fig. 17.19 In lightning protection practice, all conductors entering a building are connected to each other and to ground at the entrance to the building. Conductors that cannot be connected to ground directly are connected to it across surge protective devices (Figure created by author)

connected to ground, they are connected to ground across surge diverters or surge protective devices (Fig. 17.20).

17.10.1 Surge Protective Devices

A surge protective device (SPD) provides a high impedance between its terminals as long as the voltage between the terminals of the device is below a specific value. Thus, when a surge protector is connected between the phase conductor and the ground wire of an electricity supply to a building, under normal operation the

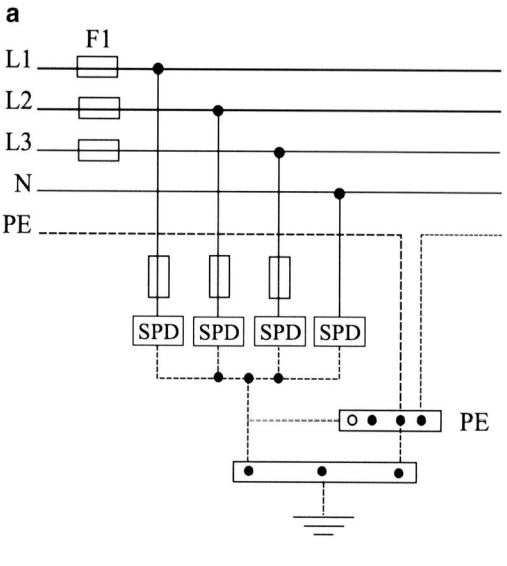

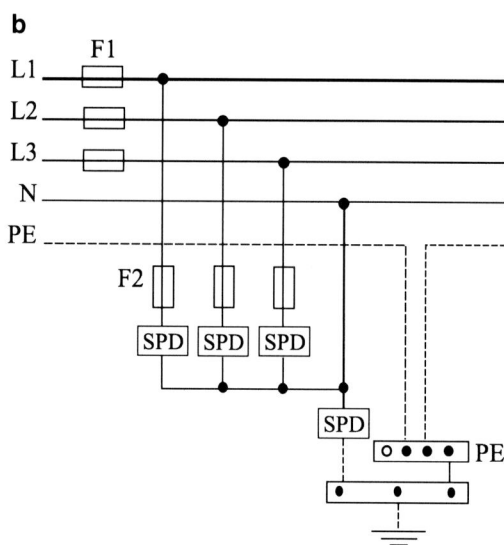

Fig. 17.20 (a) Conductors that cannot be connected to ground directly (such as those carrying electricity and data signals) are connected to ground across surge arresters. The figure shows an arrangement corresponding to a system with an isolated neutral conductor (Adapted from [1]). (b) Conductors that cannot be connected to ground directly (such as those carrying electricity and data signals) are connected to ground across surge arresters. The figure shows an arrangement corresponding to a system with neutral connected to the PE at the structure (Adapted from [1])

17.10 Internal Lightning Protection System

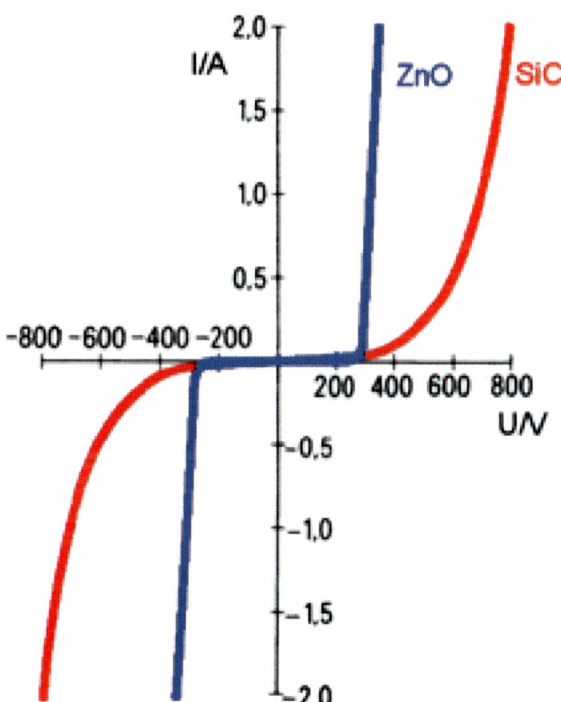

Fig. 17.21 Voltage current characteristics of a surge protective device (SPD). In the example, the V-I characteristics are shown for ZnO and SiC varistors where the nominal voltage (i.e., the voltage at which the device starts conducting) is approximately 300 V. Note that the device starts conducting when the voltage across it reaches the nominal voltage (figure from http://en.wikipedia.org/wiki/Varistor, 2014)

conductors operate as if they are not connected to each other. When the voltage between the terminals of the surge protector increases beyond the specified value or the firing voltage, the impedance reduces to a small value, creating almost a short circuit between the two terminals. Once the voltage is reduced, the impedance reverts back to the original value. Examples of SPDs are gas discharge tubes, varistors, and zener diodes. Figure 17.21 shows the voltage current characteristics of a SPD.

In an electrical power line supplying power to a building, SPDs are usually connected between the phase wires and between the phase wires and the ground (Fig. 17.20). Whenever a high voltage arises in one conductor with respect to the others, and if the amplitude of the voltage is higher than the nominal or firing voltage of the SPD, then the incoming voltage is diverted to ground when the SPD is short-circuited. In designing the internal lightning protection using a SPD it is important to have some idea of the magnitude and temporal variation of the voltages and currents entering the system so that the correct SPD can be selected for the application. The information concerning the amplitude of surges is necessary in selecting the proper nominal operating voltage of the SPD so that unwanted signals can be diverted to ground. The information concerning the temporal variation of the unwanted voltages is necessary for two reasons. First, the rapidity with which the SPD changes from a high-impedance stage to a low-impedance stage varies from one type of SPD to another. For example, if the incoming voltage has a very fast rise time, then a

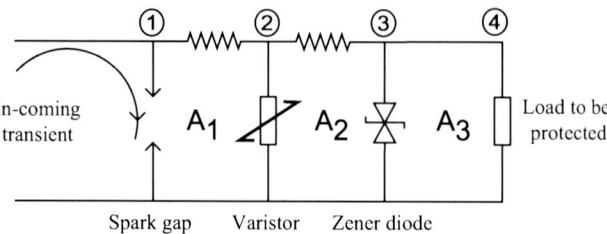

Fig. 17.22 In practice, sensitive electronics are protected by a series of surge protectors. See text for the reasons behind this arrangement (Figure created by author)

SPD that reacts very fast is necessary to divert the signal. Second, the amount of energy dissipation that can be withstood by a surge protector also varies from one type to another. The duration of the incoming voltages, together with their amplitude provides information concerning the possible energy dissipation in the SPD.

In practice, several stages of SPD are connected to reduce the unwanted voltages in a series of steps (Fig. 17.22). The first element in the series is usually a gas discharge tube. It removes a large portion of the incoming voltage. Gas discharge tubes can withstand rather high currents and therefore are suitable for this purpose. Unfortunately, the firing time of a gas discharge tube is not that short, and therefore a residual voltage is kept on moving downstream. The residual voltage that remains after the transient has passed through the gas discharge tube (voltage below the firing threshold of the gas discharge tube) is removed by a varistor. What remains after the varistor is removed by zener diodes, which are the least tolerant and most sensitive to the energy dissipation of the three types of surge protectors. In arranging the various protective elements, it is important to reduce the loop areas (marked A_1, A_2 and A_3 in Fig. 17.22) associated with various elements. During the firing of a gas discharge tube, for example, very rapid changes in currents that can cause unwanted voltages in any nearby loops are produced.

In internal lightning protection systems, SPDs are not only connected at the entrance of the structure; they are usually placed at the entrance of the power system to sensitive electronic devices. Thus, if an unwanted voltage signal were to bypass the SPD at the entrance of a building, such voltages would be removed by these SPDs protecting the sensitive electronic device.

17.10.2 Electromagnetic Zoning

General internal lightning protection can be described using the concept of zones. Assume that sensitive equipment has to be protected from unwanted voltages, currents, and electromagnetic fields. The equipment can be surrounded with several metallic shields to prevent unwanted electromagnetic fields from entering the system. However, power and other communication cables must be supplied to the equipment from outside. The procedure to do that is shown in Fig. 17.23a.

17.10 Internal Lightning Protection System

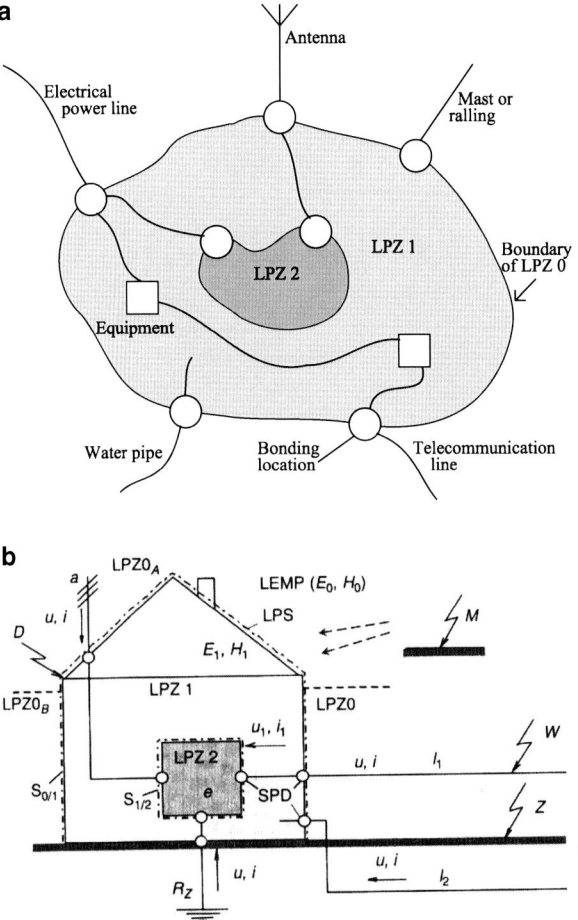

Fig. 17.23 (a) The concept of electromagnetic zoning. The idea is to screen sensitive objects with several screens that block unwanted voltages, currents, and electromagnetic fields penetrating the structure. In the method, the ground screen of any conductor penetrating through the screen is connected to the screen, and any live conductors that cannot be connected to the screen are connected to it through surge protectors (Adapted from [1]). (b) Lightning protection system of a building arranged according to the zone concept. LPZ0 is the outer zone where sources of current and voltages may directly interact with the system (i.e., a direct lightning strike). LPZ1 is the inner zone where most of the unwanted currents and voltages are removed using surge protective device (SPDs). But electromagnetic fields may exist in this region. LPZ2 is a secured zone where almost all the currents, voltages, and electromagnetic fields are removed using screens and SPDs (Adapted from [2])

Here three zones, namely, zone 0, zone 1 and zone 2 are shown. In zone 0 the direct threat of lightning current and the full exposure to the lightning generated electromagnetic fields exists. In zone 1 the impulse currents are limited by the SPD's and the electromagnetic fields are attenuated by the shield at the boundary of zone 1 and zone 0. In zone 2 the impulse currents and the electromagnetic fields are lower than in zone 1. The impulse current level and the electromagnetic fields in zone 2 can be withstood by the electrical systems in zone 2. At each entrance all the cables carrying signals are connected to the screen (at the boundary) using SPDs (shown by circles). All the screens are connected together so that they maintain the same potential. The outer screen is connected to ground. By increasing the number of screens one can drastically reduce the unwanted signals entering the equipment. Of course, in reality it may not be possible to find physical shields or screens as shown in the diagram in lightning protected buildings, but such metal shields are there in very sensitive equipment. For example, the metallic chassis of a sensitive electronic system acts as a shield. In protecting buildings from lightning, the idea is to place the conductors that divert the currents at each stage in such a way that they form an approximate shield (Fig. 17.23b). Moreover, the down conductors of the lightning protection system that represent the outer screen of the zoning procedure are placed in such a way that electromagnetic fields are reduced inside the structure.

Let us now see how a metallic screen can remove unwanted currents and electromagnetic fields.

17.10.2.1 Skin Depth

Assume that an electromagnetic field with a certain frequency (i.e., a sinusoidal signal) is incident on a metallic surface. The electromagnetic field penetrates the material and generates currents in it. However, the depth to which the electromagnetic fields penetrate can be characterized by the skin depth. As the electromagnetic field penetrates the conducting medium, its amplitude decreases exponentially with depth. The skin depth is the depth where the amplitude of the electromagnetic field is reduced to one-third the value it had at the surface. The value of skin depth is given by

$$\delta = \sqrt{\frac{1}{\pi f \mu \sigma}}, \qquad (17.7)$$

where f is the frequency of the electromagnetic field, μ is the magnetic permeability of the material, and σ is the conductivity. This equation is plotted for copper in Fig. 17.24. Since the frequencies of interest in lightning currents are a few kilohertz to several tens of megahertz, a few millimeters of copper screen can prevent unwanted lightning signals from appearing inside the enclosure. As is clear from the preceding equation, skin depth increases with decreasing frequency. Does that mean that extremely low-frequency signals can pass through a copper screen? The answer is no. When a time-varying electromagnetic field falls on a conducting medium, two processes take place. Part of the signal is reflected and the other part penetrates the medium. As the frequency decreases, the reflected part becomes increasingly higher, and only an increasingly smaller fraction enters the medium. When the frequency becomes zero, the entire signal is completely reflected. This is why a copper screen is effective in screening equipment from static fields.

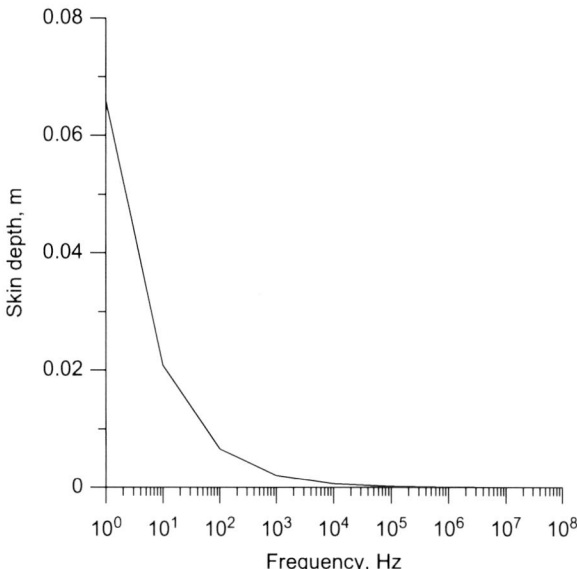

Fig. 17.24 Skin depth of copper as function of frequency of incident electromagnetic wave (Figure created by author)

17.11 Inadequacies of EGM

In EGM it is assumed that the connection between the stepped leader and the grounded structure takes place when the stepped leader comes to within striking distance of the grounded structure (i.e., when the final jump condition is reached). But in reality, the situation is more complicated. Depending on the height of the structure, before the final jump condition is reached between the tip of the leader and the grounded structure a connecting leader is issued from it. If the structure is very short, then the connecting leader is issued when the separation between the stepped leader and the grounded structure is almost equal to the striking distance. Thus, for short structures the presence of the connecting leader can be neglected. But in the case of tall structures, a connecting leader is issued long before the separation between the stepped leader and the grounded structure approaches the striking distance. The reason for this is the larger electric field that results at the top of the structure due to field enhancement. In this case, the final connection is made not between the structure and the stepped leader but between the connecting leader and the stepped leader. Again as before it can be assumed that the connecting leader will be connected to the stepped leader when the final jump condition is reached between the tip of the connecting leader and the stepped leader. When the connection between the connecting leader and the stepped leader takes place in the presence of a long connecting leader, the tip of the stepped leader will be located at a distance much greater than that stipulated by EGM. This is illustrated in Fig. 17.25. Thus, to correctly describe the attachment process, the generation of the connecting leader and its propagation and the final connection between it and the stepped leader must be taken into account.

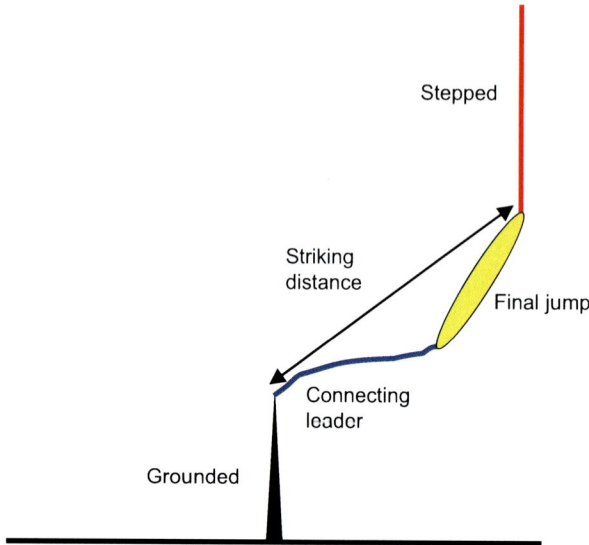

Fig. 17.25 In general, the stepped leader is met by a connecting leader issued from the lightning conductor. The striking distance in this case can be defined as the distance between the tip of the stepped leader and the point on the structure where the connecting leader is issued when the final jump condition is satisfied between the connecting leader and the stepped leader. At the final jump condition, the average electric field in the gap between the connecting leader and the stepped leader is equal to 500 kV/m (Figure created by author)

17.12 Attachment Models More Advanced Than EGM

To overcome the inadequacies of the EGM, scientists have developed several models that are capable of taking into account the presence of connecting leaders during lightning attachment. These models are described in Chap. 18.

17.13 Unconventional Lightning Protection Systems

17.13.1 Early Streamer Emission Principle

The analysis given in the previous section shows that tall structures have a longer attractive range because streamers and, hence, connecting leaders are generated when the leader is at a larger distance from its tip than in the case of short structures. Now, this will raise a question. Assume that by some other means, say by applying an external voltage, streamers are generated at the tip of a lightning conductor. Will these artificially created streamers give rise to a connecting leader, making the attractive range larger? All studies conducted to date, both theoretical and experimental, show that just by creating streamers at the tip of a lightning conductor it is not possible to enlarge the attractive range of a lightning conductor because the

17.13 Unconventional Lightning Protection Systems

generation of streamers is not a sufficient criterion for making a stable upward moving connecting leader (Chaps. 2 and 18). To create a successful connecting leader, the streamers must be transformed into a leader, and once initiated, the leader should be able to propagate continuously toward the stepped leader. The energy necessary for all these physical processes is supplied by the background electric field. Each of these processes requires the background electric field to be above a certain critical value for its completion. On the other hand, artificial generation of a streamer at the tip of a conductor is just a local effect. If these streamers are created when the leader is far away, they will just die out without leading to a connecting leader. If this is the case, how about artificially generating streamers from the conductor when all the conditions necessary for the streamer-to-leader transition and the leader propagation are satisfied? In principle, it should work, but the problem is that streamer inception from lightning rods takes place naturally long before the field conditions necessary for the successful launch of a connecting leader are satisfied. This means that when these conditions are satisfied, nature itself will provide the streamers necessary for the process. Nothing can be changed by artificially generating streamers at the tip of a rod.

Unfortunately, notwithstanding the theoretical and experimental data available, some vendors have introduced lightning rods where gadgetry is provided for the artificial generation of streamers at the tip of lightning rods. These are called early streamer emission (ESE) devices. It is claimed that these rods can generate connecting leaders long before a normal rod of similar height, and hence they have attractive radii longer than those of normal lightning rods. At present, there are no theoretical or experimental data to validate that claim. Of course, some laboratory experiments are cited in support of the claims of the ESE principle [3]. Indeed, it was a laboratory experiment that led to the creation of ESE devices. Analysis has shown that laboratory experiments cannot be used to simulate the lightning attachment process that takes place under natural conditions [4] because the electric fields used to expose the lightning rods in the laboratory are very different from the electric fields present under natural field conditions. In summary, neither theory nor experiment justifies the claims of ESE manufacturers. These conductors have no advantage over normal lightning conductors of similar heights.

17.13.2 Lightning Dissipaters

It took some time for scientists to understand how a lightning conductor works. Recall that Benjamin Franklin was of the erroneous opinion that a lightning conductor works by generating positive charges that travel upward into the charge center of a cloud and dissipating the electric charge in the cloud. In other words, he believed that a lightning conductor could dissipate the charge in a cloud, thereby preventing lightning strikes. This wrong concept is used today by manufactures of lightning dissipaters. The idea is as follows. On a tall conductor a net of sharp points is connected in the form of an umbrella. The umbrella is grounded with a ground conductor. It is claimed that the charges generated by these sharp points can

Fig. 17.26 The inception of upward initiated lightning flashes can be reduced by reducing the field enhancement at the top of the tower by placing an umbrellalike structure at the top of the tower (Figure created by author)

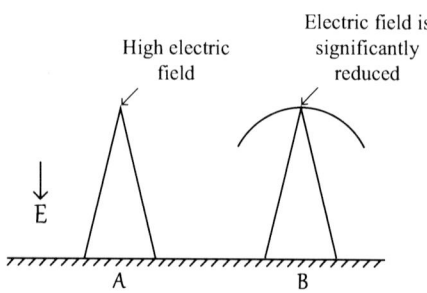

dissipate either the charge in a cloud or the charge in a downward moving stepped leader. These are called *dissipation arrays*. Calculations show that the charge generated by a dissipation array is not large enough to dissipate either the charge in a cloud (which is continuously formed) or the charge in a downward moving stepped leader. Indeed, there are many instances in which such dissipation arrays are struck by lightning directly. Thus, there is no truth to the manufacturers' claim that they can dissipate the charge in clouds and prevent lightning strikes. There are some instances where it was claimed that after connecting a dissipation array on a very tall tower, the number of lightning incidents were reduced. A possible reason for this observation is that, in the case of very tall structures, most lightning flashes are upward initiated. The ability of a tall tower to launch a connecting leader that culminates in an upward initiated lightning flash depends on the field enhancement at the extremity of the tower. By placing an umbrellalike structure at the top of a thin cylindrical tower, it is possible to reduce the field enhancement at the top, and this may lead to a reduction in upward initiated lightning flashes (Fig. 17.26). However, this effect has nothing to do with the creation of corona charges at the sharp points of the structure.

In summary, scientific claims of dissipation arrays have not been substantiated, either by theory or experiment, and they should not be used as part of structural lightning protection systems.

References

1. IEC 62305, International Standard on Protection Against Lightning, Part3: 62305–1 General principles; 62305–2: Risk assessment; 62305–3: Physical damage to structures and life hazard; 62305–4: Electrical and electronic systems inside structures, 2006
2. Flisowski Z, Mazzetti C (2010) Risk analysis. In: Cooray V (ed) Lightning protection. IET Publishers, London
3. Berger G (1992) The early streamer emission lightning rod conductor. In: Proceedings of 15th international conference on aerospace and ground: ICOLSE, Atlantic City, NJ, USA
4. Becerra M, Cooray V (2008) Laboratory experiments cannot be utilized to justify the action of early streamer emission terminals. J Phys D Appl Phys 41:085204

Chapter 18
Advanced Procedures to Estimate the Point of Location of Lightning Flashes on a Grounded Structure

18.1 Introduction

The attachment of a stepped leader to a grounded structure is mediated by a connecting leader issued by the grounded structure. In the case of short structures, the length of the connecting leader is usually small, but it could be several tens of meters long in the case of tall structures. However, in the electro-geometrical method (EGM), the presence of a connecting leader is neglected (Chap. 17). Thus, the conclusions based on EGM on the attachment of stepped leaders to tall grounded structures could be in error. This simplifying assumption in EGM motivated several scientists to create lightning attachment models that included the effect of connecting leaders. Four models are used by lightning researchers at present, and this chapter presents the basic ideas associated with these models. The four models were introduced by Eriksson [1], Dellera and Garbagnati [2], Rizk [3], and Becerra and Cooray [4, 5]. These models are referred to here as the Eriksson model, the Dellera and Garbagnati model, the Rizk model, and the Becerra and Cooray model.

A physically reasonable lightning attachment model should be able to describe the generation of a connecting leader, its subsequent propagation, and the final attachment to the stepped leader. First of all, let us consider how the initiation of a connecting leader from a grounded structure is treated in the four models. Eriksson and Dellera and Garbagnati use the *critical radius concept* (described subsequently) to determine when a connecting leader is issued by a grounded structure under the influence of a downward moving stepped leader. Rizk used his own criterion, which we refer to as the *Rizk criterion*, and Becerra and Cooray used a criterion based on the known physics of electrical discharges as outlined in Chap. 2. We refer to the latter criterion as the *Becerra and Cooray* criterion for leader inception. Let us consider these criteria individually.

18.2 Conditions Necessary for the Inception of a Connecting Leader According to Different Models

18.2.1 Critical Radius Concept

This criterion is based on observations made in several-meter-long laboratory discharges. In these experiments, sparks were created between a spherical electrode and a ground plane separated by a distance of 5–10 m. The sparks were created by applying an impulse voltage to the spherical electrode. The shape of the impulse voltage used in the study is known in high-voltage research as a *switching impulse*. A switching impulse has a rise time of approximately 200 μs and decays to 50 % of its peak value in approximately 1 ms. As the voltage rises to its peak value, the first observable physical process that takes place in the discharge gap is the generation of a streamer discharge from the high-voltage sphere. Usually two or three streamer bursts are created before a positive leader is generated from the sphere. Once initiated, the leader travels toward the grounded electrode. When it reaches the grounded electrode, a spark is created. Two important observations were made by the researchers during the experiment [6]. (1) As the radius of the high-voltage sphere increased, the voltage necessary to create a spark remained the same until its radius reached a critical value. Further increase of the radius led to an increase in the voltage necessary for breakdown. This radius was called the *critical radius* and it was approximately 38 cm. (2) When the radius of the sphere was equal to this critical radius, immediately with the creation of the first streamer burst from the electrode the positive leader was generated. The electric field at the surface of the sphere during the leader inception was approximately equal to the breakdown electric field in air, i.e., 3×10^6 V/m. This observation shows that in long gaps, when the electric field at the surface of the spherical electrode whose radius is equal to the critical radius is reached 3×10^6 V/m, a positive leader is initiated from it. In the models of Eriksson and Dellera and Garbagnati, this concept is used as follows to find out exactly when a connecting leader is issued from a grounded structure. Assume that the structure under consideration is a lightning conductor. In the model, the tip of the lightning conductor is replaced by a spherical conductor equal to the critical radius, i.e., 38 cm. In the calculations, the height of the stepped leader at which the electric field at the surface of the spherical conductor reaches 3×10^6 V/m is assumed to be the height of the stepped leader when a connecting leader is generated. To perform this calculation, it is necessary to utilize a numerical technique such as charge simulation method. Of course, in these studies it is assumed that the critical radius concept is valid also for the inception of lightning leaders.

18.2.2 Rizk Criterion

By analyzing the voltages necessary to cause electrical breakdown in long gaps, Rizk came to the following conclusion. Denote by U the electric potential at the location of the tip of a lightning conductor as a result of the presence of the charges in the cloud and the charge located on the downward moving stepped leader. In estimating this potential, the effect of the perfectly conducting ground is taken into account using the theory of images. According to the criterion developed by Rizk, a connecting leader is issued from the lightning conductor when

$$U = \frac{1556}{1 + 3.89/h}, \qquad (18.1)$$

where h is the height of the lightning conductor.

18.2.3 Becerra and Cooray Criterion

Becerra and Cooray analyzed the problem as follows. As the leader approaches the ground, they calculated the electric field at the extremity of the structure or at the tip of the lightning conductor as a function of the height of the leader tip. As the leader approaches the ground, the electric field at the extremity of the structure or at the tip of the lightning conductor increases, and a stage is reached where streamer discharges are generated from them. This generation of streamers takes place like a burst from the extremity of the structure or from the tip of the lightning rod. Utilizing the method outlined in Chap. 2, the charge in the streamer burst can be estimated. The criterion for the generation of a leader is that the charge in the streamer burst must exceed 1 µC. In many situations, the charge in the first streamer burst is less than this critical value. If this is the case, then the charge in the next streamer burst that occurs at a later time is estimated. Since the electric field increases with time (because the leader tip comes closer and closer to the structure), the charge in successive streamer bursts will be larger than the previous one. The analysis is continued until the charge in a streamer burst reaches a value of 1 µC. When a charge in the streamer burst reaches approximately 1 µC, theory shows that the stem of the streamer burst will be heated and converted into a leader. This is the condition used by Becerra and Cooray to estimate the instant of inception of a connecting leader.

18.3 Leader Propagation Models

Having described how the first inception of the leader is evaluated in each model we are now in a position to describe the various models developed to analyze lightning interception.

Fig. 18.1 Leader progression model of Eriksson [1]

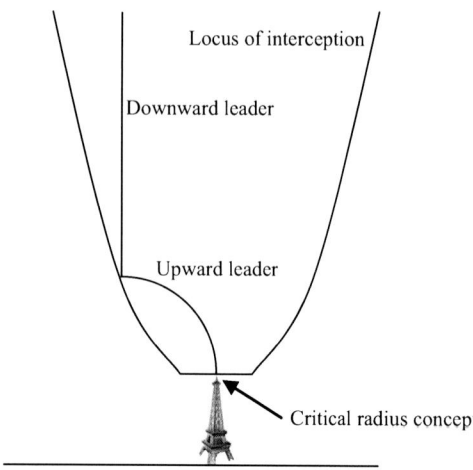

18.3.1 Eriksson's Model

Several steps of this model are illustrated in Fig. 18.1. The basic assumptions of the model are as follows:

1. The downward stepped leader is assumed to take a straight path to ground without branches.
2. The linear charge distribution on the stepped leader channel is assumed to decrease uniformly with height and reaches zero at the top of the channel. The total charge on the leader Q for a given return stroke with a peak current I_p is given by

$$I_p = 24.9 Q^{0.7}. \tag{18.2}$$

   If the height of the leader channel is known, then the linear charge density at any point on the leader channel can be estimated from the aforementioned information.
3. As the stepped leader approaches the ground, the height at which a connecting leader is field is evaluated using the critical radius concept.
4. Whether the stepped leader becomes attached to the connecting leader or whether it goes to ground depends on the following condition. If the connecting leader meets the stepped leader before it becomes attached to ground, then the flash terminates on the structure. The condition for the attachment to ground is assumed to be the following: the electric field at ground level at a point directly below the downward moving stepped leader reaches a value equal to the breakdown electric field in air, i.e., 3×10^6 V/m.
5. The path of the stepped leader is not affected by the connecting leader coming toward it.
6. The condition for the attachment of the connecting leader to the stepped leader can be analyzed as follows. Assume that the height of the stepped leader tip when the connecting leader is issued is z. Let z_m be the height of the stepped leader tip from ground when the electric field at ground level reaches the breakdown electric field.

18.3 Leader Propagation Models

Let us denote the height where the connecting leader meets the stepped leader by z_0. Then the condition for attachment of the stepped leader to the structure is $z_0 > z_m$. In other words, before the stepped leader reaches a height of z_m, the connecting leader must connect with it. The way in which the connecting leader propagates toward the stepped leader is not specified in the model. It finds the best possible path for making the connection with the stepped leader. The problem is a geometrical one, and the only parameter that is necessary to solve this problem and check whether the stepped leader becomes attached to the structure or reach the ground is the ratio between the speed of the connecting leader and the stepped leader. In the analysis, this ratio is assumed to be equal to 1.

The preceding discussion defines completely Eriksson's model. Recently the same model was introduced under the name Collection Volume Method (CVM) by several manufacturers of lightning protection equipment [7]. The only difference in CVM compared to Eriksson's model is that the charge on the leader channel is assumed to be given by $I_p = 10.7 Q^{0.7}$ instead of Eq. 18.2.

18.3.2 Dellera and Garbagnati Model

The steps of this model are illustrated in Fig. 18.2. The basic assumptions of the model are as follows:

1. As explained in Chap. 2, the stepped leader surges forward with the aid of streamer bursts. The length of the streamer region can be calculated if the charge distribution on the leader is known. In the model, it is assumed that at any given instance the stepped leader is traveling in the direction of the

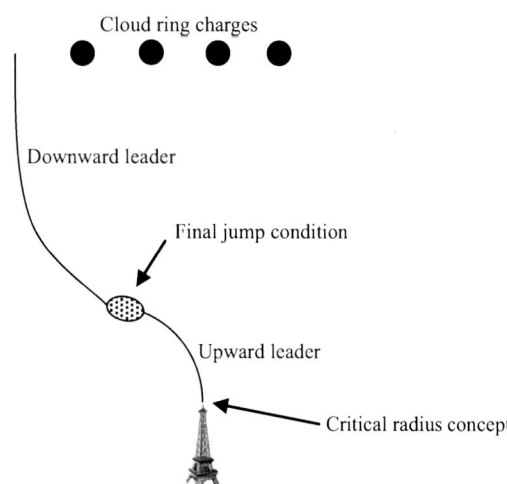

Fig. 18.2 Leader progression model of Dellera and Garbagnati [2]

background electric field that exists at the outer extremity of the streamer region. The length of the streamer region can be calculated using the technique outlined in Chap. 2.
2. The height of the stepped leader tip at which a connecting leader is incepted from the grounded structure is obtained from the critical radius concept.
3. The linear charge density on the stepped leader channel is assumed to be constant but was divided into two regions: one in the vicinity (the final tens of meters) of the leader tip, which is uncorrelated to the amplitude of the return stroke current and assumed to be equal to 100 µC/m, and the other, whose magnitude varies with the prospective return stroke current, along the rest of the leader channel. If ρ_0 (in µC/m) is the uniform linear charge density on the rest of the channel, it is given by $\rho_0 = 38 I_p^{0.68}$, where I_p is the peak return stroke current in kiloamperes.
4. Once incepted, the connecting leader moves in the direction of the maximum electric field that exists at the outer edge of its (i.e., the connecting leader's) streamer region.
5. The attachment process is complete when the streamers of the stepped leader meet the streamers of the connecting leader.
6. To perform the analysis, it is necessary to know the ratio of speeds of stepped and connecting leaders. In the model it is assumed to be equal to 4.

Once the preceding assumptions are made, it is possible to evaluate whether the stepped leader will become attached to the structure or whether the stepped leader streamer front will reach the ground (thus establishing a connection with ground) before it becomes attached to the connecting leader.

18.3.3 Rizk's Model

The steps of this model are illustrated in Fig. 18.3. The assumptions made in the model are as follows:

1. The stepped leader follows a vertical path without being affected by any other structure or by the connecting leader.
2. The charge on the leader channel is assumed to decrease linearly with height, decreasing to zero at the top of the leader channel. The total charge on the leader channel for a given return stroke peak current I_p is

$$I_p = 10.7 Q^{0.7}$$

3. The height of the stepped leader tip when a connecting leader is incepted is obtained using the Rizk criterion (Eq. 18.1).
4. Once incepted, the positive leader travels in such a way that, at any given instant of time, its direction of motion is toward the location of the tip of the stepped leader at that instant.

18.3 Leader Propagation Models

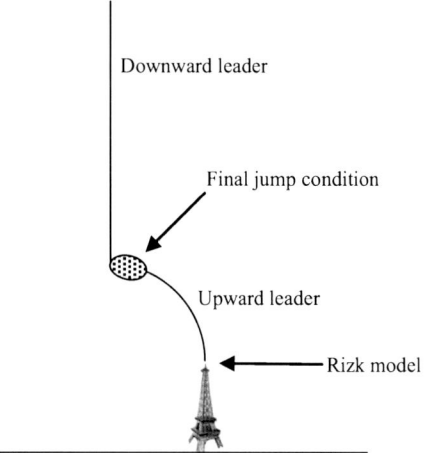

Fig. 18.3 Leader progression model of Rizk [3]

5. The attachment between the connecting leader and the positive leader takes place if they reach a critical distance so that the average electric field in the gap between the two leader tips is 500 kV/m.
6. To calculate the average electric field in the gap between the two tips of leaders, it is necessary to know the potential difference between the tips of the connecting leader and the stepped leader. The potential of the tip of the stepped leader can be calculated from its charge distribution, which was specified earlier, if the radius of the leader channel is known. Using theory, Rizk developed an equation to calculate the potential of the tip of the connecting leader [8]. According to this equation, the potential of the tip of a connecting leader of length l is given by

$$U_{tip} = lE_\infty + x_o E_\infty \ln\left[\frac{E_{str}}{E_\infty} - \frac{E_{str} - E_\infty}{E_\infty} e^{-\{l/x_0\}}\right]. \tag{18.3}$$

The numerical values of the parameters in this equation are as follows: $E_\infty = 3 \times 10^4$ V/m, $E_s = 4 \times 10^5$ V/m, $x_0 = v\theta$, where v is the ascending positive leader speed, assumed to be 1.5×10^4 m/s, and θ is the leader time constant, assumed to be 50×10^{-6} s.

The preceding description provides the basic principles used in Rizk's lightning attachment model.

18.3.4 Becerra and Cooray Model (SLIM)

The main features of the lightning attachment model introduced by Becerra and Cooray are shown in Fig. 18.4. Note that most of the assumptions and conditions used in the model are related to the basic physics of electrical discharges.

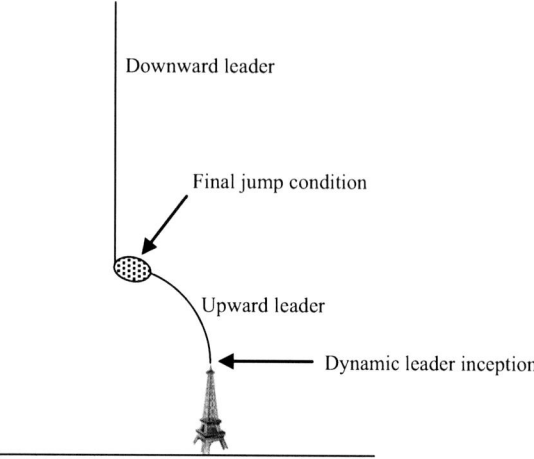

Fig. 18.4 Leader progression model of Becerra and Cooray (SLIM) [4, 5]

The reader is referred to Chap. 2 on electrical discharges for more details. What follows are the assumptions of this model:

1. The stepped leader takes a straight path to ground without being diverted either by grounded structures or by the connecting leader.
2. The charge distribution on the stepped leader is assumed to vary as the stepped leader approaches ground. The charge distribution on the stepped leader channel when the stepped leader is at any given height above ground is obtained as a function of cloud potential as described in Chap. 14.
3. The potential V of a cloud, and hence of the leader tip, is obtained as a function of return stroke peak current as described in Chap. 14.
4. As the stepped leader approaches ground, the passage of time is divided into small time steps, and at each time step whether a streamer is incepted at the extremity of the structure is checked first by calculating the electric field at the extremity of the structure in the presence of the stepped leader (a numerical technique such as charge simulation must be used to do this) and then using the streamer inception criterion as described in Chap. 2.
5. Once a streamer is incepted, the charge in the streamer zone is calculated using the procedure described in Chap. 2. If this charge exceeds 1 μC, a connecting leader is incepted.
6. Once a connecting leader is incepted, the potential along the connecting leader channel is calculated using the procedure outlined in Chap. 2.
7. The propagation of the connecting leader is mediated by streamer bursts generated at the tip of the leader. At each time step the charge in the current streamer burst is calculated, and from this charge the advancement of the leader is obtained by dividing it by the charge necessary for the creation of a unit leader channel. For example, if Q is the charge in the streamer burst, it causes an extension of the connecting leader by an amount Q/q_l. The value of q_l is on the order of 60 μC. In the later version of the model, q_l was shown to be a function of leader speed [5].

8. In the first version of the model, the connecting leader was assumed to propagate at any instant toward the tip of the stepped leader at that instant. In later versions of the model, the connecting leader tip was assumed to propagate in the direction of the maximum background electric field at the location of the tip [9].
9. The attachment between the stepped leader and the downward moving stepped leader takes place when the average electric field in the gap between the two leader tips is equal to 500 kV/m.

The preceding description defines completely the leader progression model introduced by Becerra and Cooray. Recently, the model was modified to include not only vertical leader channels but also slanted ones [9].

18.4 Comparison of Different Models

As is apparent from the preceding description, many simplifying assumptions are made in each of the lightning attachment models. Moreover, some models are empirical in nature, i.e., they use directly the results from the laboratory and plug them into the model, whereas others attempt to give a more physical picture of the event under consideration. However, no model can be categorized as perfect because of all the uncertainties inherent in the models. On the other hand, these models represent a significant advance in the study of lightning attachment in comparison to the EGM model. Hopefully, with more research and observations, the best of these models will be selected and improved further. Notwithstanding these considerations, it is of interest to compare the predictions of the different models.

One parameter that can be used to compare different models is the attractive radii of a grounded structure as a function of return stroke peak current. For comparison purposes we select a slim tower with a radius of 0.1 m. The attractive radii as predicted by the different models for different tower heights as a function of peak return stroke current is presented in Fig. 18.5. Note the significant differences in the prediction of different models. Hopefully, some of these predictions will be tested against experiments in the future.

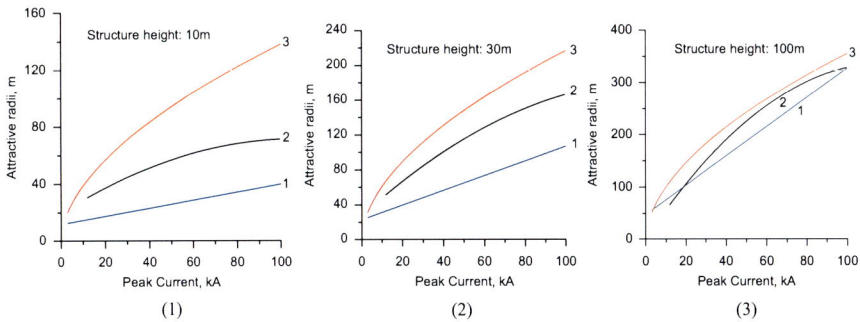

Fig. 18.5 Variation of attractive radii with structure height as predicted by (Curve 1) Dellera and Garbagnati model, (Curve 2) Becerra and Cooray model (SLIM), and (Curve 3) Rizk model (Adapted from [10])

References

1. Eriksson AJ (1987) An improved electrogeometric model for transmission line shielding analysis. IEEE Trans PWDR-2:871–877
2. Dellera L, Garbagnati E (1990) Lightning simulation by means of the leader progression model, I. Description of the model and evaluation of exposure of free standing structure. IEEE Trans Power Deliv 5(4):20009–20017
3. Rizk F (1994) Modeling of lightning incidence to tall structures Part I: Theory. IEEE Trans Power Deliv PWRD-9:162–171
4. Becerra M, Cooray V (2006) Time dependent evaluation of the lightning upward connecting leader inception. J Phys D Appl Phys 39:4695–4702
5. Becerra M, Cooray V (2006) A self-consistent upward leader propagation model. J Phys D Appl Phys 39:3708–3715
6. Carrara G, Thione L (1976) Switching surge strength of large air gaps: a physical approach. IEEE Trans PAS-95(2):512–524
7. D'Alessandro F, Gumly JR (2001) A 'collection volume method' for placement of air terminals for the protection of structures against lightning. J Electrost 50:279–302
8. Rizk F (1989) A model for switching impulse leader inception and breakdown of long air-gaps. IEEE Trans Power Deliv 4(1):596–603
9. Cooray V (2013) On the attachment of lightning flashes to grounded structures with special attention to the comparison of SLIM with EGM and CVM. J Electrost 71:577–581
10. Cooray V, Kumar U, Nucci CA, Rachidi F (2013) On the possible variation of the lightning striking distance as assumed in the IEC lightning protection standard as a function of structure height. Electr Power Syst Res 113:79–87

Chapter 19
Interaction of Lightning Flashes with the Earth's Atmosphere

19.1 Introduction

Thunderstorms and lightning flashes are processes that take place in the Earth's atmosphere and they can modify it over both the short term and the long term. Thus, understanding the effects of thunderstorms and lightning in the atmosphere is a must, and these effects must be used as inputs in any atmospheric model created to understand the effects of human-made greenhouse gases in the atmosphere. This chapter provides a brief description of the effects of lightning and thunderstorms in the atmosphere, which are of interest in climatology.

19.2 Production of Nitrogen Oxides in the Atmosphere

An assessment of the global distribution of nitrogen oxides, NO_x, (NO and NO_2 are collectively referred to as NO_x) is required for an adequate description of tropospheric chemistry and in the evaluation of the global impact of increasing anthropogenic emissions of NO_x. In the mathematical models utilized for this purpose, one needs to specify as inputs the natural and human-made sources of nitrogen oxides in the atmosphere. Lightning is one of the main natural sources of nitrogen oxides in the atmosphere, and it may be the dominant source of nitrogen oxides in the troposphere in equatorial and tropical South Pacific regions. Let us first understand how NO (Ozone gas generated by the lightning flashes oxidizes NO to NO_2) is produced in lightning and then how scientists have managed to quantify it.

To produce oxides of nitrogen in the atmosphere, it is necessary to break first both nitrogen and oxygen molecules down into their constituent atoms and then let them react with each other. However, nitrogen is a molecule with a strong bond between two atoms, and a significant amount of energy is needed to break this bond. Several processes that take place on Earth and in Earth's atmosphere break both nitrogen and oxygen molecules apart, allowing them to form nitrogen oxides and other nitrogen compounds (i.e. fixing of Nitrogen). Actually, living organisms can only use nitrogen when it is in forms other than N_2, i.e., fixed nitrogen. On Earth, nitrogen is fixed by nitrogen-fixing bacteria in soil and blue-green algae. Thus, the main natural source of nitrogen in forms other than N_2 on Earth is created by biological processes. Of course, today it is produced in mass quantities as fertilizer in the chemical industry. In the Earth's atmosphere nitrogen oxides are produced mainly by cosmic radiation and energetic particles from the Sun and electrical discharges. Nitrogen oxides (mainly NO) are produced in electrical discharges in two ways. One is due to electron impacts, and the other is due to thermal effects.

Nitrogen oxide production by electron impacts takes place in corona and streamer discharges, whereas the thermal production of NO takes place in hot electrical discharges such as leaders and return strokes. In electrical discharges, electrons are accelerated to energies in excess of the energy required to break nitrogen molecules, which is approximately 10 eV. On the other hand, the energy required to dissociate an oxygen molecule is approximately 5 eV. In electrical discharges copious amounts of electrons acquire energy that exceeds these values. The collision of nitrogen and oxygen molecules with these electrons leads to their dissociation. Thus, energetic electrons in electrical discharges such as corona and streamer discharges can create both nitrogen and oxygen atoms. These atoms combine with each other to form NO. Furthermore, since an oxygen molecule is more easily broken down, large quantities of ozone are also produced in these discharges. The NO produced in the discharges is oxidized by ozone to produce NO_2. NO is also produced in the Earth's atmosphere by cosmic radiation. The mechanism is identical to that active in corona and streamer discharges in the production of NO. NO production in cosmic rays takes place through the dissociation of O_2 and N_2 molecules by the impact of energetic electrons generated in the atmosphere by cosmic rays.

Based on theory and experimental data it is estimated that in NO production by energetic electrons each ionization collision generates approximately one NO molecule [1, 2]. Armed with this simple fact, the NO production in avalanches and streamer discharges can be easily estimated. In an avalanche the total number of ionization events can be easily estimated from theory (Chap. 2), and from that number the NO production can be estimated directly. For example, in Fig. 19.1 the NO production in avalanches is given as a function of E/p. NO production in streamer discharges can be easily estimated by estimating the number of ionizing events that a streamer will perform in moving a unit length. Let us represent the number of ions in the head of the streamer by ω. In moving a unit length the ions in the streamer head are replenished $1/2R_s$ times, where R_s is the radius of the streamer head. Thus, the total number of ionizing events taking place in the head of the streamer as it moves a unit length is $\omega/2R_s$. This number is equal to the total number of NO molecules generated by the streamer in moving a unit distance. Since $R_s \approx 10^{-4}$ m and $\omega \approx 10^8$, the number of NO molecules generated by a streamer discharge moving a unit length is approximately 5×10^{11} molecules/m. The same technique can be applied easily to evaluate the total NO production in upper atmospheric discharges (Sect. 19.4) because there, too, the discharge process is dominated by electron avalanches and streamer discharges.

19.2 Production of Nitrogen Oxides in the Atmosphere

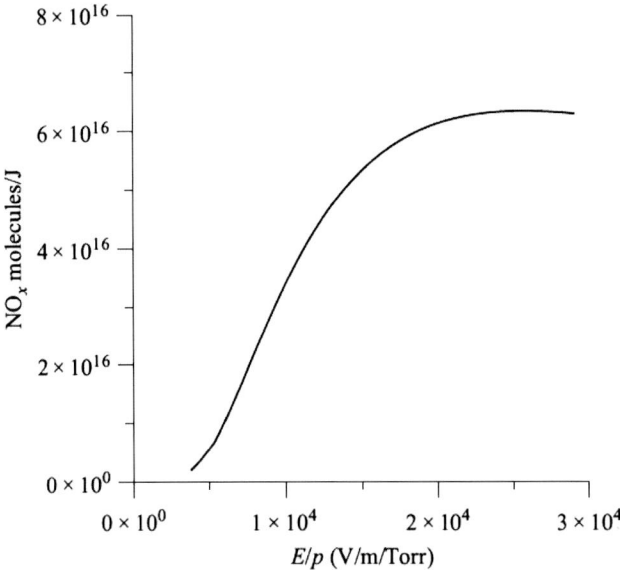

Fig. 19.1 NO production in electron avalanches as function of E/p (Figure created by author)

NO is also generated in electrical discharges through thermal effects [3]. In leader discharges and return strokes, the air in the channel is heated to more than 25,000 K. At these temperatures almost all the molecules of the gas encompassing the channel is broken down into their constituent atoms N and O. NO is produced as the air in the discharge channel cools after the cessation of the current. Let us now consider the details of this process. At any temperature there are chemical reactions that generate NO, and other chemical reactions destroy NO. The chemical reactions that produce NO are the following:

$$O + N_2 \rightarrow NO + N, \tag{19.1}$$

$$N + O_2 \rightarrow NO + O. \tag{19.2}$$

The chemical reactions that destroy NO are the following:

$$NO + N \rightarrow N_2 + O, \tag{19.3}$$

$$NO + O \rightarrow N + O_2, \tag{19.4}$$

$$NO + NO \rightarrow N_2O + O. \tag{19.5}$$

At any given temperature the NO concentration corresponds to the amount at which the NO-destroying reactions are in balance with NO-producing reactions. This is what we mean by the statement that NO is in thermal equilibrium. At around 25,000 K the species N and O are more stable than NO because NO is destroyed by the thermal collisions that take place at this temperature. As the temperature decreases, the concentration of NO starts to increase and reaches a peak value around 4,000 K. If the temperature continues to decrease extremely slowly, the concentration of NO also decreases and finally approaches the value it had just before the discharge. If this were the case, one would not expect any NO to be produced in the discharge. However, a net amount of NO is created in the discharge for the following reason: at any given temperature it will take some time for the chemical reactions to establish the equilibrium NO concentration. At 4,000 K it takes only a few microseconds to establish equilibrium. At 2,500 K several milliseconds are required, at 2,000 K the time needed is about a second, and 10^3 years are required at 1,000 K. Since the discharge cools in times on the order of milliseconds, thermal equilibrium can be established only for temperatures larger than approximately 2,500 K. The NO concentration that is present around this temperature does not get the chance to decrease further because the discharge cools too rapidly for chemical reactions to take place and establish thermal equilibrium. Thus, the concentration of NO that was in the gas around 2,500 K will remain as it is even when the temperature of the discharge channel reaches room temperature. In effect, the volume of the gas heated in the discharge channel will contain more NO after the discharge than before the discharge. This is the mechanism that generates NO in hot discharges.

This shows that the amount of NO produced in a discharge depends on how fast the temperature of the air in the discharge channel decreases after the cessation of the current. Faster cooling will give rise to more NO.

In the preceding explanation, it is assumed that the hot gas in the discharge channels cools because the energy input to the gas stops after the cessation of the discharge current. This need not be the case in every discharge. For example, in the case of long continuing currents in lightning channels, the production of NO can take place even while the current is flowing along the channel. This happens due to the chaotic influx of cold air into the channel and outflow of hot air from the channel into the colder regions. In this case, NO is produced continuously as the hot air cools rapidly from mixing with cold air, and the process of NO production continues during the lifetime of the continuing current.

A lightning discharge contains a variety of electrical discharges, namely, electron avalanches, streamer discharges, leader discharges, return strokes, continuing currents, M components, and K changes. In evaluating the production of NO by lightning discharges, scientists have assumed previously that return strokes were the most significant source of NO [3]. They also assumed that cloud discharges did not provide a significant contribution to global nitrogen oxide production. Recently, Cooray et al. [4] have questioned the validity of this procedure. Using theory and data gathered by Rahman et al. [5] from direct measurements of NOx in triggered lightning experiments conducted with triggered lightning, Cooray et al. [4] showed that all discharge processes that take place during lightning flashes should be taken into account, including cloud discharges, in evaluating the NO production from lightning flashes. Figure 19.2 shows the NO production in a single lightning discharge as a function of its length (or the total charge dissipated) separated into contributions from different events in the discharge. Note that it is the leader processes that contribute most to NO production. The global production of NO by lightning discharges, estimated taking into account all these processes

19.2 Production of Nitrogen Oxides in the Atmosphere

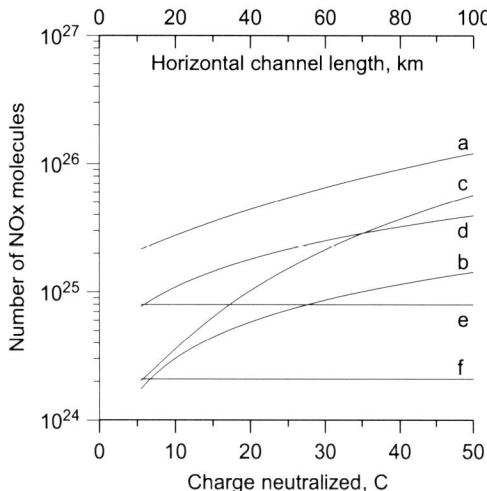

Fig. 19.2 NO production in single lightning discharge (ground flash), separated into different components, as function of its length. (*a*) Total. (*b*) Streamer discharges in corona sheath. (*c*) Core currents in leader. (*d*) M-components and K changes. (*e*) Continuing currents. (*f*) Return strokes (Adapted from [4])

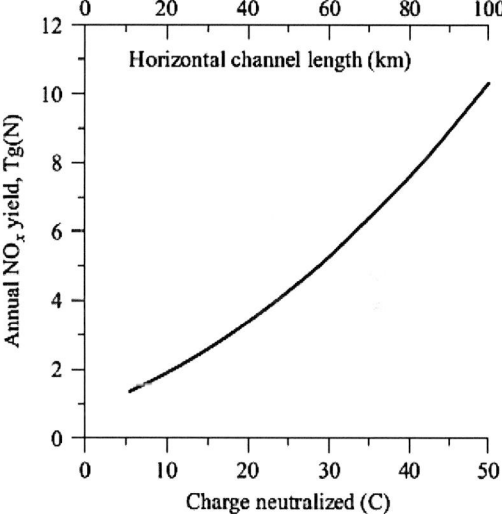

Fig. 19.3 Global annual production of NO as function of length of lightning discharges (or charge dissipated by discharge) calculated by assuming that approximately 40 lightning flashes take place in the atmosphere per second (Adapted from [4])

and assuming that approximately 40 lightning flashes take place in the atmosphere per second, is presented in Fig. 19.3. Since the total charge dissipated by a typical lightning flash is approximately 20 C, the global production of NO amounts to approximately 4 Tg of nitrogen per year due to lightning. Of course, if the charge dissipated by a typical lightning flash is larger than the value given previously, then the estimated global production of NO from lightning would be higher. Moreover, since NO production does not scale linearly with peak current or charge, an estimate of NO production based on a typical lightning flash may differ from an estimate that takes into account the real distribution of currents or charges in lightning flashes.

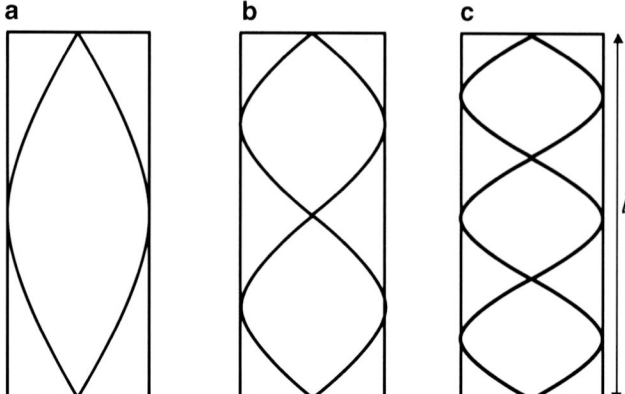

Fig. 19.4 Amplitude distribution of resonating waves in air column trapped inside cylinder closed at both ends. Note that the wavelength of the resonance frequencies are (**a**) $\lambda = 2l$, (**b**) $\lambda = l$, (**c**) $\lambda = l/2$ (Figure created by author)

19.3 Schumann Resonances

Consider an air column of length l trapped inside a cylinder closed at both ends. If the air inside the cylinder is set to vibrate, the air column will resonate with a fundamental frequency such that the length of the air column will be equal to half the wavelength. It can also resonate at higher frequencies f_n given by

$$f_n = nv_s/2l, \qquad (19.6)$$

where n is a nonzero integer and v_s is the speed of sound. The spatial distributions of these resonances are shown in Fig. 19.4. Note that the air column can resonate at many frequencies, but all are multiples of the fundamental frequency f_1 given by $v_s/2l$. Clearly, the resonance frequencies are accommodated comfortably within the cylinder with nodes at the two ends. Such resonances can also occur in electromagnetic waves if they are created inside a closed cavity. Now consider the cavity formed between the surface of the Earth and the ionosphere. If this cavity were perfect, with both boundaries perfectly conducting, then the electromagnetic resonance frequencies of the cavity would be [6]

$$f_n = (c/2\pi R)\sqrt{n(n+1)}, \qquad (19.7)$$

where c is the speed of light and R is the radius of the Earth. However, the walls of the cavity formed between the ionosphere and the Earth are not perfectly conducting. In particular, the finite conductivity of the upper boundary, i.e., the ionosphere, makes the resonance frequencies lower than that predicted by the preceding equation and the peaks wider. Resonances of the Earth–ionosphere cavity are shown in Fig. 19.5a. This figure, adapted from [6], shows the amplitude

19.3 Schumann Resonances

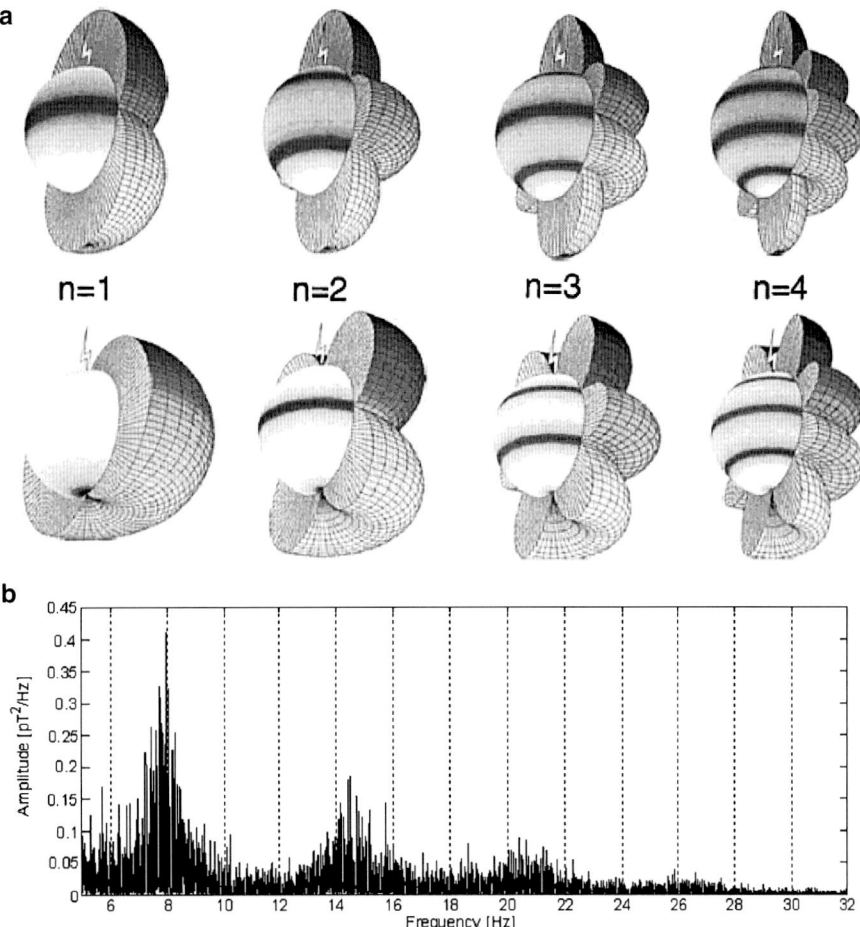

Fig. 19.5 (a) Idealized angular structure of electric and magnetic field amplitudes of lowest four modes in Earth–ionosphere cavity. The mode structure is azimuthally symmetric about the vertical dipole source located at the North Pole. In this figure, the unit sphere is split through the pole into two hemispheres, and the absolute values of the mode amplitudes are plotted in different formats for each hemisphere. In one hemisphere, the modulus of the amplitude is proportional to the radial distance above the surface of the unit sphere. In the other hemisphere, the modulus of the amplitude is shown as a *shady gray* where the nodes are *black*, and increasing absolute amplitude goes from black to white (Adapted from [5]). (b) Schumann resonances in a sample spectrum of natural electromagnetic signal. Note the peaks around 8, 14, and 20 Hz (Adapted from [6])

distribution of the lowest four TM_0 normal modes in the Earth–ionosphere cavity. In the example considered, the lightning flash is assumed to be located at the North Pole. In the figure, the absolute value of the electric field associated with each mode is presented in two different formats. These oscillations of the electric (and magnetic) fields are called Schumann resonances. Recall that each second approximately 40–100 lightning flashes take place in the atmosphere. Each of these lightning flashes taking place at different locations of the Earth gives rise to a field structure similar to that in Fig. 19.5, with the location of the lightning flash as the origin. Thus, the structure of the Schumann resonance fields at any given instant of time is much more complicated than that shown in Fig. 19.5a. However, since the frequency associated with each mode is the same for each lightning flash, they can be separated in frequency employing tuned receivers. Such measurements show peaks at frequencies around 7.83 (fundamental), 14.3, 20.8, 27.3, and 33.8 Hz. These are illustrated in Fig. 19.5b.

The amplitude of these resonances provides a measure of the total lightning activity of the globe. Thus, by monitoring Schumann resonances one can investigate the variation in global lightning activity with time.

19.4 Global Warming and Lightning

The global increase in temperatures due to human-generated emissions of greenhouse gases is a burning problem in modern society. Since lightning occurrence is higher in warmer regions in comparison with colder regions of the Earth, scientists have raised the question of whether the number of lightning flashes can be used as a global thermometer [7]. This has been investigated by scientists and they have found that lightning activity is a measure of the Earth's surface temperature. For example, Fig. 19.6 shows the relationship between surface temperature and lightning activity. There is a clear tendency toward increased lightning activity with increasing surface temperature.

Do the preceding statements mean that lightning activity increases with increasing global temperature? The jury is still out on this question [8]. We really do not know yet how lightning and thunderstorm activity will react when global temperatures increase. The model results are still contradictory. Global warming models show that the greatest warming due to greenhouse gases will take place not at the

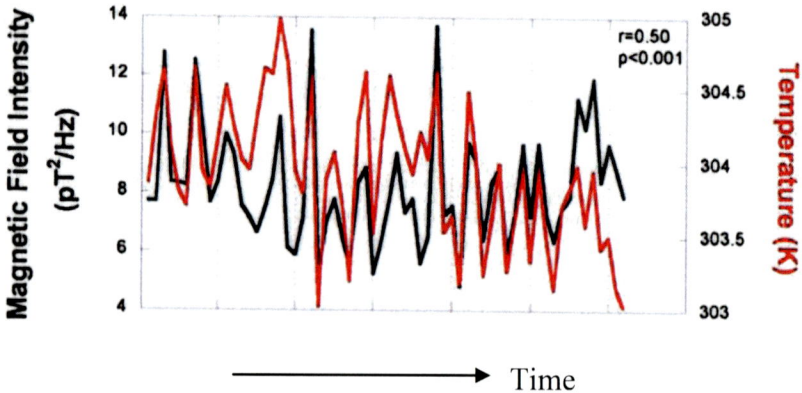

Fig. 19.6 Relationship between surface temperature of Earth and amplitude of magnetic field of Schumann resonances. Recall that this amplitude is directly related to global lightning activity. With this interpretation note that global lightning activity is directly related to the surface temperature of the Earth (Adapted from [7])

19.5 Upper Atmospheric Electrical Discharges

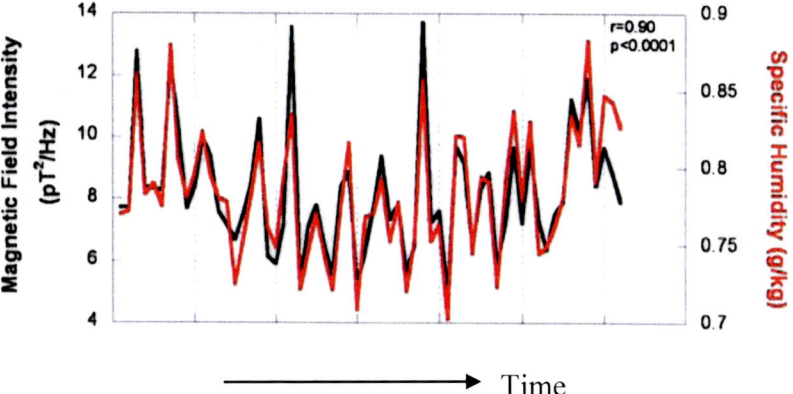

Fig. 19.7 Relationship between water vapor content in upper atmosphere and amplitude of magnetic field of Schumann resonances. Recall, as before, that this amplitude is directly related to global lightning activity. With this interpretation note that the vapor content of the upper atmosphere is directly related to global lightning activity (From [7])

surface but in the equatorial upper atmosphere. Heating of the upper atmosphere will stabilize the atmosphere, reducing convection. This means fewer occurrences of thunderstorms. On the other hand, climate model simulations show that lightning activity will increase in a warmer climate. An increase of approximately 10 % of lightning activity is expected for each 1 K increase in global temperature. More recent studies show that global warming can reduce the number of thunderstorms but the strongest or severe storms will increase by 25 %. This means that we can expect severe thunderstorm activity more often as the global temperature increases.

Another important parameter related to the greenhouse effect is the content of water vapor in the upper atmosphere. Water vapor absorbs the infrared radiation going out from the Earth's surface and therefore traps the heat leaving the Earth, amplifying global warming. Thus, the climate is very sensitive to the water vapor in the upper troposphere. Experiments show that thunderstorms deposit large amounts of water vapor in the upper troposphere. Figure 19.7 shows the relationship between water vapor in the upper atmosphere and lightning activity. In fact, lightning activity is a measure of the amount of water vapor transported to the upper troposphere [8] (Chap. 4). This shows that thunderstorms play an important role in the regulation of the heat balance of the Earth's atmosphere.

19.5 Upper Atmospheric Electrical Discharges

Until recently, lightning researchers believed that lightning flashes were confined to a thin layer of the atmosphere that extends from ground level to the tropopause, which is located at 10–15 km height. However, since 1886 eyewitness accounts of lightning flashes taking place in the upper atmosphere (in the region between cloud tops and the ionosphere) appeared in various articles. In 1925, C. R. T. Wilson [9], a Noble Prize winner, also suggested the possibility of the existence of lightning flashes that strike from cloud tops to the upper atmosphere. With the increase in air travel after the Second World War, many observations by pilots and air travelers where lightning was seen to strike from cloud tops to the upper atmosphere were published in various magazines. However, these reports were not taken seriously by the scientific community. The direct evidence for their existence did not appear until July 6, 1989. On that day, a thunderstorm was brewing on the horizon as Winckler and his associates from the University of Minnesota were testing a highly sensitive video camera by directing it toward the stars on the horizon [10]. By pure chance, they recorded a luminous event that took place between the cloud tops and the upper atmosphere. This discovery showed that lightning flashes are not confined to the troposphere but extend far into the upper atmosphere. These lightning flashes that take place in the upper atmosphere are now known as *ionospheric* or *upper atmospheric lightning flashes*. Following this initial discovery many recordings of these ionospheric lightning flashes were made by television observations from the Space Shuttle. Since then, many ground-based observations of upper atmospheric lightning flashes have been made [11]. According to the knowledge gathered so far by research groups working in different parts of the world, these ionospheric lightning flashes can be divided into three types: sprites, blue jets, and elves [12]. These activities are summarized in Fig. 19.8.

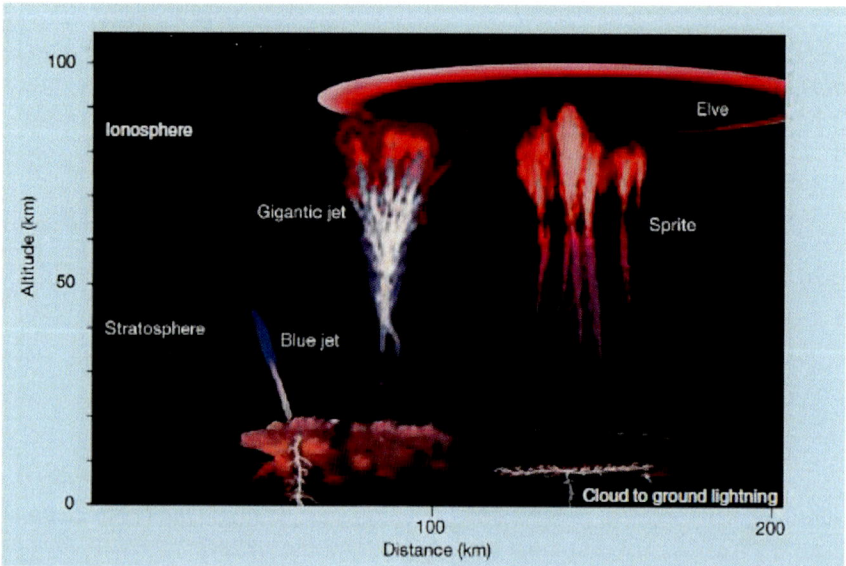

Fig. 19.8 Various types of electrical discharges taking place above thunderclouds (Adapted from [13])

19.5.1 Sprites

Sprites have the appearance of a jellyfish with an oval-shaped body and numerous tendrils. The main body of a sprite is located approximately 70–90 km above the ground and the tentacles extend downward to a height of approximately 40 km. The main body of the sprite is red in color, with tendrils becoming increasingly bluish as they come closer to the Earth. The total length of a sprite is approximately 30–60 km, and they have a width of approximately 5–30 km. One can compare this with a normal lightning flash, which has a vertical length of approximately 5–10 km and a width of approximately 1–5 cm. The individual tendrils of a sprite may be on the order of 10 m across. The total volume of air illuminated by a sprite can be approximately 10,000 km^3. The lifetime of a sprite is approximately 10–100 ms, but the details of the events that take place during its lifetime are still unknown to researchers.

One interesting feature of sprites is that they have no physical connection to thunderclouds active in the troposphere, but researchers now know that sprites are created by the action of thunderclouds. As described in Chap. 4, the atmosphere close to the ground is a good insulator, but with increasing height it gradually turns into a good conductor. The reason for this is the increase in the number of free electrons in the atmosphere with increasing height. The creation of a sprite takes place as follows (Fig. 19.9). As the charge in a thundercloud builds up, it creates an

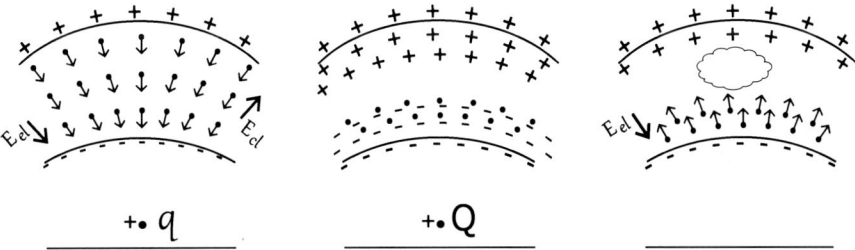

Fig. 19.9 Physical mechanism leading to creation of sprites during thunderstorms. The charge in the thundercloud creates an electric field, marked E_{cl}, in the upper atmosphere. In the upper atmosphere, there are free electrons, and these electrons experience a force from this electric field and start moving downward. Of course, positive ions move upward, but their movement can be neglected in the analysis because they are much heavier than electrons and, therefore, move very slowly. Since the amount of free electrons decreases with decreasing altitude (i.e., the conductivity of the atmosphere decreases with decreasing altitude), this polarization of the atmosphere gives rise to a layer of negative charge in the lower regions of the upper atmosphere. This displacement of electrons gives rise to an electric field, E_{el}, that opposes the electric field produced by the cloud. This concentration of electrons takes place until the electric field produced by the electrons counterbalances the electric field produced by the cloud. As the charge in the thundercloud increases, the separation of electrons and the concentration of the negatively charged layer increase. When the concentration of the charge in the cloud reaches a critical value, a lightning discharge is created, removing the thundercloud electric field from the upper atmosphere. The removal of the thundercloud electric field causes the electrons to recoil back under the influence of the electric field E_{el} and move into their original place. This rapid movement and acceleration of electrons due to their collective electric field E_{el} causes ionization that rapidly leads to the formation of streamers. The light emitted by this ionization phenomenon gives rise to the visual observations in the upper atmosphere known as sprites (Figure created by author)

electric field in the upper atmosphere. The free electrons and the positive ions in the upper atmosphere experience a force due to this electric field. As a result of this force, the electrons and positive ions move in opposite directions, leading to a separation of charge and, hence, the creation of an electric field. Of course, since ions are much heavier than electrons they move very slowly and their displacement can be neglected when compared to electrons. The electric field created by the displacement of electrons and ions is opposite to that created by the cloud. Consequently, the effect of the displacement of electrons and ions in the ionosphere is to reduce the net electric field in the ionosphere. The buildup of the charge in the thundercloud takes place rather slowly (in several tens of seconds to minutes), and the response of the electrons in the upper atmosphere also follows this slow pace. The action of electrons at this stage could be compared to the movement of a spring that is extended slowly by an external force. This continues slowly until the charge in the thundercloud increases to such a level that it gives rise to a normal lightning discharge. During the lightning discharge most or a large fraction of the charge accumulated in the cloud will be neutralized, and this takes place in approximately a few milliseconds. As the charge disappears from the cloud, the electric field produced by this charge in the upper atmosphere also disappears. Since the cloud field was balanced and neutralized by the electric field generated by the displacement of electrons, the removal of the cloud field leaves behind an electric field (the field produced by the displacement of electrons) in the upper atmosphere that will force the electrons to return to the original position they had before the charging of the cloud took place. This is similar to the sudden removal of force from the extended spring. In the same way that the spring recoils violently, the electrons in the upper atmosphere recoil, moving rapidly into the places they occupied before the charging of the thundercloud. This movement and acceleration of electrons cause excitation and ionization in the air. The light emitted by these processes is observed optically as sprite discharges. In general, only very strong lightning flashes that can remove a large amount of charge from a thundercloud can cause sprites to flash up in the upper atmosphere. Today we know that sprites are triggered by very strong positive ground flashes and only rarely by negative ground flashes. The reason for this could be the large charge moment (charge times displacement) associated with positive lightning discharges. It is estimated that approximately ten sprites occur every second around the globe.

One interesting question is this: why do sprites not extend all the way to the clouds even though the electric field generated by thunderclouds is larger close to their top than at several tens of kilometers higher up above the clouds? It is true that the electric field generated by the charging of a thundercloud is high close to its top and the electric field decreases as one goes higher up. On the other hand, the atmospheric density δ decreases with height z as (Chap. 4)

$$\delta(z) = \delta(0) e^{-z/\lambda_d}, \tag{19.8}$$

19.5 Upper Atmospheric Electrical Discharges

where $\delta(z)$ is the air density at height z, $\delta(0)$ is the air density at ground level (1.208 kg/m³), and λ_d is approximately equal to 7×10^3 m. Since the breakdown electric field, E_b, in air is proportional to the air density, it varies with height as

$$E_b(z) = E_b(0)e^{-z/\lambda_d}. \tag{19.9}$$

Note that the breakdown electric field decreases with height exponentially. In the preceding equation, $E_b(0)$ is the breakdown electric field at ground level. Now, let us consider the spatial variation of the electric field generated by the thundercloud. Let us denote the positive charge dissipated by a lightning flash by Q. The height at which this charge is located is denoted by h. Let us assume that R is the radius of the region where this charge is located. The electric field at the surface of this region must be less than the breakdown electric field in air at that altitude. Assuming that the electric field at the surface of this region is equal to the breakdown electric field in air at that altitude, one can write

$$R = \sqrt{\frac{Q}{4\pi\varepsilon_0 E_b(h)}}. \tag{19.10}$$

The way in which the electric field generated by this charge varies as a function of height for values of z larger than $h + R$ is depicted in Fig. 19.10 for

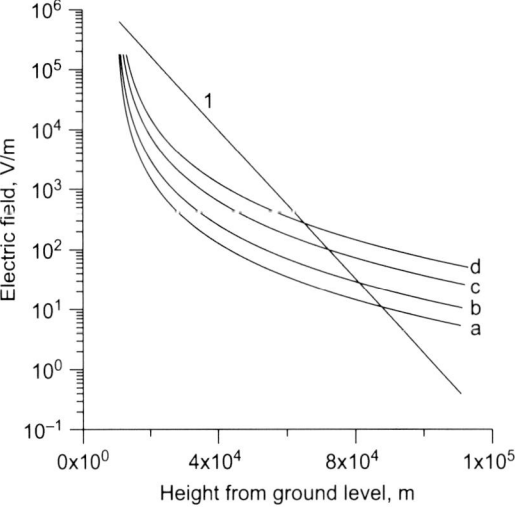

Fig. 19.10 The way in which the electric field generated by a lightning flash varies as a function of height above a thundercloud. The results are presented for four values of charge dissipation in the lightning flash: (a) 20 C, (b) 40 C, (c) 100 C, and (d) 200 C. The curve marked *1* shows how the electrical breakdown electric field varies with altitude. Note that if the charge transferred by a lightning flash is 200 C, then the electric field produced by the lightning flash will exceed the electrical breakdown field at altitudes greater than approximately 60 km (Figure created by author)

four values of Q. The same figure shows the breakdown electric field in air as a function of height. In the calculation, the height of the location of the charge dissipated by the lightning flash is assumed to be 10 km (i.e., $h = 10$ km). Consider, for example, the case $Q = 100$ C. During the discharge the electric field change produced by the discharge exceeds the breakdown electric field around 70 km. For 200 C the height where the electric field change becomes equal to the breakdown electric field is reduced to 60 km. In the case of a 200 C discharge, electrical breakdown may take place around 60 km, and this initial breakdown can give rise to a streamer-type discharge that propagates in electric fields smaller than the breakdown value. These streamer discharges create the tendrils of the sprite. The data given in Fig. 19.10 show that for a given height of charge location the charge transfer to ground by a lightning flash must be larger than a critical value to create a sprite. If the charge brought to ground is smaller than this critical value, the electric field at all heights remains less than the breakdown electric field in air, and such lightning flashes do not give rise to sprites.

19.5.2 Blue Jets and Gigantic Jets

Blue jets are electrical discharges that emanate from cloud tops (starting altitude approximately 15–18 km) and travel approximately 40 km toward the ionosphere. They have a blue conical shape and travel upward at a speed of approximately 10^5 m/s. The cone angle is approximately 15° and they last for approximately 200–300 ms. Usually they are produced at a time when the lightning flash rate is high. But they are not connected directly to a given lightning flash. Giant jets are similar in appearance to blue jets but they travel all the way from the cloud top to the ionosphere. The speed of propagation of these discharges is also on the order of 10^5 m/s. Blue jets are positive discharges, that is, they transport positive charge toward the upper atmosphere, whereas giant jets are negative discharges. These discharges are not triggered by cloud-to-ground lightning flashes.

   Several theoretical models attempt to explain the occurrence and the nature of these discharges. Initially, they were considered to be gigantic positive or negative streamers or a beam of runaway electrons. The latest models suggest that they are the result of positive and negative leaders penetrating through the thundercloud and extending upward with the aid of very long streamer systems [14]. It is suggested that blue jets start as leaders originating at the upper region of a positive charge center and travel toward the negative screening layer at the top of the cloud (Fig. 19.11a). The leader penetrates through the screening layer and gives rise to a blue jet that travels toward the upper atmosphere. Giant jets originate as leaders traveling from the upper part of a negative charge center toward the upper positive charge center (Fig. 19.11b). They penetrate through the positive charge center and travel upward as giant jets toward the upper atmosphere. In order for a leader to penetrate a charged region of opposite polarity, the charge density in the parent charge center of the leader (i.e., the charge center in which the leader is generated)

19.5 Upper Atmospheric Electrical Discharges

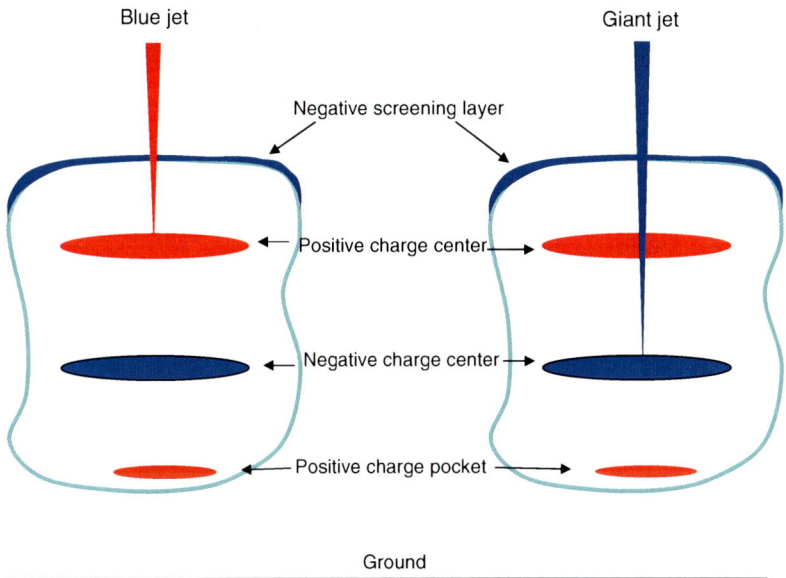

Fig. 19.11 (a) (diagram to the left) Blue jets start as a leader originating at the upper region of the positive charge center and travel toward the negative screening layer at the top of the thundercloud. (b) (diagram to the right) The giant jets start as a leader originating at the upper region of the negative charge center and travel toward the positive charge center. The leader penetrates the positive charge center and propagates toward the upper atmosphere as a giant jet (Figure created by the author)

must be significantly greater than the charge density of the opposite polarity. Otherwise, the leader will not gain enough potential to create its own electric field to penetrate a charge center of opposite polarity.

19.5.3 Elves

As described in previous chapters, a lightning flash gives rise to an electromagnetic wave that spreads out in all directions at the speed of light. We have also seen that whenever this electromagnetic wave hits a conductor, it generates a current in the conductor. Now, as the electromagnetic wave from a lightning flash travels upward, it encounters the conducting ionosphere, which is located at a height of approximately 60–90 km. The distance to the ionosphere from the lightning flash varies with the direction of travel. It is shortest directly above the thundercloud and increases as the direction of travel moves closer and closer toward the horizon. As the electromagnetic wave spreads out from the lightning flash like a hemisphere, it encounters the ionosphere first directly overhead, and the point of encounter moves outward like that of an expanding ring (Fig. 19.12). This encounter between

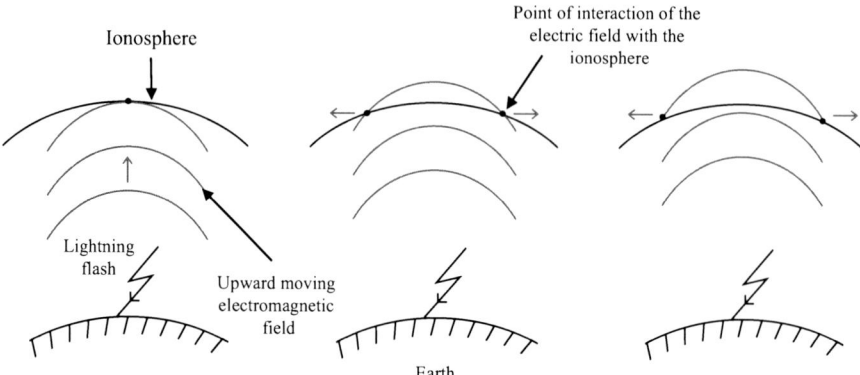

Fig. 19.12 The interaction of a lightning-generated electromagnetic field with the upper atmosphere gives rise to a current leading to further excitation and ionization in the upper atmosphere that spread out as a ring as the electromagnetic field interacts with ever more distant areas of the upper atmosphere. Since the ionization and excitation of the upper atmosphere lead to the emission of light, a rim of light seems to move outward as the circle on which the crossing points of the upper atmosphere with the electromagnetic field are located moves outward (Figure created by author)

electromagnetic waves and the ionosphere gives rise to a current in the ionosphere, and this current spreads out like an expanding ring as the point of interaction between the radio waves and the ionosphere expands outward. As we know, an electric current is simply the movement of electrons. As the electrons move, they collide with atoms of nitrogen and oxygen in air, and some of these collisions can excite the atoms, which leads to the emission of light. Thus, the interaction of the radio waves produces a ring of light that expands outward in the upper atmosphere. These bright rings are called elves. Elves are very brief events and are over in approximately 1 ms.

19.6 Energetic Radiation Production from Lightning and Thunderstorms

In January 2011 NASA reported that the Fermi Gamma-ray Space Telescope had detected beams of antimatter produced by thunderstorms on Earth, a phenomenon never seen before. This is actually the culmination of a process anticipated by C. R. T. Wilson in 1924 [15]. Wilson realized that the electric fields in thunderclouds are large enough and extend spatially long enough to accelerate electrons to relativistic speeds. This is called the *electron runaway process* (Chap. 7). These runaway electrons, when suddenly stopped during collisions with air atoms and molecules, dissipate their energy by creating more electrons, positrons (antimatter), and X-rays. This is the source of X-rays and antimatter from thunderclouds.

Bursts of X-rays and gamma rays generated by thunderstorms were first discovered in 1994 in an experiment carried out onboard the Compton Gamma Ray Observatory orbiting the Earth. These bursts were named terrestrial gamma-ray flashes, or TGFs. Initially, it was assumed, based on theoretical calculations, that the energy of TGFs follows a power-law spectrum with a cutoff near 10 MeV. Later measurements indicated that the energy spectrum of these flashes extended up to 100 MeV [16]. Recently, scientists have also observed gamma-ray flashes generated by thunderstorms with energies as high as 10 MeV at ground level [17]. Scientists now believe that approximately 500 TGFs are produced daily by thunderclouds. Since the initial study onboard Compton Gamma Ray Observatory, many studies have documented X-ray generation inside thunderstorms [18–21]. More recently, scientists have observed thundercloud-generated X-ray bursts even at ground level [22]. Studies have also discovered X-ray emissions not only by thunderclouds but also from lightning and electrical sparks [23–25]. For example, X-ray bursts have been observed within several hundred meters of stepped leaders, dart stepped leaders, and dart leaders. Experiments show that in sparks X-rays are generated just before the final breakdown during which time leaders and streamers are active in the discharge gap. X-ray bursts measured from lightning leaders and sparks contain energies extending up to approximately 1 MeV.

However, the exact mechanism of X-ray generation in lightning and sparks is not yet known, and X-ray flashes directed toward the Earth by thunderstorms have not been explored. Even though it is understood that an electron runaway mechanism is responsible for X-ray generation, the exact mechanism that leads to the seed electrons necessary for the formation of runaway electron avalanches is not completely understood. It is believed, however, that in thunderclouds the relativistic seed electrons are generated by energetic gamma rays. On the other hand, theoretical calculations demonstrate that in dart leaders they could be generated by the acceleration of thermal electrons to relativistic energies by very high electric fields existing only for a short time in the discharge; in the case of sparks, the thermal electrons may be accelerated to relativistic energies by the strong electric fields created during the meeting of streamers of opposite polarity [26, 27]. Of course, once leaderlike discharges are formed inside thunderclouds, they can also inject relativistic electrons into the thundercloud environment, which may result in relativistic electron avalanches.

References

1. Nicolet M (1975) On the production of nitric oxide by cosmic rays in the mesosphere and stratosphere. Planet Space Sci 23:637–649
2. Cooray V, Rahman M (2005) Efficiencies for production of NO_x and O_3 by streamer discharges in air at atmospheric pressure. J Electrost 63:977–983
3. Chameides WL (1986) The role of lightning in the chemistry of the atmosphere, the earth's electrical environment. National Academy Press, Washington, DC

4. Cooray V, Rahman M, Rakov V (2009) On the NOx production by laboratory electrical discharges and lightning. J Atmos Sol Terr Phys 71:1877–1889
5. Rahman M, Cooray V, Rakov VA, Uman MA, Liyanage P, DeCarlo BA, Jerauld J, Olsen RC III (2007) Measurements of NOx produced by rocket-triggered lightning. Geophys Res Lett 34:L03816. doi:10.1029/2006GL027956
6. Colin Price C, Pechony O, Greenberg E (2007) Schumann resonances in lightning research. J Light Res 1:1–15
7. Williams ER (1992) The Schumann resonance: a global tropical thermometer. Science 256:1184–1186
8. Price C (2008) Thunderstorms, lightning and climate change. In: Proceedings of the 20th international conference on lightning protection, Uppsala, Sweden
9. Wilson CTR (1925) The electric field of a thundercloud and some of its effects. Proc R Soc Lond 37:32D
10. Winckler JR, Lyons WA, Nelson TE, Nemzek RJ (1996) New high-resolution ground- based studies of sprites. J Geophys Res 101:6997–7004
11. Franz RC, Nemzek RJ, Winckler JR (1990) Television image of a large electrical discharge above a thunderstorm system. Science 249:48–51
12. Füllekrug M, Mareev EU, Rycroft MJ (eds) (2006) Sprites, elves and intense lightning discharges. Springer, Dordrecht, The Netherlands
13. Heavner MJ, Sentman DD, Moudry DR, Wescott EM, Siefring CL, Morrill JS, Bucsela EJ (2000) Sprites, blue jets, and elves: optical evidence of energy transport across the stratopause. In: Siskind DE, Eckermann SD, Summers ME (eds) Atmospheric science across the stratopause. American Geophysical Union, Washington, DC
14. Krehbiel PR, Riousset JA, Pasko VP, Thomas RJ, Rison W, Stanley MA, Edens HE (2008) Upward electrical discharges from thunderstorms. Natl Geosci 1(4):233–237. doi:10.1038/ngeo162
15. Wilson CTR (1925) The acceleration of β-particle in strong electrical fields of thunderclouds. Proc Camb Philos Soc 22:534
16. Tavani M et al (2011) Terrestrial gamma-ray flashes as powerful particle accelerators. Phys Rev Lett 106:018501
17. Tsuchiya H et al (2007) Detection of high-energy gamma rays from winter thunderstorms. Phys Rev Lett 99:165002. doi:10.1103/PhysRevLett.99.165002
18. Shaw GE (1967) Background cosmic count increases associated with thunderstorms. J Geophys Res 72:4623–4626. doi:10.1029/JZ072i018p04623
19. Parks GK, Mauk BH, Spiger R, Chin J (1981) X-ray enhancements detected during thunderstorm and lightning activities. Geophys Res Lett 8:1176–1179. doi:10.1029/GL008i011p01176
20. McCarthy M, Parks GK (1985) Further observations of X-rays inside thunderstorms. Geophys Res Lett 12:393–396. doi:10.1029/GL012i006p00393
21. Eack KB, Beasley WH, Rust WD, Marshall TC, Stolzenburg M (1996) X-ray pulses observed above a mesoscale convection system. Geophys Res Lett 23:2915–2918. doi:10.1029/96GL02570
22. Dwyer JR et al (2004) A ground level gamma-ray burst observed in association with rocket-triggered lightning. Geophys Res Lett 31:L05119. doi:10.1029/2003GL018771
23. Dwyer JR et al (2003) Energetic radiation produced during rocket triggered lightning. Science 299:694–697. doi:10.1126/science.1078940
24. Dwyer JR et al (2004) Measurements of X-ray emission from rocket triggered lightning. Geophys Res Lett 31:L05118. doi:10.1029/2003GL018770
25. Rahman M, Cooray V, Ahmed NA, Nyberg J, Rakov VA, Sharma S (2008) X rays from 80-cm long sparks in air. Geophys Res Lett 35:L06805. doi:10.1029/2007GL032678
26. Cooray V, Becerra M, Rakov V (2009) On the electric field at the tip of dart leaders in lightning flashes. J Atmos Sol Terr Phys 71:1397–1404
27. Cooray V, Arevalo L, Rahman M, Dwyer J, Rassoul H (2009) On the possible origin of X-rays in long laboratory sparks. J Atmos Sol Terr Phys 71:1890–1898

Chapter 20
Unusual Forms of Lightning Flashes

20.1 Introduction

Based on various experiences and observations several phenomena associated with thunderstorms and lightning have become part of folklore. These are forked lightning, bead lightning, ribbon lightning, and ball lightning. As outlined subsequently, evidence, based on experimental observations, for some of these are well documented in the literature whereas others are based on personal accounts of observations. A brief description of each of these phenomena is given in this chapter.

20.2 Forked Lightning

In some lightning strikes to ground, the channel at the lowest section is divided like a fork and apparently strike the ground simultaneously at two different points (Fig. 20.1). Lightning flashes having this feature are called *forked lightning*. Two physical scenarios give rise to forked lightning. The first one is as follows. As the stepped leader comes down toward the ground, it gives rise to several branches. Thus, at any given time several branches could be traveling toward the ground. One of these branches (let us call it the main branch) touches the ground first and initiates a return stroke. The return stroke travels along the main channel and along the branches and neutralizes the charges located on the branches, thereby stopping their continuous propagation. However, in very rare cases immediately after the main branch touches the ground another branch may touch the ground before its charge is neutralized by the return stroke generated by the main branch. Since the speed of a return stroke is very fast, the two branches should touch the ground within a few microseconds of each other, so that the return stroke of the main branch does not have enough time to travel the distance to the tip of the second branch. In this case, another return stroke is generated when the second branch touches the ground. This return stroke travels along

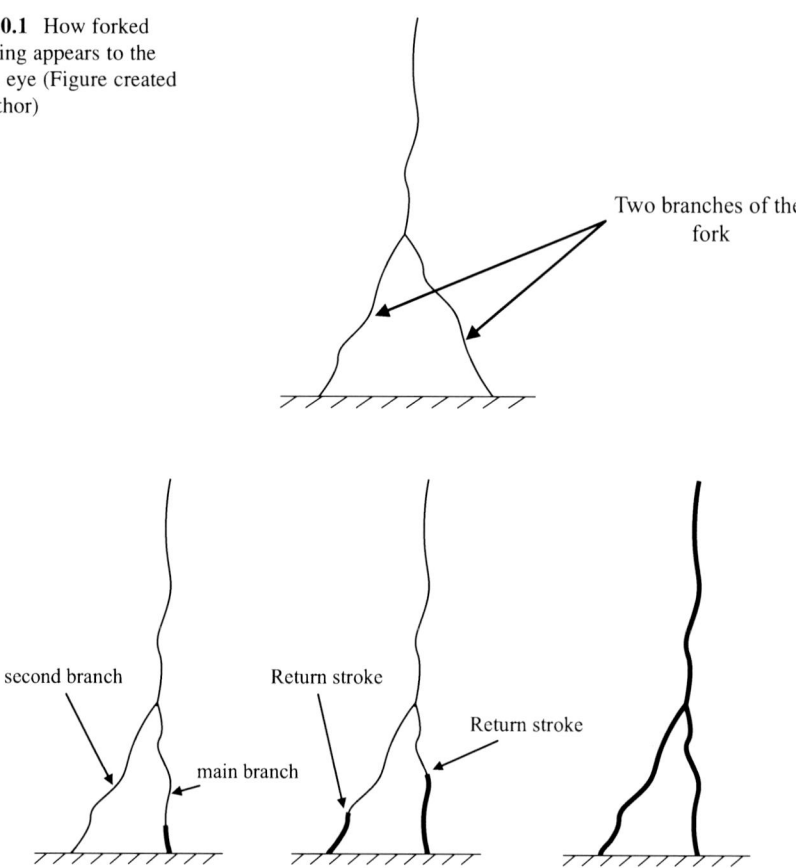

Fig. 20.1 How forked lightning appears to the naked eye (Figure created by author)

Fig. 20.2 The first mechanism that can give rise to forked lightning. In this case, two branches of the stepped leader strike the ground almost at the same time, giving rise to two return strokes that share the same channel at higher locations along the stepped leader. In order for this to happen the two branches must touch the ground within a few microseconds. Otherwise, the return stroke of the first event would flow along the other branch, neutralize its charge, and kill it before it touches the ground (Figure created by author)

the branch and merges into the return stroke created by the main branch. To the naked eye the complex process cannot be resolved and the lightning channel seems to be striking two points on ground simultaneously. This process is illustrated in Fig. 20.2. The second mechanism that can generate forked lightning is as follows. In general, after the first return stroke the subsequent strokes are created by dart leaders traveling along the same channel. However, in some cases the dart leader may deviate from this path and travel to ground as a stepped leader along a new path. When the new stepped leader makes the connection with ground, the resulting return stroke travels along this new path to the cloud. It is only the lowest section of the new return stroke path that differs from the previous one. If the deviation of the paths happens at a point close to ground, then a still photograph may show a forked-shape channel at the ground end, even though the one part of the so-called fork was created many milliseconds after the first one. This is illustrated in Fig. 20.3.

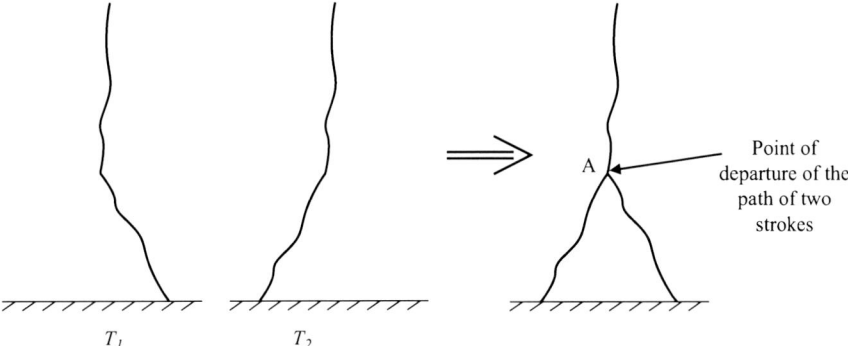

Fig. 20.3 The second mechanism that can give rise to forked lightning. In this case, two strokes of the same flash (separated in time by the amount $T_2 - T_1$) take two different paths to ground from a point located below the thundercloud (point marked A). To the naked eye it appears as a fork, even though the two paths of the fork are created by events several tens of milliseconds apart (Figure created by author)

20.3 Bead Lightning

In several time-resolved lightning photographs published in the literature it is clear that a decaying lightning channel consists of several periodically changing dark and bright regions (Fig. 20.4). The length of the dark and bright regions can range from a few meters to several meters. These types of lightning flashes are called *bead lightning*. Several theories have been advanced to explain the occurrence of bead lightning [1, 2]. The first one is that either rain or clouds covers the lightning channel periodically so that the lightning channel can be seen as consisting of bright and dark regions. However, it is doubtful whether this is the correct explanation because in time-resolved photographs the bead nature appears at the later stages of the lightning flash when the channel is decaying. The second explanation of bead lightning is that the section of the channels slanted away from the observer generates a brighter section in the image than the section located perpendicular to the field of view (Fig. 20.5). This, of course, explains the bright spots that sometimes appear in photographs. However, to explain bead lightning, this change of geometry should happen periodically, and there is no physical reason for a natural process to behave in a periodical manner. However, the lightning channel geometry is very complex, and occasionally very strange geometries can be observed. The third explanation for bead lightning attributes its cause to the pinching of the channel by magnetic forces. As a current flows along a channel, it generates a magnetic field that forms closed loops around the channel. This magnetic field excert a force on the moving charge carriers in the channel (electrons in our case) forcing them to move toward the center of the channel (Fig. 20.6). If the channel is very thin and the current is very large, a considerable force could act on the channel (or the charge carriers) to compress it and make the channel thinner. This is called the *magnetic pinch effect*. As the channel becomes thinner, the current density flowing through it increases, leading to a higher illumination (bright channel).

Fig. 20.4 Sometimes, the decaying stage of a lightning channel may appear as bright and dark sections or as beads in a string. This phenomenon is called bead lightning (Figure created by author)

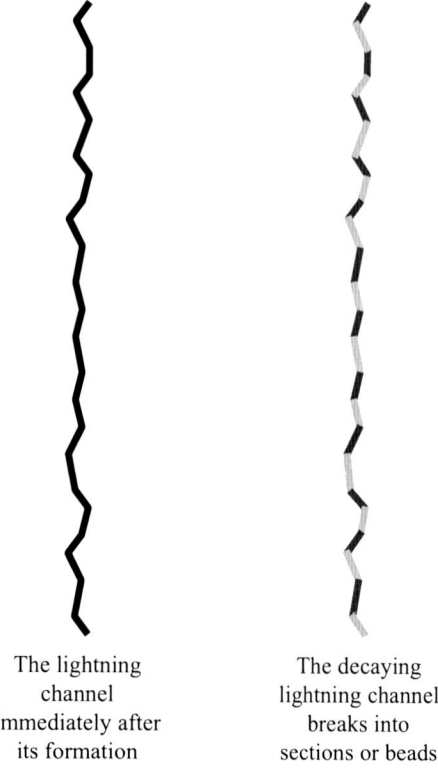

The lightning channel immediately after its formation

The decaying lightning channel breaks into sections or beads

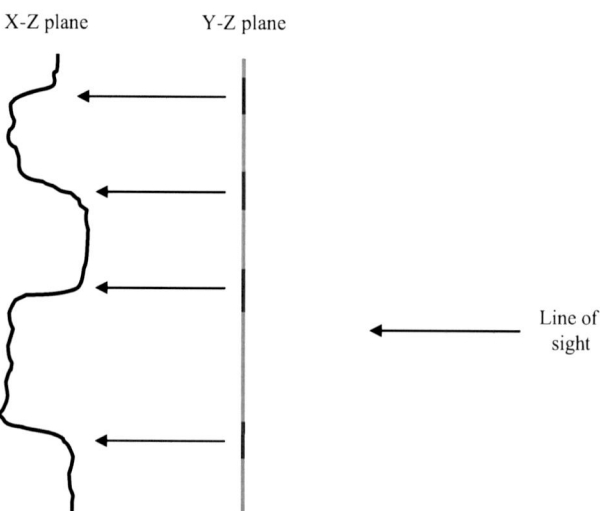

Fig. 20.5 The sections of a channel slanted away from an observer may appear as a bright spot when the slanted lightning channel is located along the line of sight of the observer (Figure created by author)

20.3 Bead Lightning

Fig. 20.6 The magnetic field generated by the current flowing in a lightning channel exerts a force and causes moving electrons to move toward the center of the channel. For example, if **v** is the velocity of the electron, then the force experienced by it from the magnetic field **B** is $-e\mathbf{v} \times \mathbf{B}$, where e is the electronic charge. This leads to the shrinking of the section of the channel transporting current. This effect is called the magnetic pinch effect (Figure created by author)

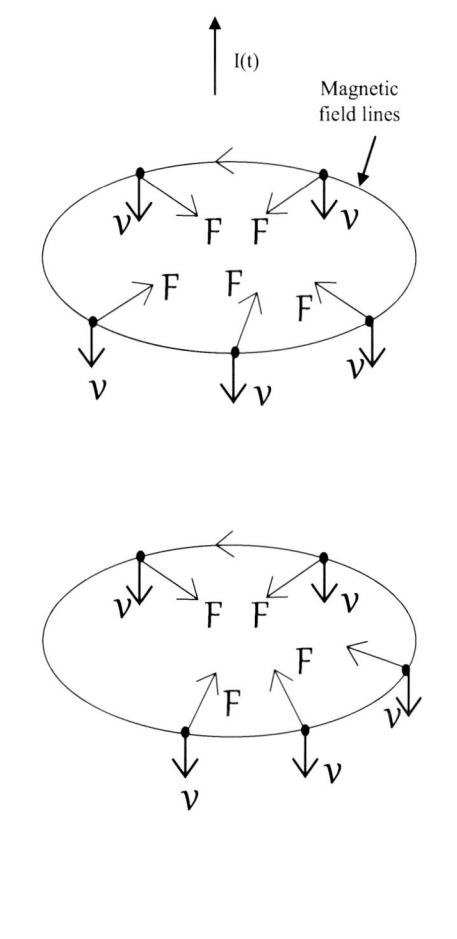

However, for this process to generate bead lightning, the pinch effect should be active alternatively along the channel so that some sections of the lightning channel becoming thinner (and brighter) while other sections remain thicker and, hence, less luminous. Why it should happen periodically remains an unsolved question.

In connection with bead lightning, a possible explanation that has not yet been considered in the literature is the following. A lightning channel is not straight; it has a rather complicated and tortuous structure. When a strong return stroke current propagates through it, the channel at the kinks experiences a force that in principle attempts to make the channel straighter. This imparts a momentum to the charge particles flowing in the channel, and this momentum is transferred to the neutral gas molecules through collisions. Thus, the channel at the kinks moves in space relative to other sections, and this movement not only displaces the channel but also cools it. The reason for the cooling is that as it moves through colder air, the hot air mixes with the cold air through turbulent mixing. The effects are apparent especially at the

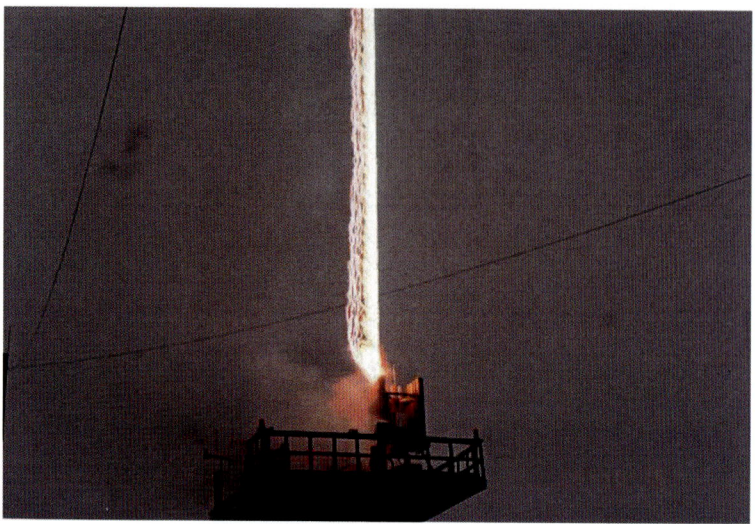

Fig. 20.7 Different strokes of triggered lightning flash separated in space due to movement of channel in wind during interstroke time interval (Courtesy Prof. V. Rakov, University of Florida at Gainesville, Florida, USA)

decaying stage of the channel, where the forces have enough time to displace and cool the channel sections. This rapid cooling of parts of the channel with respect to the other parts may give rise to the bright and dark sections giving the appearance that the channel has broken into bright and dark sections at the decaying stage. This mechanism could also explain the formation of bead lightning.

20.4 Ribbon Lightning

Sometimes a lightning channel appears to be separated into several ribbons displaced slightly from each other in space (Fig. 20.7). Such lightning is called *ribbon lightning*. As described in Chap. 7, a single lightning flash consists of many return strokes that take place in the channel at time intervals of 40–60 ms. If there is strong wind, the lightning channel (the hot gas) might move slowly in the wind, and therefore each subsequent stroke would appear to be displaced from the previous one spatially, even though the strokes travel along the same hot channel. Lightning with such an appearance is called *ribbon lightning*.

20.5 Lightning without Thunder

Sometimes when lightning flashes are taking place at a considerable distance, they are observed without any accompanying thunder. Such lightning flashes gave rise to the notion that some lightning flashes do not produce any thunder. In reality, all lightning

20.6 Ball Lightning

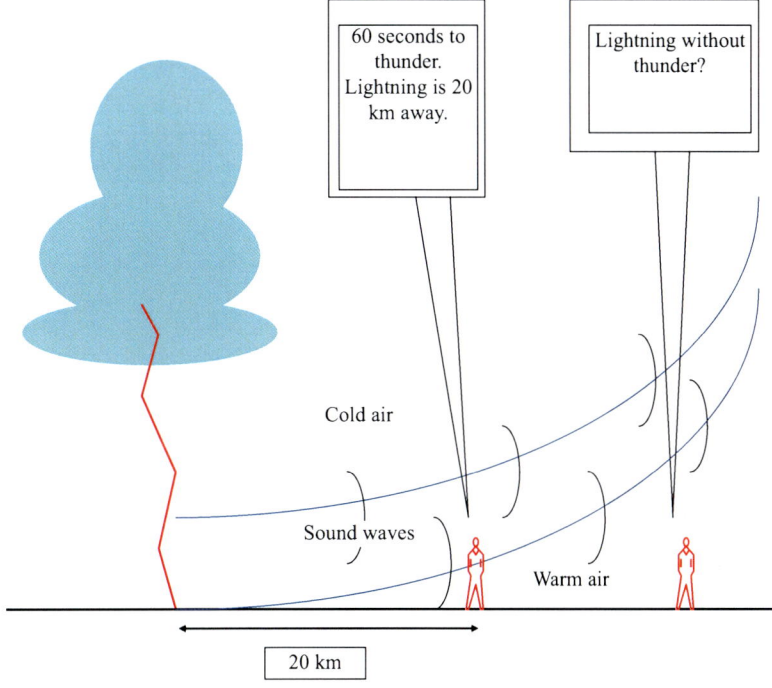

Fig. 20.8 Lightning flashes always produce thunder. Usually thunder can be heard only if observers are located within approximately 20 km of the lightning flash. This is because the air is warm near the surface and cold aloft. This causes sound waves to bend upward and pass over distant observers without reaching their ears (Figure created by author)

flashes generate thunder as a consequence of the rapid heating of the channel during the return stroke. The thunder is not heard because of the diffraction of sound waves. Sound waves travel faster in hot air. Since the air layer close to ground is warmer than the air at higher altitudes, sound waves propagating in air are diffracted upward, preventing them from reaching observers on the ground. Thus, a thunder signal that has propagated for some distance may pass over the heads of observers, making them believe that a lightning flash has occurred without thunder (Fig. 20.8).

20.6 Ball Lightning

Ball lightning has been seen and described since antiquity and recorded in many places around the globe [2, 3]. Because ball lightning has not been produced in the laboratory and the authenticity of the available photographs is questionable, the properties of ball lightning must be extracted from eyewitness accounts. According to eyewitness accounts, ball lightning is described as being spherical in shape, although other shapes such as teardrops or ovals have also been reported. On rare occasion, ball lightning shaped like rods has also been observed. The diameter of ball lightning is usually 10–40 cm, but ball lightning as large as 1 m has been reported. The lifetime of ball lightning is reported to be approximately 10 s, but occasionally ball lightning lasting as long as 1 min has been observed. Ball lightning may manifest in different colors such as red, reddish yellow, yellow, white, green, and purple. In some cases, the intensity of ball lightning may increase with time and becomes a dazzling white before it disappears explosively. The structure of the ball lightning may vary from one report to another. Sometimes a solid core surrounded by a translucent envelope is reported, whereas in other cases it is a rotating structure or a structure that emits sparklike phenomena. Most ball lightning phenomena are said to move horizontally; however, descriptions that include motionlessness, a zigzag movement, or moving from a cloud toward the ground can also be found. The speed of movement is reported to be walking speed, i.e., approximately 1–2 m/s, and ball lightning that moves against the wind has also been reported. Some reports describe sharp or acrid odors, and 10 % of observers report sounds associated with the manifestation. The ball may disappear silently or explosively. However, in many cases where the ball explodes in the vicinity of various objects, no movement of objects or damage to them has been reported.

Ball lightning is usually observed during thunderstorms, but a significant number of sightings have also been reported during fine weather with no connection to thunderstorms or lightning. In addition, they have been reported under stressful natural conditions, such as tornadoes, storms, and earthquakes. Most ball lightning has been sighted indoors, and at least in one case it was observed inside a commercial aircraft. Ball lightning has also been reported by sailors under stormy conditions but without any lightning. According to observations, balls of lightning can enter or depart from closed rooms and pass through solid walls and closed windows without any apparent change in its structure and without causing any damage.

There is a wide variety of theories on ball lightning. Some theories suggest that the ball lightning is internally powered by energy coming from chemical, electrical, or nuclear sources. Other theories suggest external energy input from electromagnetic radiation generated by lightning or the electric fields caused by current flowing along the ground. These theories attempt to describe how the structure of the ball is maintained without dispersion and to explain the energy output similar to the descriptions given by observers. Unfortunately, none of these theories is testable since almost all ball lightning descriptions are subjective descriptions and do not provide ways to test the theories objectively. Moreover, the descriptions given may vary from one observation to another, and so single theory explains all the observed features. For example, a recent theory suggests that ball lightning is burning silicon that has been vaporized by a lightning strike to soil [4]. If this is the correct theory, how does it explain, for example, the observations that ball lightning can penetrate into rooms through walls and closed windows. The great difficulty of encompassing all observed features of ball lightning in a single theory makes it highly probable that many observations and experiences having no connection to ball lightning are also categorized as ball lightning experiences.

It is possible that some of the observations are really the effects of direct lightning strikes misinterpreted as ball lightning. Moreover, some of the observations could be experiences with other causes more related to the brain than

Fig. 20.9 Elementary visual hallucinations as perceived and drawn by patients with occipital epilepsy [5]

lightning. For example, visual experiences very similar to ball lightning are described by patients undergoing epileptic seizures of the occipital lobe [5, 6]. This is especially the case with idiopathic occipital epilepsy, a benign condition affecting predominantly children. During a seizure, the patient remains conscious with visual sensations similar to ball lightning appearing in the field of vision. Visual hallucinations in the form of luminous objects that are either circular or ball-like are not uncommon. Several sketches made by patients undergoing seizures adapted from [5] are shown in Fig. 20.9. In particular, luminous balls of various colors – red, yellow, blue, and green – moving horizontally from the periphery of vision to the center have been described that may appear to be rotating or spinning. Sometimes the ball also appears to have a solid structure surrounded by a thin glow or, in other cases, to generate sparklike phenomena. When the ball is moving toward the center of vision, it may increase in intensity, and when it reaches the center, it can "explode," illuminating the whole field of vision. During hallucinations, the vision is obscured only in the area occupied by the apparent object. The hallucinations may last for 5–30 s or, rarely, up to 1 min. Occipital seizures may spread to other regions of the brain, inducing auditory, olfactory, and sensory sensations. These sensations can take the form of buzzing sounds, the smell of burning rubber, pain with thermal sensation, especially in the arms and face, and numbness and a tingling sensation. In some cases, people may experience only one or a few seizures during their lifetime and may not be aware of the cause of the experience. If they read descriptions of ball lightning given in the literature, they might categorize their own experience also as ball lightning. Epileptic seizures can be triggered by the intermittent lightning generated by lightning flashes in intense thunderstorms, giving a possible explanation for the association of the observations with lightning [7]. Recent studies also show that the strong magnetic field from lightning flashes and the energetic radiation, such as γ-rays and X-rays, generated by lightning flashes could also generate visual excitations known as *phosphenes* [8–11]. Sketches of phosphenes as seen by the Apollo astronauts due to cosmic rays adapted from reference [10] are shown in Fig. 20.10. Such experiences could also be categorized as ball lightning. All these observations and experiences, which have nothing to

Fig. 20.10 Schematic diagrams of phosphenes observed by astronauts in Apollo missions (Adapted from Fuglesang [10])

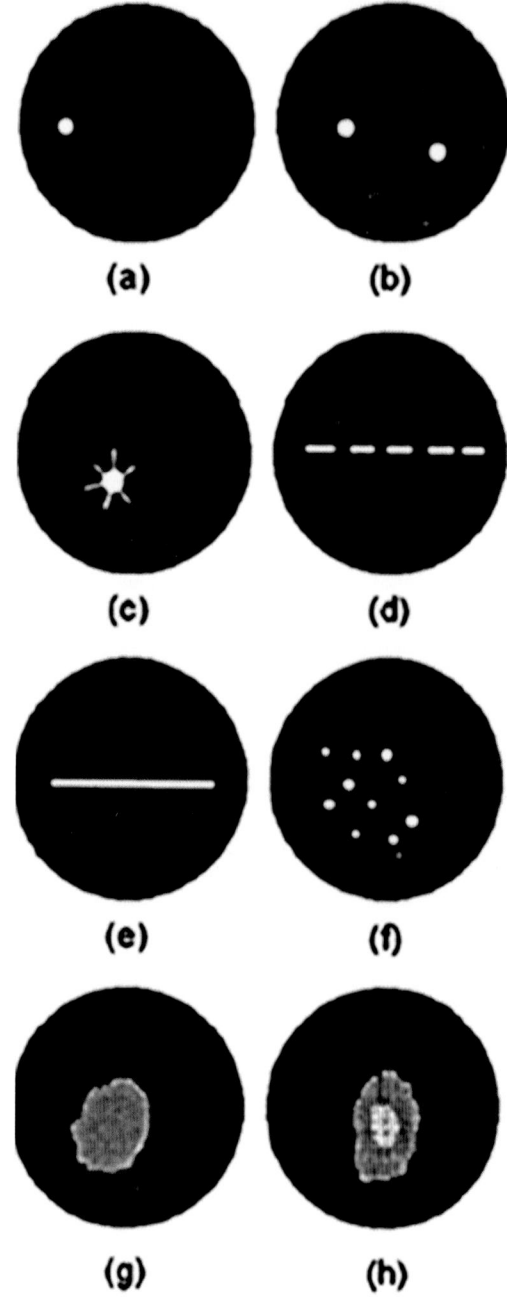

do with the real phenomenon of ball lightning, may contaminate the literature to such an extent that there is no gain to the scientific community by creating more theories to explain the phenomenon. What is needed are well-documented video records of ball lightning that can be analyzed and scrutinized for scientific research. Given the vast amount of mobile cameras equipped with video possibilities that are being used at any given time, such records should be available to scientists in the near future if ball lightning is a real phenomenon. Very recently, scientists from China recorded a ball of fire that rose to approximately 5 m from the strike point of a lightning flash. The ball lasted approximately 1.5 s [12]. This was hailed in the media as the first record of ball lightning. The ball consists of burning dirt. Even if this really is ball lightning, it is still very different from what has been described in ball lightning encounters.

What we can conclude is the following. It is possible that a real physical phenomenon called ball lightning exists, but we need more experimental evidence for it. However, the ball lightning descriptions available in the literature are probably contaminated by visual disturbances caused by electromagnetic fields, hallucinations caused by epileptic seizures of the occipital lobes, which are probably triggered by intermittent light bursts generated by lightning flashes during thunderstorms, and direct effects of lightning that are wrongly attributed to ball lightning.

References

1. Uman MA (1968) Decaying lightning channels, bead lightning and ball lightning. In: Coroniti SC, Huges J (eds) Planetary electrodynamics, vol 2. Gordon and Breach, New York, p 199
2. Barry JD (1980) Ball lightning and bead lightning. Plenum Press, New York
3. Stenhoff M (1999) Ball lightning – an unsolved problem in atmospheric physics. Plenum Press, New York
4. Abrahamson J (2002) Ball lightning from atmospheric discharges via metal nanosphere oxidation: from soils, wood or metals. Phil Trans R Soc A 360(15):61–88
5. Panayiotopoulos CP (1999) Elementary visual hallucinations, blindness, and headache in idiopathic occipital epilepsy: differentiation from migraine. J Neorol Neurosurg Psychiatry 66:536–540
6. Cooray G, Cooray V (2008) Could some ball lightning observations be optical hallucinations caused by epileptic seizures? Open Atmos Sci J 2:101–105
7. Cooray V, Cooray G (2010) Could the intermittent light generated by lightning flashes trigger epileptic seizures? Letter to the Editor, J Neorol Neurosurg Psychiatry http://jnnp.bmj.com/letters?first-index=31&hits=10
8. Peer J, Kendl A (2010) Transcranial stimulability of phosphenes by long lightning electromagnetic pulse. Phys Lett A 374:2932–2935
9. Peer J, Cooray V, Cooray G, Kendl A (210) Erratum and addendum to Transcranial stimulability of phosphenes by long lightning electromagnetic pulse. [Phys Lett A 374 (2010), 2932], Phys Lett A 374:4797–4799
10. Fuglesang C (2007) Using the human eye to image space radiation or the history and status of the light flash phenomena. Nucl Instrum Phys Res A 580:861–865
11. Cooray V, Cooray G, Dwyer J (2011) On the possibility of phosphenes being generated by the energetic radiation from lightning flashes and thunderstorms. Phys Lett A 375:3704–3709
12. Cen J, Yuan P, Xue S (2014) Observation of the optical and spectral characteristics of ball lightning. Phys Rev Lett 112:035001

Appendices

Appendix A: How to Avoid Lightning Injuries

A.1 How to Avoid Lightning Injuries

In this section, let us consider what one should do to avoid injuries from lightning. If you get caught in a thunderstorm in an area where there are no buildings, the first thing that enters your mind is to run to the nearest tree and find shelter from the rain. This is a very dangerous thought, especially if the tree is an isolated one. There is a high probability that an isolated tree, being the highest structure around, will be struck by lightning. If you happen to be near a tree, you could experience a side flash from the tree, and you will also be exposed to a strong step voltage. Moreover, during a lightning strike the tree can explode, exposing you to flying debris. So never stay close to a tree during a thunderstorm. Furthermore, if caught in a thunderstorm outside, do not make yourself a lightning conductor by projecting yourself up artificially with umbrellas, ladders, or golf clubs or by being on a bicycle or a tractor. The best thing to do in such a situation is to find a ravine, valley, or depression in the ground and make yourself as low as possible in relation to the surrounding objects. The best posture you can take is to squat down so that your feet are in contact with the smallest possible area on the ground; this will reduce the step current that will flow through your body if lightning happens to strike your vicinity. If you are caught in a thunderstorm in a forest, then you cannot avoid being close to trees. In such a situation, locate yourself between trees and squat down. Leave at least 2 m of space between you and both the trunks and foliage of the trees. In this way you may avoid being struck by side flashes either from a tree trunk or its branches if a nearby tree is struck by lightning (Fig. A1). By squatting down you will make yourself as low as possible, while at the same time, since the separation between your feet is small, you will not be exposed to a large step current.

Based purely on geometrical considerations, it may be inferred that the overhead ground wire of a power line may provide an area of protection around the power

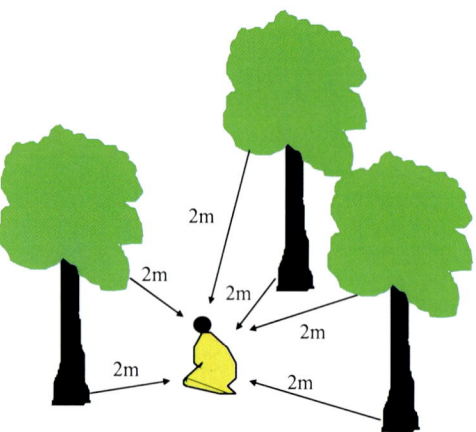

Fig. A1 What to do if caught in a thunderstorm while in a forest. If one happens to be in a forest during a thunderstorm, it is impossible to avoid being close to trees. The best posture to assume in such a situation is to squat down at least 2 m from the trunks and foliage of nearby trees. This will reduce the possibility of injury from side flashes and step currents if a nearby tree is struck (Figure created by author)

line. However, the secondary effects from a lightning strike to a power line may make the space below the power line unsafe during thunderstorms. First, if the power line is struck by lightning, the resulting current is brought to ground at the poles. Thus, in the vicinity of poles one may be exposed to large step currents. Moreover, if for some reason the grounding system of the overhead ground wire happens to be faulty, one could also receive a side flash from the pillion. Furthermore, the pole-mounted transformers could explode if a strong lightning flash becomes attached to the phase conductors of the power line. In addition, some of the rare positive ground flashes may transport a huge amount of electric charge to ground. If such a lightning flash becomes attached to a conductor of a power line, the resulting melting of the conducting material may lead to the breaking of the conductor, which is already under mechanical tension. This can lead to the electrocution of a person located under the power line.

Avoid touching or standing close to fences with metal wires during thunderstorms. Even if such a fence is struck by lightning at a point far away from you, the current may travel along the metal wires for a long distance. This current can pass directly through your body if you happen to be holding the fence during the lightning strike. Or you might get a side flash if you happen to be standing close to it (Fig. A2).

Avoid open spaces such as rivers and lakes during thunderstorms. In such spaces, you become the highest object in the area and you will serve as a lightning conductor. It is also dangerous to swim in lakes during thunderstorms. If there is a lightning strike to water in your vicinity, a large current may flow through your body. This current may knock you unconscious or temporarily paralyze your limbs. The result could be drowning. If you happen to be in a small boat in a lake or sea during a thunderstorm, remain seated inside the boat; under no circumstances should you stand on the boat and fish. Also, avoid putting your hands or legs in

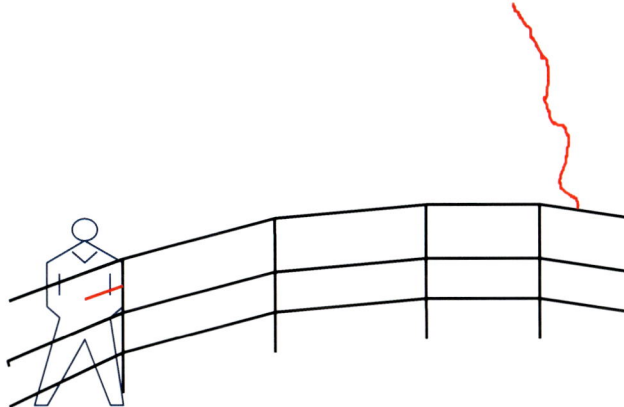

Fig. A2 Avoid touching or standing close to metal fences during thunderstorms. If a metal fence is struck by lightning, the current may flow along it and could pass to ground through the body of a person holding it. The fence could also direct a side flash to a person standing close to it, as shown here (Figure created by author)

the water because if the boat is struck by lightning, the current, seeking the least resistive path, can pass through your body to the water.

It is also a good practice to avoid picnicking in mountainous regions and climbing mountains during thunderstorms. If you are caught in a thunderstorm while climbing a mountain, try to avoid large metal objects such as axes, tent poles, and metallic picks; crouch down in a ravine or in a dip. In mountainous regions, the ground is not very conducting, and the lightning current may travel a considerable distance in search of proper ground. The current may enter any metallic structures such as water pipe systems and railings. Therefore, stay away from these metallic objects during thunderstorms.

If you are caught in a thunderstorm in an urban area, immediately find shelter in a building, especially one with a metal frame and preferably with a lightning conductor. Modern-day concrete buildings provide reasonable protection, because of their metal frame, to people inside the building. If such buildings are not in the vicinity, seek shelter in any building. However, remember that small sheds with metal roofs without proper connection to ground may produce side flashes inside them and do not provide complete protection. An automobile (not a convertible) is a very safe place to be during thunderstorms. The reason for this is not, as is sometimes asserted, the car's rubber tires. A lightning stepped leader that has traveled several kilometres in air cannot be stopped by a few inches of rubber. The reason why an automobile protects occupants from a lightning strike is related to a physical principle discovered by the British scientist Michael Faraday in the seventeenth century. Faraday showed that if you inject a current onto a metal cage, all of it flows on the outer surface of the cage; no current flows inside and no electrical discharges take place inside the cage. Automobiles, being metal, are good Faraday cages. If lightning strikes an automobile, the current will flow along the outer metal surface of the car and jump to ground across the rubber wheels. Remember, however, that the lightning flash can cause a short circuit in the batteries leading to a fire, and the electronics of the car can be damaged or destroyed including the door-locking mechanism. Moreover, the windscreen can shatter due to the shock wave. Furthermore, some modern cars contain a lot of plastic material and may not act as a Faraday cage during a lightning strike. If you happen to be driving during a thunderstorm, drive slowly. If a lightning strike occurs in front of you, the bright light and the frightening display of fireworks at the point of strike may confuse you and cause you to lose control of the vehicle.

When a lightning flash strikes an unprotected house, the lightning current, as usual, tries to find the least resistive path to ground. Therefore, the current may enter metal parts such as electricity, telephone, gas, and water installations. If you happen to be holding parts of these installations or standing beside them, you might be subjected to a touch current or side flash from these installations. Therefore, you should avoid using electrical appliances, bathtubs, sinks, and telephones during thunderstorms. There are many instances where people have suffered injuries, sometimes fatal, by being in contact with land-based telephones and water installations during lightning strikes. During thunderstorms, stay near the middle of the room away from metal objects. Lightning currents may also enter a house through television antenna masts, telephone lines, and power lines. It is the responsibility of the telephone and power companies to provide proper protection to divert these currents to ground and prevent them from entering the house. Television antenna masts should be provided with a proper path to ground. Otherwise, if there is a strike to the television mast the current will flow along the signal cable and may lead to an explosion of both the cable and the television (Fig. A3). In such a situation, anyone

Fig. A3 Explosion caused by passage of a current similar to that of a lightning return stroke through a metal wire (Photograph courtesy of Division of Electricity, Ångström Laboratory, Uppsala University, Uppsala, Sweden)

sitting close to the television may be subjected to a side flash. Even if the television mast is provided with a proper grounding system, a lightning strike may generate enough current in the signal cable to damage the television. Thus, at the entrance of the signal cable to the television it is necessary to connect a protecting device to divert this unwanted current to ground. Indeed, modern-day electrical appliances are so sensitive to overvoltages and currents that it is not necessary to have a direct lightning strike to destroy them. A nearby lightning strike may generate, through the interaction of its strong electromagnetic fields, currents in power cables and signal antennas that are strong enough to destroy them. It is a good habit to remove electrical appliances from their power sockets and, in the case of television sets, from signal cables during thunderstorms. Large unwanted currents flowing in power and antenna systems can cause enough damage, even if, for example, the television is switched off at the moment of strike.

The injuries caused by lightning to human beings can be mild in some cases or fatal in others. Some people believe that lightning victims are burned to death, and what remains after the lightning flash is just a pile of ash. This is not true. In most cases, the burns inflicted by lightning are not severe. The injuries caused by lightning can be divided into mild, moderate, and severe. A victim who has experienced a mild injury may still be able to answer questions but may be confused and amnesiac with regard to the event. He may have problems with short-term memory for hours or days. In the case of moderate injuries, the person may be disoriented or paralyzed in the lower limbs. The person can also lose consciousness during the event. Burns may appear on the victim, who might also suffer a cardiac arrest but recover spontaneously. In the case of severe injuries, the person may suffer respiratory arrest and cardiac arrest. If the respiratory or cadiac arrest continues for a sufficient time the person may also suffer brain damage. However, in some cases the person can be brought back to life by artificial respiration and cardiac resuscitation.

A.2 Most Commonly Asked Questions

(1) *How can you tell when there is about to be a thunderstorm?*

On clear days, the approach of a thunderstorm is noticeable from the appearance of dark cumulonimbus clouds. However, during frontal storms the sky can be overcast, and it is difficult to notice the approach of a thunderstorm. You may also observe gust winds as the thunderstorm approaches, and if you observe hail or snow pellets, the thunderstorm is over your head. If you hear thunder, then the thunderstorm is very close. If you have a small AM radio, the approach of a thunderstorm can be recognized by the loud and frequent static coming from the receiver. The approach of a thunderstorm can also be recognized by thunder. Usually thunder can be heard only when the thunderstorm is less than 20 km away, in which case you should seek proper shelter almost immediately.

(2) *What are the ways in which a person can be injured by lightning?*

There are several: (1) direct strike (2) side flash, (3) connecting leader, (4) contact potentials, (5) step voltages, (6) shock from apparatus connected to power lines or through telephones, and (7) shock wave of a lightning flash or by flying debris generated by a tree or other object exploding during a lightning strike.

(3) *Which of those scenarios are most dangerous?*

Injuries from a direct strike are the most dangerous. However, injuries from a side flash or contact potential can be just as severe as a direct strike. Usually the shocks from step voltages are usually mild; however, the victim may still receive burn injuries (on the foot where the current entered into the body) and in the worst case may have difficulty walking for some time.

(4) *How do I reduce the chances of receiving a direct strike?*

There are three ways to avoid a direct strike. First, stay in a space, such as a concrete building or motor vehicle with a metal chassis, where a lightning channel cannot reach. Remember that in a motor vehicle that can be considered as a Faraday cage you are safe from lightning currents but secondary effects due to a direct strike may cause injuries. Second, stay within the protective zone of a tall object. In this case, make sure that you are far enough from the possible path of a lightning current to ground to avoid step voltages. Squat down (sit in a crouching position with knees bent and the buttocks on or near the heels) so as not to be the tallest object in a area within a circle of about 30 m. This position will also reduce the effects of step voltages. However, squatting down does not always provide protection from a lightning strike.

(5) *What kind of spaces can a lightning channel not reach?*

Spaces surrounded by metal boards and frames such as inside concrete buildings, trains, buses, cars (with complete metal chassis), steel boats, etc. However, you should not stick out any part of your body or something you are holding from an opening (such as a window). Note that open cars and uncovered boat decks are not safe spaces.

(6) *How do I avoid a side flash or contact potential from an object?*

Maintain a distance of at least 2 m from an object, such as a tree, that is likely to be struck by lightning.

(7) *How do I reduce the risks from step potential?*

Minimize bodily contact with the ground. Keep the feet close together. Shoes will reduce the current flowing through the body when exposed to a step current.

(8) *How do I avoid shocks and side flashes from electrical apparatus?*

Maintain a distance of at least 1 m from electrical lines and electrical apparatus connected to power or communication networks. Try not to use a land-based telephone during thunderstorms. Cell phones pose no threat. You are also advised to keep away from TVs because there could be a direct strike to the TV antenna and the lightning current may enter the building through connecting cables. In the worst case, the TV may explode.

Appendices

(9) *Is it safe inside a building?*

Usually armored concrete buildings and buildings equipped with lightning protection systems are safe. However, even in these buildings you should avoid handling electrical apparatus, telephones, or any other metal objects during thunderstorms. Stay at least a meter away from these items while inside the building.

(10) *Are wooden buildings with no lightning protection system safe?*

During a lightning strike to a wooden building, side flashes may occur inside the building, injuring the occupants. If the wooden building has electricity, most probably the lightning current will flow along these conductors. Thus, if you happen to be inside a wooden building, stay a few meters away from the walls, roof, and electrical apparatus to avoid side flashes.

(11) *Are wooden buildings with a metal roof a safe place?*

If the metal roof is not properly connected to ground with conductors, it is no better than a wooden house with a wooden roof. Occupants of such buildings can receive side flashes from the roof.

(12) *Are tents safe places during lightning strikes?*

No. The inside of a tent is more risky than open spaces. Tent poles are higher than humans and so have a higher probability of receiving a lightning strike than a person crouching down in an open field. If there is a lightning strike to the tent, there is a high probability of receiving a side flash from the tent poles.

(13) *Why do computers and TVs suffer damage during thunderstorms if the house is not struck by lightning?*

During a lightning strike to power lines, large surges of voltage and currents can enter a building along the power lines. Such surges can destroy electrical apparatus connected to the system. Moreover, electromagnetic waves from a lightning flash that strikes several hundred meters can induce high voltages, large enough to cause damage in electrical apparatus. So it is advisable to have proper surge protection at the service entrance of these systems or to remove them from power sockets during thunderstorms.

(14) *Do trees provide a safe space for human beings?*

No. Staying near a tree puts you at risk of receiving a side flash. Moreover, during a lightning strike, the tree can explode and you could be injured by flying debris. If you happen to be in a forest, find a place where trees are not very dense and squat down, leaving a space of at least 2 m from the trunk and the foliage.

(15) *What are the riskiest places in terms of lightning strikes?*

Open fields, open decks of boats, mountaintops, and mountain ridges. In addition, inside tents and huts located in such places are risky.

(16) *Is it dangerous to carry pieces of metal and long metal objects during thunderstorms?*

Wearing metal objects such as metal jewelry on your body has no effect on the probability of a direct lightning strike. However, metal objects can modify the flow of lightning current around your body. Do not carry long objects such as, for example, iron golf clubs, fishing rods, or ladders during thunderstorms.

(17) *Will rubber boots and raincoats protect me from lightning strikes?*

No. Any insulating effects of rubber boots and raincoats will not protect you from a lightning strike. However, rubber boots can reduce the risk of injuries due to step voltages.

(18) *How do I recognize the outbreak or approach of a thunderstorm?*

The growth of cumulus clouds and loud peels of thunder (at night you can also see clouds illuminated by distant lightning) are the signs to watch for. Often thunder clouds are obscured by other clouds and it is not possible to detect them when the sky is covered with thick cloud layers. Wind gusts, hail pellets, and strong rain are also signs that you are probably close to the thunderstorm center.

Using the time to thunder you can estimate the distance to the lightning flashes. If you have an AM radio, listen carefully to the static. When the static becomes more frequent and louder, the storm is headed your way. Remember that thunderstorm conditions change very rapidly. Cumulus clouds can grow into thunderclouds in around 15 min, and thunderstorms can move at speeds as fast as 40 km/h. Moreover, lightning strike points can be located outside the thunderstorm area.

(19) *What should I do if I am in an open space and thunderstorms are approaching?*

Whenever you notice thunderstorm activity (distant rumbling, for example) move to a safe place. If you cannot find a safe place, squat down. Try to find a depression in the ground and squat down in it. Wait there until the lightning activity diminishes.

(20) *What should I do if I am walking down a road and thunderstorms are approaching?*

Go inside any nearby buildings. If there are none, squat down and watch for lightning activity.

(21) *What should I do if I'm caught in a thunderstorm while riding a bicycle or a horse?*

Get off and move to a safe place.

(22) *What should I do if I am on a yacht or a small open boat?*

When you notice signs of thunderstorms, go back to shore. If you are caught in the storm, stay as low as possible, and stay away from the mast of the boat.

Do not stick your hands or legs out of the boat and touch the water. A bridge can provide a protective zone. If you do not have time to go to the shore, move your boat just under the bridge.

(23) *What should I do if I am on a mountain top?*

Leave as soon as possible. If there is no time, find a depression and lower yourself into it. If you cannot find a depression, stay on a slope and squat down. If a forest is nearby, go to it and search for a place where trees are most sparse and squat down. If the mountain trail contains railings for climbers, do not hang onto them.

(24) *Is paragliding dangerous during thunderstorms?*

Avoid paragliding under thunderstorm conditions. The gliders do not protect you from lightning currents, and if there is a lightning strike, the consequences can be disastrous.

(25) *Can I use a mobile phone during a thunderstorm?*

It is safe to use a mobile phone during a thunderstorm. However, avoid listening to music with earphones from mobile phones during thunderstorms. If there is a direct strike, the lightning current may pass into your body through the ears following the wires of the earphones. Even in the case of a nearby strike, large currents could be induced in the earphone wires and generate a spark inside the ear.

(26) *How are people injured by lightning?*

The worst case is death. Most deaths are caused by cardiorespiratory arrest, and many victims can be saved by cardiopulmonary resuscitation (CPR). In the long term, the most likely consequences are muscle soreness, difficulty hearing, problems with balance, and possibly vision and neuropsychological changes.

(27) *How should I treat someone struck by lightning?*

First aid is urgently required. First assist the most injured, that is, those who appear to be dead should be assisted first. The required first aid is CPR and it must be continued until the arrival of skilled medical personnel. If a person has a pulse but is not breathing, then mouth-to-mouth resuscitation should be applied.

Appendix B: Advanced Books on Lightning for Further Reading

E. M. Bazelyan and Y. P. Raizer, Lightning Physics and Lightning Protection, IOP Publishing, Bristol, UK, 2000.
V. A. Rakov and M. A. Uman, Lightning: Physics and Effects, Cambridge University Press, Cambridge, UK, 2003.

Cooray, V. (Editor), The Lightning Flash, Institution of Engineering and Technology (IET, formerly IEE), London, 2003.

Cooray, V. (Editor), Lightning Protection, Institution of Engineering and Technology (IET, formerly IEE), London, 2010.

Cooray, V: (Editor), Lightning Electromagnetics, Institution of Engineering and Technology (IET, formerly IEE), London, 2012.

Index

A
Action integral, 40, 125, 129–130, 252, 253, 266
Active region of the streamer, 15
Aerosols in air, 70
Air conductivity, 70
Alternative explanations of ball lightning
    hallucinations cased by epileptic seizures, 369
    phosphenes created
        by lightning magnetic fields, 367
        by X-rays and gamma rays, 367
Amperes law, 35–36, 141
Amplitude of continuing currents, 131, 176, 318
Analytical expression for channel base current, 183–184
Angle of protection, 308–311
Atmosphere variation
    of air density with height, 60–61
    of air pressure with height, 60–61
    of air temperature with height, 59–60
    of electron density with height, 62–64
    of ion concentration with height, 62
Atmospheric density, 9, 91, 93–94, 352
Atmospheric sign convention, 40, 142–149, 152, 156, 158–160, 162
Attempted leaders, 25, 107
Attraction of two current carrying conductors, 252–253
Attractive radii, 304–306, 313, 314, 329, 339
Attractive range of a lightning conductor, 302–305, 314, 328
Automobiles and lightning strike, 373
Avalanche to streamer transition, 11–16, 24

B
Ball lighting, 365–369
Bannister theory, 205
Bead lightning, 359, 361–364
Becerra and Cooray criterion of leader inception, 331, 333
Bidirectional leader propagation, 95, 96, 264
Bipolar lightning flashes, 86
Branch component, 25, 26, 105, 109
Breakdown electric field, 10, 16, 96, 171, 332, 334, 353, 354
Burns caused by lightning, 297

C
Cardiac and respiratory arrest caused by lightning, 295
Carnegie curve, 66, 67
Central core, 99, 105, 171, 172
Chaotic pulse trains, 147
Charge
    associated with the first return stroke, 115
    associated with the subsequent strokes, 115
    generation in thunderclouds, 79–88
    of streamer discharges, 17–18
Charge on the dart leader, 233–244
Charge per unit length on the stepped leader (close to ground end), 115
Charge structure of clouds in different geographical regions, 83–84
Classical explanation of initiation of lightning flashes, 92–93
Cloud flashes, 26, 86, 95, 157, 158, 160, 162–164, 230

Cloud potential as a function of first return stroke peak current, 234–239
Coconut trees struck by lightning, 249, 261, 286
Collection volume method (CVM), 335
Compact cloud flashes, 160–162
Cone of protection, 310–313
Connecting leader, 20, 21, 25, 97, 100–104, 112, 118, 124, 171, 172, 249, 259, 261, 262, 269, 281, 283, 285, 292–294, 301, 302, 327–339, 376
Continuing current, 25, 97, 109, 110, 112, 113, 120–124, 130–132, 142, 144, 145, 147, 258, 259, 262, 263, 266, 344, 345
Convective Available Potential Energy (CAPE), 72, 73
Convective thunderstorms, 75
Corona discharges, 11, 16–17, 269
Corona sheath, 99, 100, 105, 106, 171–174, 178, 194, 345
Critical electric field necessary for electrical breakdown, 9–12, 91
Critical field necessary for streamer propagation, 15–16, 92
Critical radius concept of leader inception, 332
Cumulonimbus, 71, 72, 375
Cumulus cloud, 71, 72, 378
Cumulus congested, 71, 72
Cumulus stage, 73, 74
Current above thunderclouds, 67
Current generation type return stroke models, 170, 172–179, 183
Current propagation type return stroke models, 170, 179–183
Currents in lightning flashes, 117–132, 135, 312
CVM. *See* Collection volume method (CVM)

D

Dart leader, 25, 26, 97, 107–109, 112, 113, 120–124, 142, 144, 147, 149, 194, 233–244, 262, 357, 360
Dart stepped leader, 25, 107–108, 149, 357
Descartes, 3
Different types of clouds, 84–85
Dipole, 56, 58, 84, 195, 205–207, 220, 347
Direct lightning strike to humans, 292, 296
Direct strike
  to an airplane, 263–268
  to a power line, 271–273
  to a tree, 259–263
  to a windmill, 268–271

Disability caused by lightning, 299
Distribution of charge brought to ground by return strokes and lightning flashes, 128–129
Distribution of negative charge
  on the dart leader channel, 239–241
  on the stepped leader channel, 239–241
Distribution of positive charge deposited by return stroke, 241–242
Downdrafts, 74, 75
Drag force, 93
Drift speed, 8
Dry adiabatic lapse rate, 72, 73
Duration
  of continuing currents, 115, 262
  of a lightning flash, 115, 266
  of the M component current, 115

E

Early streamer emission (ESE), 329
EGM. *See* Electro-geometrical-method (EGM)
Electrical runaway mechanism of initiation of lightning flashes, 93–94
Electric conductivity of the atmosphere, 63–66
Electric field, 4, 7–21, 24, 25, 29–37
  generated by a tripolar thundercloud, 40–42
  inside and on surface of a perfect conductor, 32
  mill, 135, 138–140
  of a point charge, 29–30
  of a point charge over a perfect conductor, 32–35
Electric field change
  caused by a stepped leader, 44–47
  caused by leader return stroke combination, 48–49
  due to a cloud flash, 42–44
  due to a ground flash, 43–44
Electric potential of a point charge, 30
Electro-geometrical-method (EGM), 233, 239, 302–308, 327–328, 331, 339
Electromagnetic fields
  from cloud flashes, 155–160
  of a current element in time domain, 56
  of a dipole, 53–55
  of a dipole over a perfectly conducting ground, 54–55
  of a return stroke, 56–58

Index 383

Electromagnetic shielding using skin depth, 326–327
Electromagnetic spectrum of return strokes, 162, 163
Electromagnetic zoning, 324–327
Electron attachment, 8, 15
Electron avalanche, 7–14, 16, 17, 24, 66, 91, 94, 160, 342–344, 357
EMF, 37, 38
Energetic radiation produced by lightning, 356–357
Energy dissipated by subsequent return strokes, 244
Energy dissipation, 128, 129, 198, 199, 243–244, 252, 257–258, 324
Energy dissipation in first return strokes, 243
Environmental lapse rate, 73
External lightning protection system, 314–317, 319–320

F
Fair weather electric field in the atmosphere, 66–68, 70
Faraday cage, 32, 267, 316, 317, 373
Faraday's law, 36–38, 140
Fast transition of return stroke radiation fields, 152–153
Final jump condition, 102, 103, 302, 303, 327, 328, 335, 337, 338
Fine structure of radiation fields, 148–155
First and subsequent return stroke currents, 124–130, 153, 154, 183, 184, 230, 237
First return stroke, 25, 26, 97, 98, 107–109, 112, 124–126, 129, 130, 143, 144, 147, 149, 153, 154, 163, 183, 184, 188, 197, 198, 200, 212, 230, 234, 236–243, 257, 258, 266, 284, 293, 303, 312, 360
Force between current carrying conductors, 38–40
Forked lightning, 359–361
Formation
 of cumulonimbus, 71, 72
 of thunderclouds, 71–77
Franklin, Benjamin, 3, 4, 301, 329
Frontal thunderstorms, 75
Function of a lightning conductor, 301–302

G
Gamma ray flashes generated by thunderstorms, 357
Gauss's law, 30–32, 52
General features
 of currents in triggered lightning flashes, 120–124
 of upward initiated lightning flashes, 112, 113, 121–123
Generation of thunder, 247–249
Global distribution
 of lightning flashes, 86–87, 227
 of thunderstorm days, 87, 88
Global electrical circuit, 67–70
Global warming and lightning, 348–349
Graupel, 73–75, 79–85, 91

H
Hail, 11, 74, 369, 375, 378
Heating of material by lightning flashes, 252
Herodotus, 2
High-frequency (HF) radiation, 146
Horizontal electric field, 54, 203, 204, 275–277, 279
How to avoid lightning injuries, 371–379
Hydration of fast ions, 70

I
Ice crystals, 73, 74, 79–83, 91
Ice-Graupel collision, 79–85
Illustration of propagation effects, 210–214
Indirect effects of lightning on power lines, 274–279
Induction
 of voltages by electric fields, 255–257
 of voltages by magnetic fields, 254–255
Initiation of lightning flashes, 91–94
Injuries caused by shock waves, 299
Intercloud lightning flashes, 85
Internal lightning protection system, 320–327
Intracloud lightning flashes, 85
Ionospheric reflections, 149, 154–156, 219
Isolated electrical breakdown, 160

J
J processes, 26, 109
Junction processes, 26, 109
Jupiter, 2

K

K changes, 25, 26, 97, 107, 111, 112, 142, 145–147, 157, 159–160, 344, 345
Khan, Genghis, 2

L

Leader discharge, 18–22, 92, 94, 95, 100, 160, 264, 343, 344
Leader propagation model SLIM, 337–339
Leader propagation models, 333–339
  of Becerra and Cooray, 331, 337–339
  of Dellera and Garbanati, 331, 332, 335–336, 339
  of Eriksson, 331, 332, 334–335
  of Rizk, 331, 336–337, 339
Length of the dart of the dart leader, 115
Length of the steps of stepped leaders, 99, 115
Lightning conductor, 10, 16, 17, 20, 21, 129, 254, 268, 270, 301–311, 314, 315, 318, 328, 329, 332, 333, 371–373
Lightning dissipaters, 329–330
Lightning injuries
  by connecting leader, 281, 283, 285, 292–294
  to the ear, 296, 299
  to the eye, 296, 299, 360, 361
  to the nervous system, 296–297, 299
  by shock waves, 281, 292, 293, 296, 299, 376
  by side flash, 281, 284–286, 371–373, 376, 377
  by step voltage, 281, 287–291, 371, 376, 378
  by touch voltage, 281, 286–287
Lightning localization systems in practice, 229–230
Lightning without thunder, 364–365
Localization of lightning flashes, 217–230
  using Interferometric method, 224–227
  using Magnetic direction finding, 217–219
  using peak electric field, 228–229
  using satellites, 227
  using thunder ranging, 227
  using time of arrival method, 220–224
  using time reversal method, 229
Lorentz law, 51, 93
Low pressure electrical discharges, 24

M

Magnetic field
  of a current element, 36, 56
  of a long conductor, 35–36
Mathematical expression for step voltage, 290–291
Mature thundercloud, 75, 119
Maxwell's equations, 49, 168, 169, 189, 196
M-component current, 110, 115, 130–131
M-components, 25, 97, 109–110, 115, 121, 123, 124, 130–131, 142, 144, 145, 159–160, 345
Measurement of lightning currents
  Rogowski coil method, 118, 119
  Shunt method, 118, 119
  using instrumented towers, 117–120, 124
Measurement of magnetic field, 140–142
Mechanism of cloud flashes, 110–112
Melting of material by lightning flashes, 251
Mesh method of lightning protection, 317–319
Mobile phones and lightning strike, 379
Modern day myths, 4–5
Most commonly asked questions, 375–379

N

Narrow band electromagnetic spectrum of lightning flashes, 163, 164
Narrow bipolar pulses, 160, 162
Negative charge center of tripolar thundercloud, 40, 41
Negative ground flashes, 24, 26, 67, 85, 86, 112–114, 123, 124, 132, 157, 160, 217–218, 352
Negative streamer, 13–15, 22, 23, 172, 354
Norton theory, 210
Number of return strokes per flash, 115

O

Orographic thunderstorms, 75, 76

P

Paralysis caused by lightning, 292, 297
Parameters of downward lightning ground flashes, 126
Peak current
  derivative in return strokes, 125, 128–129
  distribution in return strokes, 126–127
  in the first return stroke, 115
  of M components, 115
  of the subsequent strokes, 115

Peak distribution of return stroke radiation fields, 154, 219
Percentage of flashes
  with continuing currents, 109, 115
  with single strokes, 115
Physics sign convention, 40, 146, 156–158
Pilot system, 22, 23
Positive charge center of a tripolar thundercloud, 40, 41
Positive charge pocket of tripolar thundercloud, 40
Positive ground flashes, 24, 26, 67, 84–86, 112–114, 352, 372
Positive polarity discharge, 21
Positive streamer, 11–16, 20, 22, 23, 25, 92, 105, 106, 176, 178, 354
Potential gradient along the leader channel, 21, 22, 194, 195, 235, 236
Potential of leader channel, 21–22
Preliminary breakdown, 24, 97–99, 142–147, 157, 160, 230
Production of nitrogen oxides
  by cloud flashes, 344
  by continuing currents, 344, 345
  by ionization collisions, 342
  by K changes, 344, 345
  by leaders, 342–345
  by lightning, 341–346
  by M-components, 345
  by return strokes, 342–345
  by streamers, 342, 344, 345
Production of NOx by lightning, 341
Propagation effects, 153, 203–214, 224
Propagation of leader discharges, 19–21, 94
Psychological injuries caused by lightning, 298–299, 379

R
Radiation field from stepped leaders, 148–155
Radiation fields, 54–56, 58, 143, 147–149, 151–154, 156, 157, 159, 160, 162, 163, 181, 207–209, 211–213, 218–220, 223, 228
  of cloud flashes, 157–160, 163
  produced by K changes and M-components, 159–160
  from return strokes, 148–155, 228, 230
Rate of change of current
  of first return strokes, 115
  of subsequent strokes, 115
Recoil leaders, 96–97, 107–110, 113, 114, 121
Relaxation time, 51–53, 63–65

Return stroke, 25, 48, 65, 97, 120, 137, 167, 187, 204, 217, 234, 247, 283, 303, 334, 342, 359, 374
Ribbon lighting, 359, 364
Rise time
  of the current of first return strokes, 115
  of the current of subsequent strokes, 115
  of M component current, 115
Rizk criterion of leader inception, 331, 336
Rolling sphere method of lightning protection, 307, 319
Runaway electron, 93–94, 354, 356, 357

S
Schumann resonances, 346–349
Secondary avalanches, 12, 13, 15
Sentry box experiment, 4
Slow front of return stroke radiation fields, 152–153
Socrates, 3
Space leader, 22–23
Space stem, 22, 23
SPD. *See* Surge protective devices (SPD)
Speed of dart leaders, 115
Speed of stepped leaders, 115
Standard atmospheric condition, 7, 9, 15
St. Elmo's fire, 11
Stem of streamer discharges, 18
Stepped leader, 16, 20–26, 44–47, 97, 99–109, 112, 113, 118, 126, 142–144, 147–155, 162, 171–173, 180, 188, 233–244, 259–263, 283, 292, 293, 301–308, 313, 318, 319, 327–339, 357, 359, 360, 373
Streamer discharges, 11–18, 24, 92, 94, 95, 99, 103, 105, 171, 172, 298, 332, 333, 342, 344, 345, 354, 357
Streamer head, 11–15, 22, 23, 342
Streamers, 12–25, 92, 95, 97, 102, 105, 106, 171–174, 176, 178, 302, 328–329, 332, 333, 335, 336, 338, 342, 351, 354, 357
Streamer to leader transition, 99, 329
Striking distance, 102–103, 126, 233, 234, 239, 292, 302–308, 327, 328
Subsequent return strokes, 25, 26, 97, 108–109, 114, 124–130, 147–149, 153, 154, 168, 179, 183, 184, 188, 189, 196, 230, 237, 238, 240–242, 244, 257, 266, 272–273
Super cooled water droplets, 79
Surge protective devices (SPD), 320–326

T

Temperature of the lightning channel, 249–250
Theories of return stroke speed, 181–201
Thermal ionization, 19
Thermalization, 19, 24, 94, 192, 342–344, 357, 367
Thickness of lightning channel, 250
Thor, 2
Thunderstorm days, 88, 281
Thunderstorm hours, 88
Tiberius, 2
Time derivative of return stroke radiation fields, 153
Time interval
  between steps of stepped leaders, 99, 115
  between strokes, 105, 115
Time varying electromagnetic fields, 49–51, 326
Townsend discharges, 24
Translational energy, 19
Triggered lightning, 117, 119–124, 128, 130, 344, 354, 364, 367, 369
Tripolar structure of cloud charge, 81–83
Tripolar thundercloud, 40–42

U

Unstable atmosphere, 72, 73
Updrafts, 67, 74, 75, 82, 84, 87

Upper atmospheric electrical discharges, 63, 349–356
  blue jets, 350, 354–355
  elves, 350, 355–356
  gigantic jets, 354–355
  sprites, 350–354
Upward initiated lightning flashes, 26, 85–86, 112–114, 118, 120–124, 128, 129, 330

V

Vertical antenna, 135–138
Vibrational energy, 19
Voltage induced in a loop, 36–38, 119, 127, 140, 141, 254, 255
Volume of protection, 308, 311–312

W

Wet adiabatic lapse rate, 72, 73

X

X-rays generated by lightning, 357, 367

Z

Zero crossing time of return stroke radiation fields, 212, 213
Zeus, 2, 3